Pervasive Communications Handbook

Pervasive Communications Handbook

Edited by
Syed Ijlal Ali Shah
Mohammad Ilyas
Hussein T. Mouftah

CRC Press
Taylor & Francis Group
Boca Raton   London   New York

CRC Press is an imprint of the
Taylor & Francis Group, an **informa** business

CRC Press
Taylor & Francis Group
6000 Broken Sound Parkway NW, Suite 300
Boca Raton, FL 33487-2742

© 2012 by Taylor & Francis Group, LLC
CRC Press is an imprint of Taylor & Francis Group, an Informa business

No claim to original U.S. Government works

Printed in the United States of America on acid-free paper
Version Date: 20111031

International Standard Book Number: 978-1-4200-5109-4 (Hardback)

Library of Congress Cataloging-in-Publication Data

Pervasive communications handbook / editors, Syed Ijlal Ali Shah, Mohammad Ilyas, Hussein T. Mouftah.
    p. cm.
    "A CRC title."
    Includes bibliographical references and index.
    ISBN 978-1-4200-5109-4 (hardcover : alk. paper)
    1. Ubiquitous computing--Handbooks, manuals, etc. 2. Wireless communication systems--Handbooks, manuals, etc. I. Shah, Syed Ijlal Ali. II. Ilyas, Mohammad, 1953- III. Mouftah, Hussein T.

QA76.5915.P44 2012
004--dc23
                        2011034287

Visit the Taylor & Francis Web site at
http://www.taylorandfrancis.com

and the CRC Press Web site at
http://www.crcpress.com

Contents

SECTION I   Technology

SECTION II Architecture

SECTION III Applications

Preface

The field of pervasive communication represents the next logical step in the evolution of communication networks. It essentially means that in a pervasive communication environment, we will all be able to communicate with others, whenever, wherever, and with whatever communication device(s) we carry. We will be connected all the time and have the ability to take our personal and corporate information with us wherever we go. This ubiquitous communication ability includes one-to-one, one-to-many, and many-to-one exchange of information.

This pervasive communication is being enabled by continuous advancements in multiple areas of research and development. First and foremost are developments in the device and process technologies that are able to drastically reduce the size and power requirements of sensing and communication devices, while doubling the processing capacity every couple years. This reduction in size and power, while more than an equal increase in the processing capacity, is enabling these technologies to be embedded in every aspect of our daily lives. So much so that it is almost impossible to avoid sensors even in our very basic mundane activities. These days sensors are embedded in the automobiles we drive, the cooking ranges we cook our food on, air conditioners, and the communication devices we use.

Equally important are the advancements in application-level software and firmware, and communication technology that seamlessly connect these sensors to the rest of the world, in a manner that makes sense. These advances are making it possible for us to develop appropriate human machine interfaces. Standardization of communication protocols such as 802.17 (Zigbee) and 802.11n are further helping pervade the 24/7 online concept.

Examples of how the pervasive computing paradigm is changing our world are plenty; however, we will mention only two possible examples. We all drive on highways; however, in a pervasive communication paradigm, the drivers will have a different user experience than what they currently have. In a pervasive communications paradigm, users will start receiving location- and need-specific information about the neighborhood they are driving through. For example, information about local restaurants (including discount coupons) could be presented to the driver if it is time for dinner. Similarly, users may start receiving offers on hotels that are within, say, 5 miles of radius of their location on the highway, if they are looking for hotels.

In the field of medicine, a pervasive communication environment, for example, will enable hospitals to gather all information about a patient while they are being transported to the hospital. By the time the patient arrives at the hospital, information about the medical history of the patient, medicines being administered, vital signs, etc. will not only be already available to medical professionals but may also have been analyzed and decisions made about the most likely treatment options. The pervasiveness of this environment will enable many more applications, which will, undoubtedly, change the way in which we conduct our daily lives.

The aim of this book is to cover all these topics, contributed by experts in this field. The book has 21 chapters that tackle three aspects of pervasive communications: (1) technology; (2) architecture; and (3) applications. In the technology section we have contributions on quality of service routing in mobile networks, a survey and categorization of MAC protocols, low power design for Smart Dust Networks, security and reliability in wireless sensor networks, and more covering other aspects of the technology. In the architecture section we have contributions on P2P-based VoD architecture, and interoperability in a pervasive environment. And finally in the application section we discuss how pervasive communication is helping health care providers and energy companies better manage their vital assets.

The *Handbook* should be useful for anyone who deals with or intends to become involved with the field of pervasive communication. The targeted audience for the *Handbook* includes professionals who are designers and developers for pervasive communication, researchers (faculty members and graduate students), and those who would like to learn about this field.

Many people have contributed to this *Handbook*: first and foremost are the researchers and technical professionals who have contributed 21 chapters. These people deserve our appreciation for taking the time out of their busy schedules to contribute to this book. We would also like to thank Andrea Dale, Ashley Gasque, and Jessica Vakili for their patience and hard work.

We also extend our very special thanks to our families for their unconditional love and support throughout this project.

Syed Ijlal Ali Shah
Mohammad Ilyas
Hussein Mouftah

Editors

Dr. Syed Ijlal Ali Shah has more than 15 years of experience in the telecommunications and datacom industry. He has contributed to leading-edge technology through academic papers, industrial contributions (products), and patents. He was a key member of the team that worked on ATM switches and networks when the technology was first introduced. He was also a key member of the team that worked on a new generation of devices called Network Processors in early 2000. Network Processors like ATMs created a paradigm shift on reconfigurable multiprotocol processors. He has worked and contributed heavily to the definition and creation of new interconnect and switching technology called RapidIO (RIO). RapidIO is an alternative backplane interconnect technology to Ethernet and is the preferred interconnect technology in base stations.

Dr. Shah is currently with Freescale Semiconductor as a data path Systems Architect. At Freescale Semiconductor, he has defined and worked on the architecture of Freescale's traffic management co-processor, the RapidIO Fabric, and has been a key contributor to several other network-related projects.

Prior to Freescale/Motorola, Shah was an associate professor at the Lahore University of Management Sciences (LUMS). And prior to LUMS, he was at Nortel as senior member of the technical staff, where he helped define and architect packet switching products, and Quality of Service (QoS) and network dimensioning methodologies.

Dr. Shah holds a PhD and an MS in electrical engineering from Columbia University and a master's in engineering management from the University of Ottawa. He holds four patents on call admission control algorithms for switches/routers and dynamic IP/ATM congestion management, with several more pending. He was recently awarded the "Engineering Award" by Freescale for his contributions. He has presented on IP/ATM congestion control and traffic management at various conferences and invited sessions, and has published over 35 industrial and academic papers in conferences and journals on various aspects of QoS, network design and architecture, and all-optical networks, wireless sensor networks, and Multicore processors.

He was the editor for the Traffic Management specification for the RapidIO Trade Association from 2004 to 2006. He was also the Technical Committee co-chair for Power.Org from 2007 to 2009, where he was instrumental in streamlining the technical committee activities and aligning them with membership interests. He was the guest editor for *IEEE Communication Magazine* on Fabric and Interconnect Standards and was the associate technical editor for the magazine from 2000 to 2002. His research interests include sensor network design, QoS in IP networks, optical networks, wireless networking, and pervasive communications.

Dr. Mohammad Ilyas received his BSc in electrical engineering from the University of Engineering and Technology, Lahore, Pakistan, in 1976. From March 1977 to September 1978, he worked for the Water and Power Development Authority, Pakistan. In 1978, he was awarded a scholarship for his graduate studies, and he completed his MS in electrical and electronic engineering in June 1980 at Shiraz University, Shiraz, Iran. In September 1980, he joined the doctoral program at Queen's University in Kingston, Ontario, Canada. He completed his PhD in 1983. His doctoral research was about switching and flow control techniques in computer communication networks. Since September 1983, he has been with the College of Engineering and Computer Science at Florida Atlantic University, Boca Raton, Florida, where he is currently associate dean for research and industry relations. From 1994 to 2000, he was chair of the Department of Computer Science and Engineering. From July 2004 to September 2005, he served as Interim Associate Vice President for Research and Graduate Studies. During the 1993–1994 academic year, he was on his sabbatical leave with the Department of Computer Engineering, King Saud University, Riyadh, Saudi Arabia.

Dr. Ilyas has conducted successful research in various areas, including traffic management and congestion control in broadband/high-speed communication networks, traffic characterization, wireless communication networks, performance modeling, and simulation. He has published 1 book, 8 handbooks, and over 150 research articles. He has supervised 10 PhD dissertations and more than 37 MS theses to completion. He has been a consultant to several national and international organizations. Dr. Ilyas is an active participant in several IEEE technical committees and activities.

Dr. Ilyas is a senior member of IEEE and a member of ASEE.

Hussein T. Mouftah received his BSc and MSc from Alexandria University, Egypt, in 1969 and 1972, respectively, and his PhD from Laval University, Quebec, Canada, in 1975. He joined the School of Information Technology and Engineering (SITE) of the University of Ottawa in 2002 as a Tier 1 Canada Research Chair Professor, where he became a *University Distinguished Professor* in 2006. He has been with the ECE Department at Queen's University (1979–2002), where he was prior to his departure a full professor and the department associate head. He has six years of industrial experience, mainly at Bell Northern Research of Ottawa (now Nortel Networks). He served as editor-in-chief of the *IEEE Communications Magazine* (1995–1997) and IEEE ComSoc director of Magazines (1998–1999), chair of the Awards Committee (2002–2003), director of Education (2006–2007), and member of the Board of Governors (1997–1999 and 2006–2007). He has been a distinguished speaker of the IEEE Communications Society (2000–2007). He is the author/coauthor of 7 books, 48 book chapters and more than 1000 technical papers, 12 patents, and 140 industrial reports. He is the joint holder of 12 best paper and/or outstanding paper awards. He has received numerous prestigious awards, such as the 2007 Royal Society of Canada Thomas W. Eadie Medal, the 2007–2008 University of Ottawa Award for Excellence in Research, the 2008 ORION

Leadership Award of Merit, the 2006 IEEE Canada McNaughton Gold Medal, the 2006 EIC Julian Smith Medal, the 2004 IEEE ComSoc Edwin Howard Armstrong Achievement Award, the 2004 George S. Glinski Award for Excellence in Research of the University of Ottawa Faculty of Engineering, the 1989 Engineering Medal for Research and Development of the Association of Professional Engineers of Ontario (PEO), and the Ontario Distinguished Researcher Award of the Ontario Innovation Trust. Dr. Mouftah is a fellow of the IEEE (1990), the Canadian Academy of Engineering (2003), the Engineering Institute of Canada (2005), and the Royal Society of Canada RSC Academy of Science (2008).

Contributors

Ishfaq Ahmad
University of Texas
Arlington, Texas

Rana Ejaz Ahmed
Department of Computer Science and
    Engineering
American University of Sharjah
Sharjah, United Arab Emirates

Khaled A. Ali

Jamal N. Al-Karaki
Department of Computer Engineering
The Hashemite University
Zarka, Jordon

Ibrahim Al-Oqily
Department of Computer Engineering
The Hashemite University
Zarka, Jordon

Arny Ambrose
Department of Computer Science and
    Engineering
Florida Atlantic University
Boca Raton, Florida

Sami Saleh Al-Wakeel
College of Computer and Information
    Sciences
King Saud University
Riyadh, Saudi Arabia

Ahmed Badi
Department of Computer Science and
    Engineering
Florida Atlantic University
Boca Raton, Florida

Guruduth Banavar
IBM India Research
    Laboratory
IBM India/South Asia
Bangalore, India

Djamel Belaïd
Insitut Telecomm
Paris, France

Mihaela Cardei
Department of Computer Science and
    Engineering
Florida Atlantic University
Boca Raton, Florida

Norman Cohen
Thomas J. Watson Research Center
IBM
Hawthorne, New York

Muhammad K. Dhodhi
Ross Video Limited
Ottawa, Ontario, Canada

Melike Erol-Kantarci
School of Information Technology
and Engineering
University of Ottawa
Ottawa, Ontario, Canada

Marwan Fayed
Department of Computer Science
and Math
University of Stirling
Stirling, Scotland, United Kingdom

Eduardo B. Fernandez
Department of Computer Science and
Engineering
Florida Atlantic University
Boca Raton, Florida

Joel I. Goodman
MIT Lincoln Laboratory
Lexington, Massachusetts

Zhihai (Henry) He
University of Missouri
Columbia, Missouri

Chi-Fu Huang
National Chiao-Tung University
Hsincho, Taiwan

Zdravko Karakehayov
Technical University of Sofia
Sofia, Bulgaria

Sheng-Po Kuo
National Chiao-Tung University
Hsincho, Taiwan

Imen Ben Lahmar
Insitut Telecomm
Paris, France

Victor Yongfang Liang
Apple, Inc.
Cupertino, California

Imad Mahgoub
Department of Computer Science and
Engineering
Florida Atlantic University
Boca Raton, Florida

David R. Martinez
MIT Lincoln Laboratory
Lexington, Massachusetts

Claire Maternaghan
Computing Science and Mathematics
University of Stirling
Stirling, Scotland, United Kingdom

Hussein T. Mouftah
School of Information Technology and
Engineering (SITE)
University of Ottawa
Ottawa, Ontario, Canada

Hamid Mukhtar
National University of Science and
Technology
Islamabad, Pakistan

Huangmao Quan
Department of Computer and
Information Sciences
Temple University
Philadelphia, Pennsylvania

Sk. Md. Mizanur Rahman
University of Ottawa
Ottawa, Ontario, Canada

Albert I. Reuther
MIT Lincoln Laboratory
Lexington, Massachusetts

Jahangir H. Sarker
School of Information Technology and
Engineering (SITE)
University of Ottawa
Ottawa, Ontario, Canada

Syed Ijlal Ali Shah
Freescale Semiconductor
Austin, Texas

Yuan Shi
Department of Computer and
    Information Sciences
Temple University
Philadelphia, Pennsylvania

Michael Slavik
Department of Computer Science and
    Engineering
Florida Atlantic University
Boca Raton, Florida

Danny Soroker
Thomas J. Watson Research Center
IBM
Hawthorne, New York

Yu-Chee Tseng
National Chiao-Tung University
Hsincho, Taiwan

Kenneth J. Turner
Computing Science and Mathematics
University of Stirling
Stirling, Scotland, United Kingdom

Jie Wu
Department of Computer and
    Information Sciences
Temple University
Philadelphia, Pennsylvania

Xuexue Zhang
Beijing Jiaotong University
Beijing, China

Junhui Zhao
Beijing Jiaotong University
Beijing, China

I

Technology

1

Privacy-Preserving Anonymous Secure Communication in Pervasive Computing

Sk. Md. Mizanur Rahman
University of Ottawa

Hussein T. Mouftah
University of Ottawa

1.1 Security and Privacy Preservation for Pervasive Communication

When someone is asked to use his/her office, what goes through his/her mind? Is it the probability that may be it is for an overseas call? Is it the fear that may be it is for browsing his/her profile or confidential documents lying on the table? Or in the case of a total stranger, how it is known about the office? Certainly, it is not easy to deal with all these questions at once in everyday life, but somehow it should be handled if it is arisen. As the reality of pervasive computing becomes more and more apparent, these types of requests become more subtle, frequent, and potentially impacting [1]. Considering recent technology and ongoing advances, devices embedded in smart environments and worn on our bodies will communicate seamlessly about any number of different things. In such kind of interactions, huge amounts of information are shared and exchanged [2]. Even though this may be the means of enjoying context-based and other enhanced services, there is an increased risk involved in some of these interactions and collaborations, if collaborators are about to use our private possessions [3]. This further illustrates how combined assessment of the interrelationships between trust, security, privacy, and context aid in confident decision-making [4]. In everyday life we do not treat these concerns in isolation; we actually make spontaneous decisions that are based on maintaining a "comfortable" balance. Although we do not completely understand these basic building blocks, the potential trade-offs are intuitively understood, even if technically underexplored [1].

A user who does not wish to be tracked by an application will want to use different pseudonyms on each visit. Many applications can offer better services if they retain per-user state, such as personal preferences; but if a set of preferences were accessed by several different pseudonyms, the application would easily guess that these pseudonyms map to the same user [1]. We therefore need to ensure that the user state for each pseudonym looks, to the application, different from that of any other pseudonym. There are two main difficulties. First, the state for a given user (common to all that user's pseudonyms) must be stored elsewhere and then supplied to the application in an anonymized fashion. Secondly, the application must not be able to determine that two sets of preferences map to the same user, so we might have to add small, random variations. However, insignificant variations might be recognizable to a hostile observer, whereas significant ones could adversely affect semantics and therefore functionality [1].

In the following sections, at first we discuss privacy and security parameters for pervasive computing, then we discuss some existing protocols for different pervasive environments. In Section 3, we discuss a protocol on position-based privacy-preserving secure communication for pervasive mobile ad-hoc communication. In Section 4, we discuss a user controllable privacy-preserving secure communication in pervasive computing.

1.2 Privacy and Security Parameters

The key notions on privacy and security associated with pervasive communication are summarized in the following sections.

1.2.1 Privacy Parameters

Identity Privacy: Identity privacy means no one knows the real identity of the nodes in the network. We are especially interested in discussing identity privacy of entities involved in packet transmission, namely the source, intermediate nodes, and the destination [5].

Location Privacy: Requirements for location privacy are as follows: (a) no one knows the exact location of a source or a destination, except themselves; (b) other nodes, typically intermediate nodes in the route, have no information about their distance, that is, the number of hops, from either the source or the destination. It is said that a protocol satisfying (a) achieves weak location privacy and a protocol satisfying both (a) and (b) achieves strong location privacy [5].

Route Anonymity: Requirements for route anonymity are as follows: (a) adversaries either in the route or out of the route cannot trace packet flow back to its source or destination; (b) adversaries not in the route have no information on any part of the route; (c) it is difficult for adversaries to infer the transmission pattern and motion pattern of the source or the destination [5].

1.2.2 Security Parameters

Passive Attacks: Passive attack typically involves unauthorized "listening" to the routing packets or silently refusing execution of the function requested. This type of attack might be an attempt to gain routing information from which the attacker could extrapolate data about the positions of each node in relation to the others. Such an attack is usually impossible to detect, since the attacker does not disrupt the operation of a routing protocol but only attempts to discover valuable information by listening to the routed traffic.

Active Attacks: Active attacks are meant to degrade or prevent message flow between nodes. They can cause degradation or a complete halt in communications between nodes. Normally, such attacks involve actions performed by adversaries, for example, replication, modification, and deletion of exchanged data.

The traffic analysis is usually passive. After performing traffic analysis, an adversary can set a target node and conduct an intensive attack against the node. We call such an attack as "target-oriented." Such attacks are often active. The following are examples of active attacks:

DoS: Multiple adversaries in co-operation or one adversary with enough power can set a specific node as a target in order to exhaust the resource of that node. That is to identify a node and make a target to that specific node.

Wormhole Attack: In wormhole attack, an attacker records a packet in one location of the network and sends it to another location through a tunnel [6] made between the attacker's nodes. Afterwards, the packet is retransmitted to the network under his control.

Rushing Attack: Existing on-demand routing protocols forward a request packet that arrives first in each route-discovery. In the rushing attack, the attacker exploits this property of route discovery operation. If the route requests forwarded by attackers arrive at a target node earlier than other route requests, any route discovered by this route discovery includes a hop via the attacker. In general, an attacker can forward a route request more quickly than legitimate nodes can, so he can enter a route. Such a route cannot be easily detected.

Since nodes in pervasive environment move dynamically, adversaries cannot conduct active attacks without knowing the location or name of nodes. It thus often happens that traffic analysis is conducted passively at first and active attacks are conducted later.

1.3 Anonymous On-Demand Position-Based Secure Communication in Pervasive Communications

Sk. Md. Mizanur Rahman et al. [5,7] proposed an on-demand position-based routing protocol for pervasive mobile *ad hoc* environment, where position information of a node was kept secret during the communication among the nodes in the network. A detailed description is given in the following section.

1.3.1 Fundamental Properties of the Protocol

To understand the operating principle of the protocol, we need to clarify the following properties of the protocol.

1.3.1.1 Position Management

Known position-based routing protocols [8–11] use a position/location management scheme, called a virtual home region (VHR)-based distributed secure position service (DISPOSER). In this scheme, each node has its own VHR, which is a geographical region around a center specific to the node. The center is a fixed center and anyone can identify it by taking a concatenation of two publicly known values, namely the node's ID and position information regarding the center of the whole network, as input to a publicly known hash function. There are position servers (PSs) for each node in the network. PSs of a node N exist only inside the VHR of N and manage position information of N as follows. To report the position of N to its PSs, N executes a region-based broadcast [10] in the VHR if N stays inside its VHR. If N stays out of its VHR, N sends a packet containing position information of N and the center of N. The latter position information is used for determining which node forwards the packet. Once the packet reaches a node in the VHR, the node executes a region-based broadcast. After the region-based broadcast, the PSs can store the latest position information of N. To retrieve position information of N, a source sends a request packet in the direction of the center of N. When the packet reaches a node in the VHR of N, the node executes a sequential searching method [10] and finally the packet reaches one of the PSs. The source authenticates itself to the PS, and then the PS provides the required position information. Using this position information, the source can establish a path from him to the destination. PSs are determined from the node density, the size of the VHR, the robustness of the system, and so on, and the number of the PSs is set in an appropriate value that makes the sequential search more cost-effective than the region-based broadcast and the management cost of the position information low enough. More details on the VHR are described in [8,10].

The PS of this scheme has an additional property: a PS provides a source with additional information to enhance the authenticity and secrecy of services provided by the PS. Before describing this scheme, we define two notations: *Position information* denotes a pair composed of position and time, and legitimate nodes denote nodes that have registered with PS and received a common key (C_K) from PS.

In contrast to ordinary PS, our PS provides a source with a C_K for all legitimate nodes, public key (PK) of the destination, *position information* of the destination, authentication information *Auth*, and a *Token*.

When a node joins a network, it is registered the PS and gets a C_K and a pair of PK/SK from the PS.

When a node updates its *position information* and sends it to the PS, it generates a *random number* and sends it together with its *position information* to the PS. This *random number* is used for generating *Auth*, where *Auth* = [H_1 (Destination's *Position*, Destination's *random number*)] and H_1 is a global hash function. The notation A = [B, C, …, Z] means variable A is substituted by the concatenation of B, C, …, Z. Later, at the route discovery phase, *Auth* is used for authenticating the destination to the source.

To obtain the *position information* of the destination from the PS, the source has to send a signed position request to PS with a *route-request sequence number* (RRQSeqNo). After verification of the signature, the PS responds to the source's request with the *position information* of the destination, *Auth*, public key PK_{Dest} of the destination, and a *Token* defined as *Token* = [H_{PS}(Sender Temp ID, Receiver Temp ID), Time, RRQSeqNo], where H_{PS} is a local hash function defined by the PS. *Position information* is used for generating the temporary ID, Temp ID. In contrast, *Position* is used only for routing, and it is encrypted by C_K in the route-request phase.

A sender keeps *Auth* received from the PS for a session of communication. At the last phase of the route-discovery procedure, destination will reply with a *route-reply message* (RRPMsg) for its authentication to the sender: RRPMsg = [$Sig_{SK_{Dest}}$ (*Auth*)], where Sig_{SK} is a digital signature function under secret key (SK) and SK_{Dest} is the SK of the public/secret key pair of the destination. With this RRPMsg, the sender authenticates the destination.

A *Token* is sent in the last phase of data transmission to the destination. At the end of the communication, the destination sends this *Token* to PS, so that PS can determine whether the communication between the source and the destination is valid. If a node takes the *position information* of the destination and does not make a data transmission, then PS will not supply any further position information to that node.

1.3.1.2 Dynamic Handshaking

A type of handshaking, called dynamic handshaking, which is established from the ending point to the beginning point, is defined here as shown in Figure 1.1. At first, node A sends a signal for node D via B. B will response to A after getting a response from C. That means A will wait for a certain time. The whole handshaking process is performed from the ending to the beginning.

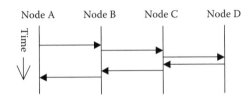

FIGURE 1.1 Dynamic handshaking.

1.3.1.3 Control Packets of the Protocol

Three control packets are used for route discovery of AODPR: *Route Request Packet* (RRQ), *Route Reply Packet* (RRP), and Fail Packet (*Fail*). These packets are described as follows:

C_K is used for encryption and decryption by all legitimate nodes. E_{C_K}: means encryption with C_K.

1.3.1.3.1 Route Request Packet

Sender Temp ID	E_{C_K} (RRQSeqNo)	E_{C_K} (PD)	E_{C_K} (NH)	E_{C_K} (TempNH)	E_{C_K} (EM)

For construction purposes, when senders or forwarders forward any packet, they generate a large bit random number and make parts of that random bit corresponding to the number of fields of the packet. And they specify all the fields with a specific bit number. They then encrypt these fields by padding with random bits. When a packet reaches a node, the node first decrypts and extracts the random bits from the fields and pads their own random bits. As all the fields of a packet are changed, when a packet moves from node to node, it appears new to the network. This procedure is applicable to all the encrypted fields of all the packets. Encryption/decryption is performed as necessary when a packet moves from node to node.

> *Sender Temp ID:* For every session of communication, a source generates its temporary ID Temp ID, computed as Temp ID = [H (Position, Time, PK)], where H is a global hash function known to all legitimate nodes in the network, Position is the position of the source, Time is the present time, and PK is a public key of the source. Temp ID uniquely identifies the source in each session of communication and is dynamically changed from session to session and from hop to hop. When nodes staying within the sender's radio range receive the *RRQ* packet, they will become new senders or forwarders and update the Temp ID into their own Temp ID, which is generated in a similar way as mentioned above.
> For successful identification, the Temp ID should be unique for each session of communication. To this end, H should be collision resistant. Theoretically proven collision-resistant hash functions are slow; thus, in practice, hash functions that are expected to be collision-resistant, such as Message Digest algorithm 5 (MD5) [12] and Secure Hash Algorithm 1(SHA-1) [13], are used instead. The probability of finding a collision for MD5 with respect to 128-bit output and that for SHA-1 with respect to 160-bit output have been estimated as, on average, 2^{64} and 2^{80}, respectively. As long as these probabilities hold, it is difficult to find the same Temp ID for different nodes in each session of communication.
> *RRQSeqNo:* It is generated by the source uniquely, for the uniqueness of a session.
> *Position of Destination* (PD): The geographical position (X_T, Y_T) of the destination, taken from PS and encrypted by C_K.
> *Number of Hops* (NH): NH is the minimum number of hops that an *RRQ* packet travels to find a route from the source to the destination. NH is estimated by the *source*. It is changed by the source when the source tries to find a route with a new estimation. It is also encrypted by C_K.

Temporary Number of Hops (Temp NH): At the beginning of route discovery, Temp NH is initiated as NH by the source, Temp NH = NH and it is encrypted with C_K by the source. After receiving the RRQ packet by legitimate nodes, it is updated. Update means decrementing by *one*, that is, Temp NH = Temp NH − 1. When the RRQ packet travels from node to node it is updated each time by each node. Moreover, the nodes perform encryption/decryption operations and vice-versa by C_K.

Ensure Message (EM): This examines the genuineness of the destination. The source generates an EM when it receives the destination's position. EM = [H_2 (position of destination, time)], where H_2 is the global hash function.

1.3.1.3.2 Route Reply Packet

E_{C_K} (RRQSeqNo)	Sender Temp ID	Receiver Temp ID	RRPMsg

Receiver Temp ID: For every session of communication, an intermediate node or the destination generates its Temp ID in the same procedure as the sender Temp ID. Temp ID is the only identification of a node in one session of communication. It is dynamically changeable from session to session. When packets are forwarded, this field is updated by nodes according to their own Temp ID.

1.3.1.3.3 Fail Packet (Fail)

E_{C_K} (RRQSeqNo)	Sender Temp ID	Receiver Temp ID	E_{C_K} (NH)

1.3.2 Protocol Description

1.3.2.1 Parameters of the Protocol

In certain environments, such as stadiums, classrooms, disaster areas, and battle fields, node placements and their corresponding density can be defined as follows.

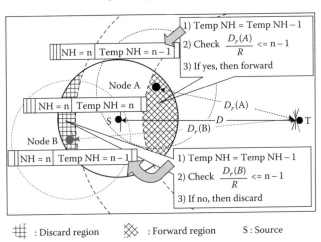

FIGURE 1.2 Quad-placement-connected network.

Quadratic placement means that a node is connected in its radio range with its neighbors in all four compass directions from its center (Figure 1.2): thus, their corresponding
densities are approximated as $\mu_{quad} \approx \sqrt{n} / [\{\pi + (\sqrt{3}/2) + 1\} \times R^2]$, where n is the number
of nodes to make the connection and R is the radius of the maximum radio-range cover-
age of each node of the ad-hoc network. When any node sends a packet within its radio
range, the other nodes within its radio range can receive the packets. *Line placement*
means that a node can be connected to any node in a line via intermediate nodes. *Least
placement* means that a node can reach another node with just one connection to its
neighbor (Figure 1.3).

At first, we describe the estimation of NH by the source for different placements of the
nodes in the network. The source estimates NH on the basis of the density of the nodes
in the network, and NH is the highest when node density is the lowest and vice-versa.
NH is thus proportional to $1/\mu$, where μ is the density of the nodes.

For *line placement*, NH = D/R, where R is the radius of the maximum radio-range
coverage of each node of the ad-hoc network, D is the distance from the source to the
destination, $D = \sqrt{(X_T - X_S)^2 + (Y_T - Y_S)^2}$, where (X_S, Y_S) and (X_T, Y_T) are the source's
and destination's positions, respectively. In this placement, NH is the minimum number
of hops, from the source to the destination, estimated by the source.

For μ_{qaud} it is assumed that NH = $f(L,B)/R$, where f is a linear function in L and B,
where length L is the horizontal distance from the source to destination and breadth B
is the vertical distance from the source to destination.

For *least placement*, it is assumed that NH = $(k \times g(C))/R$, where k is a constant and
a function of L/R or B/R; and g is an exponential function in circumference C of the area
of the network. In this placement, NH is the maximum number of hops, from the source
to the destination.

1.3.2.2 Overview of the Protocol

The protocol is described in detail with respect to the functionalities of the nodes.

Source: The source sends a request to the PS for the position information of the destination when it wants to communicate with the destination. AODPR is thus an on-
demand protocol. The source generates its own Temp ID, RRQSeqNo, and estimates NH
and the maximum number of hops.

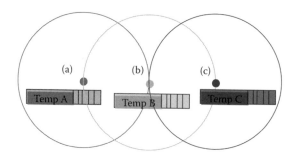

FIGURE 1.3   Least-placement-connected network.

After receiving the destination's position, the source estimates NH and assigns this NH to Temp NH. It then source sends an *RRQ* packet within its radio range and waits to receive a *response*, which is either *RRP* or *Fail* during time 2 × TTL, where TTL denotes *time to leave* and is estimated by the source from *TTL = (traveling time for one hop) × (number of hop).*

- If the source receives RRP, by decrypting RRPMsg of RRP, it tries to find a match with *Auth*. If a match is found, it stores the corresponding RRQSeqNo, NH, receiver's Temp ID, and status (i.e., "yes") in its routing table. It then sends data encrypted by the destination's PK. Lastly, sender sends *Token* to the destination so that destination can inform the PS of this communication.
- If it receives a *Fail* packet, it stores the corresponding RRQSeqNo, NH, and status ("no") to its routing table, and again tries with a new estimated NH.
- If it does not receive any *response* and TTL is exceeded, it stores RRQSeqNo, NH, and status ("no") in its routing table, and again tries with a new estimated NH.

As a result of this procedure, if the source fails to find the destination with an estimated NH, it tries with the next estimated NH until it finds the route. In this way, it can try with the minimum to the maximum estimated NH. Moreover, the maximum number of hops can be varied for different placements.

Intermediate Nodes or Forwarders: If a node receives a packet RRQ, but it is not the destination, it is a *forwarder* and becomes a new sender. Forwarder *F* generates its own Temp ID and calculates distance $D_r(F)$ between *F* and its destination *T* by $D_r(F) = \sqrt{(X_T - X_F)^2 + (Y_T - Y_F)^2}$ from the forwarder's position (X_F, Y_F) and destination's position (X_T, Y_T). F then updates Temp NH by Temp NH = Temp NH − 1. It compares this updated Temp NH with $D_r(F)/R$ and makes the following decision, as shown schematically in Figure 1.4.

- If $D_r(F)/R \leq$ Updated Temp NH, forwarder F forwards the packet to its radio region and keeps the route information.
- If $D_r(F)/R >$ Updated Temp NH, forwarder F discards the packets.

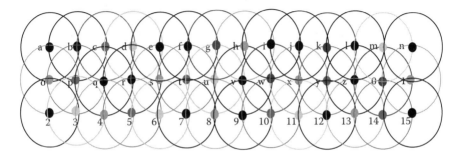

FIGURE 1.4   Packet forwarding or discarding in intermediate nodes.

After forwarding a packet, the forwarder waits to receive a *response* for time $2 \times TTL_1$, where TTL_1 is computed from TTL_1 = *(traveling time for one hop)* × *(updated number of hops)*.

- If the forwarder receives RRP, it just forwards it on the reverse path and keeps the route information.
- If the forwarder receives *Fail*, it also forwards it on the same reverse path and keeps the route information.
- If it does not receive any *response* and its waiting time exceeds TTL_1, it generates *Fail* and forwards it on the reverse path.

Destination: The destination checks EM of RRQ to confirm the destination of RRQ. Finally, it replies by RRP and keeps the route information.

1.3.2.3 Working Principle

Carrier sense multiple access with collision avoidance (CSMA/CA) [14] is used as the channel-access mechanism for control messages. A sender (a source or a forwarder) of an RRQ transmits the RRQ after sensing the channel and finding idle time for a distributed inter frame space (DIFS). When there is a collision, the sender retransmits the RRQ after a short inter frame space (SIFS). The same procedure is applicable for any node for the RRP as well as *Fail*.

> *Initial Procedure:* A source makes a signed position request to the PS, and receives required information C_K, destination's *position information, Auth, Token,* and PK of the destination from the PS.
>
> *Source's Working Procedure:* The source generates an RRQ and sends it to its radio region and waits to receive a *response* for time $2 \times TTL$.

If it receives the following *response:*

- If source receives RRP, then it compares *Auth* with RRPMsg by decrypting it.
  - If it matches, then source sends data in the path and at last sends the *Token*.
  - If it does not match, then source discards this *RRP* and estimates a new NH and again tries this procedure until it receives a valid *RRP*.
- If source receives a *Fail* packet within time $2 \times TTL$, it estimates a new NH and again tries this procedure until it receives an *RRP that does not exceed the maximum number of hops for that environment.*

If the source does not receive any response and the waiting time exceeds $2 \times TTL$, the source estimates a new NH and again tries the above procedure until it receives an RRP. The source repeats this procedure as long as the NH of its packet is smaller than the *maximum number of hops for that environment.*

Forwarder's or Destination's Working Procedure: On receiving an RRQ, a forwarder checks whether it is the destination or not.

If it is the destination, then it generates an RRP and sends this RRP on the reverse path.

If it is not the destination, then it forwards the RRQ and waits for time $2 \times TTL_1$.

- If the forwarder receives an RRP, it keeps the route information and sends it on the reverse path.
- If the forwarder receives *Fail*, then it keeps the route information and sends it on the reverse path.

If the waiting time for the forwarder exceeds $2 \times TTL_1$ time, then the forwarder generates *Fail* and sends it on the reverse path.

1.3.3 Anonymity Achievement and Security Analysis

When senders or forwarders forward any packets, they generate a large bit random number and use parts of that random bit sequence corresponding to the number of encrypted fields of the packet, that is, RRQ and RRP. The packets are described in the Appendix. They also specify all the fields with a specific bit number. They then pad the fields with random bits and encrypt these fields. When a packet reaches a node, the node first decrypts it, extracts the random bits from the fields, and pads these fields with its own random bits. All the fields of a packet are thus changed. As a result, when the packet moves from node to node, it appears new to the network. This procedure is applicable to all the encrypted fields of all the packets. Encryption and decryption are performed as necessary when a packet moves from node to node.

In an ad-hoc security routing protocol, the most expensive operation is the PK operation [15]. To guarantee the anonymity in the AODPR, every node generates its Temp ID, which is a hash computation, and a random bit corresponding to the fields of the packets, and it finally performs symmetric encryption/decryption of the fields. These computations are not more computationally complex than those of some other *ad-hoc* security routing protocol [16].

Identity Privacy: In AODPR, the identities Temp ID of the nodes are changing in each hop as a packet is forwarded. Location of destination is encrypted and padded with random bits. Also the Temp ID is changed in each session of communication. The Temp ID depends on not only the position of the node and the PK but also on time, so it is changeable within a hop range. So, AODPR ensures identity privacy.

Location Privacy: The general concept of the current attacks on the location privacy is to observe the route request and route response packets and to estimate the distance between the source and the destination from the traveling information added to the packet, that is, how many hops it travels. In contrast to existing anonymous ad-hoc routing protocols, there is no extra traveling information added to the packets in our scheme, and estimating the distance between the source and the destination is not possible in a straightforward way. No node knows anything about the location and identity of the other nodes, including the source, and it does not know from where a packet starts to travel in the network. Even though all legitimate nodes can determine the distance from themselves to the destination and also know the Temp ID of other nodes in the neighboring region, no one except the source can determine the

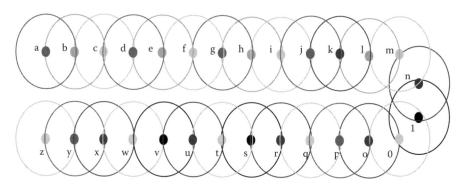

FIGURE 1.5   Route anonymity model.

distance from the source to the destination by using this information. Location privacy is thus achieved.

Route Anonymity: Current attacks on route anonymity are based on traffic analysis [17]. The general theory behind these kinds of attacks is to trace or to find the path in which the packets are moving. For this purpose, a malicious node mainly looks for unchangeable information, that is, common information in a packet, so that it can trace the movement of control packets. As a result, the adversaries can find or estimate the route from the source to the destination. In AODPR, all the control packets appear new (Figure 1.5) in the network when packets move from node to node. So, no one can trace the path of the route. Route anonymity is thus achieved. A detailed description is given in the Appendix.

DoS: Multiple adversaries cooperatively or one adversary with enough power can exhaust the resource of a specific target node. To this end, adversaries need to identify a node and set that specific node as a target. In AODPR, identity privacy is achieved as discussed above and *DoS* can be protected.

Wormhole Attacks: In wormhole attack, there could be a long distance for a packet to travel for finding the route from the source to the destination. In AODPR, the source and the forwarders wait for a *limited time*, TTL or TTL_1, for getting a response based on the estimated NH. If an attacker's response exceeds a *limited time*, it cannot be a forwarder within a routing path. If the attacker is a forwarder within a path limit and does not reply properly, this path no longer remains valid. The sender will try another path. A wormhole attack is therefore not effective in the case of AODPR.

Rushing Attack: Many existing on-demand routing protocols forward only the request that arrives first from each route discovery. In a rushing attack, the attacker exploits this property of the operation of route discovery and establishes a rushing attack. A more powerful rushing attacker may employ a wormhole to rush packets. By using the tunnel of a wormhole attack, the attacker can introduce a rushing attack. As shown above, AODPR can prevent a wormhole attack. It is thus also robust against a rushing attack.

1.3.4 Theoretical Analysis

In the case of AODPR, the source can determine the direct distance from him to any node connected in the network. Let the distance from the source to a node be D, so the number of hop given by $h = D/R$, where R is the radio-range coverage around a node. For route discovery, when a control packet travels from hop to hop, h is decremented by one. When a packet is forwarded to a specific node, the values of h will thus converge to a smaller value than its previous value. Let t be the time a packet needs to travel h number of hops, within time $2 \times t$, the source will receive a response. If the source does not receive any response, it will estimate new hop h_1 and will wait for a corresponding traveling time $2 \times t_1$. Thus, by consecutive estimation of new hops, the source reaches to the goal, as long as there is at least one path to reach the goal. If the density of the network is more than the quadratic-placement density μ_{quad} (Figure 1.2), it can reach the goal directly. If there is a shield on the path, it is also informed to the source by sending a fail packet after a certain amount of time. The source therefore estimates a new hop number by increasing its value more than in the previous attempts. If the source fails again, it will try as previously with a new estimate. If there is at least one path from the source to a node, then it can be found out, and successful communication is accomplished.

If the nodes in the network are at least-connected as shown in Figure 1.3, the maximum hop count to reach the goal is $n - 1$, where n denotes the number of nodes in a network. Let us consider a path from node **a** to **z**, as an example path. At first **a** will calculate $D_r(a)$ from node **a** to **z** and also estimates NH, so that the packet travels according to the protocol for this NH. If the relation $D_r(F)/R >$ Updated Temp NH holds for a node in the path, that node discards the packet. After that, node **a** sends the packet with a new estimated NH, which is a value greater than the old NH. Either with current estimated NH or a new estimated NH on the consecutive estimation of NH, the relation $D_r(F)/R >$ Updated Temp NH does not hold for that node anymore and the node finally forwards the packet. As long as the NH of a packet from **a** is smaller than the maximum hop count, this procedure will continue. By taking an appropriate value for the maximum hop count, the packet can reach from node **a** to node **z**. The simulation results of least-connected nodes in a network are given in the next section with a different estimation of NH.

1.3.5 Simulation Result

The reach ability in a network with least placement was simulated by varying the number of nodes, as shown in Figure 1.6, under a C++ programming environment. The graph shows the number of trials with respect to the number of nodes, in different estimation. For all the estimation methods, the source at first initializes NH = D/R. With this initial value, the source tries to reach the destination. If the source fails, it estimates a new NH value and tries to reach to the goal with this value. Each time the source tries to reach the goal, the trial number is counted. For estimating the NH value, we experimented with seven estimation functions. For all the estimation functions, the *estimation value* is initialized by NH = D/R. These functions are mainly defined in two ways such as (i) linear and (ii) exponential, which are described as follows.

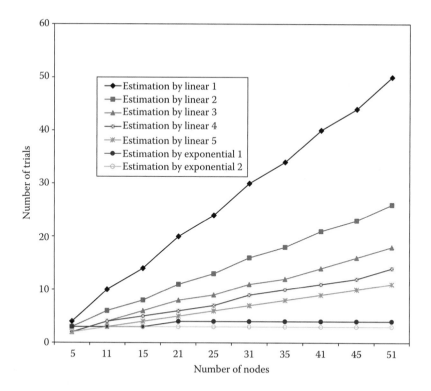

FIGURE 1.6 Number of trials for different estimation methods to find a route for different numbers of nodes in a least-placement-connected network.

Estimation by linear I ($I = 1–5$): After initializing the estimation value, it is incremented by I, so *estimation value = estimation value + I*. Detailed results for various I ($I = 1–5$) are shown in Figure 1.6.

Estimation by exponential I ($I = 1, 2$): After initializing the estimation value, it is incremented as a power, so *estimation value = (estimation value)$^{I+1}$*. When $I = 1$, the source tries four times for 21 numbers of nodes to reach the goal and for 51 numbers of nodes, it tries four times, but the trial value for 5–15 numbers of nodes differs from the previous value and it is 3. When $I = 2$, the source tries three times to reach the goal for 21 numbers of nodes, and for 51 numbers of nodes, it also tries three times and it remains constant from any number of nodes from 5 to 51. Exponential 2 is thus the best estimation for a least-placement-connected network.

1.4 User-Controllable Security and Privacy System for Pervasive Computing

Jason et al. proposed an user-controllable security and privacy system in [18], where they developed and evaluated three different applications, that are (1) a contextual instant messenger, (2) a people finder application, and (3) a phone-based application for access control. In this work, the authors describe their work with respect to three pervasive-computing scenarios and then drew out themes for the applications. The applications are as follows:

I. *Contextual Instant Messaging:* Users can inquire about each other's context (e.g., interruptability, location, and current task) through an instant messaging service.

II. *People Finder Application:* Users are equipped with location-aware smart phones. They interact with their devices to inquire about the locations of others subject to privacy policies.

III. *Access Control to Resources:* Smart phones are used to access both physical and digital resources. Users can use their smart phones to create and manage their *security policies*, and to give others credentials to access different resources.

1.4.1 Contextual Instant Messaging

Privacy controls and feedback mechanisms for *imbuddy411* have been iteratively designed, which is a contextual Instant Messenger (IM) service that lets any AOL Instant Messenger (AIM) users query for three types of information: interruptibility, location, and current task (abstractly represented as the name of the current window being viewed). Currently, AIM users can only query information of AIM users who are running client software, which collects and reports their contextual information.

To configure the contextual IM privacy settings, a group-based approach is used which is based on the works of Patil and Lai [19]. Users can modify their privacy control setting via a web browser. All buddies are first classified under a "default" privacy group that denies all disclosures. Users can create as many groups as they want and move buddies from the default group to any other group. Other AIM users who request information from *imbuddy411*, but are not part of the user's buddy list are dynamically added to the default group.

Three feedback mechanisms are developed: (a) a notification letting users know when their information is being seen, (b) a grounding and social translucency mechanism that facilitates conversation by letting users know what others know about them, and (c) a history letting users know what information has been disclosed to others.

Imbuddy411 is implemented as an AIM robot that could answer queries, such as "HowBusyIs alice" and "WhereIs Bob," and a Trillian plug-in that can sense contextual information such as interruptibility (using the SUBTLE toolkit [20]), location (using PlaceLab [21]), and current task. To introduce *imbuddy411* to the participants' buddies, a short blurb was included in each participant's profile. Trillian plug-in also advertised the *imbuddy411* service whenever a conversation starts between a user and their buddies.

To evaluate their research, a two-week study was conducted with 10 IM users. There were 193 queries not counting users querying themselves, including 54 interruptibility requests, 77 location requests, and 62 active window requests. Also, 63 queries were hits to the database (i.e., when users were not online). There were 46 distinct users who queried *imbuddy411* and 9 of those were repeat users and all the participants agreed that the three information types being disclosed were all potentially sensitive interruptibility: 3.6, location: 4.1, active window: 4.9, all out of 5). Participants informed that they were comfortable with their privacy settings (4.1/5). The result is particularly interesting, since as the part of the experiment. However, most of the participants' settings were set up not to reveal anything by default, and so they were unconcerned and did not mention this issue at a debriefing at the end of the study.

1.4.2 People Finder Application

The emergence of cell-phone-based location-tracking opens the door to a number of new applications, including recommendations, navigation, safety, enterprise applications, and social applications. Experiments conducted with some of these applications in the context of MyCampus show that adoption of these services often depends on whether users feel they can adequately control when their location is shared [22]. To better understand the privacy preferences users have in the context of these applications, as well as what it takes to capture these preferences, a series of experiments were conducted involving a cell phone-based people finder that lets users inquire about the location of their friends, family members, and colleagues.

In the first set of experiments of the research, 19 participants were presented with situations simulating queries from others. The queries were customized to capture elements of their daily activities involving friends, family, and colleagues. Each participant was asked to specify rules indicating the conditions under which she would be willing to share her location information with others (e.g., "My colleagues can only see my location on weekdays and only between 8 a.m. and 6 p.m."). The experiments involved presenting each participant with a total of 30 individualized scenarios (45 scenarios for each of the last 4 participants). Each individualized scenario included asking the participant whether she felt comfortable disclosing her location, showing her what her current policies would do, and offering her a chance to refine her policies.

Experiments show that users often have fairly sophisticated privacy preferences, requiring over 5 minutes just to specify their initial rules and nearly 8 minutes if one adds time spent revising these rules as they get confronted with new situations. Several users ended up with eight or more rules by the end of the experiments. More surprisingly, despite the time and effort spent specifying and refining their policies, participants were generally unable to achieve high levels of accuracy. Rules specified at the beginning of the experiments only captured their policies 59% of the time. When given a chance to revise their rules over time, that percentage only went up to 65%. Even when using the rules that users ended up with at the end of the experiments and re-running these rules on all 30 (or 45) scenarios, decisions were only correct 70% of the time.

The results with case-based reasoning (CBR) suggest that it is possible to train a system to learn a user's policies that can be more accurate than those specified by users— 82% accuracy using CBR. While additional experiments are required to validate statistical significance, these preliminary findings suggest that requiring users to fully specify their policies may be unrealistic. Instead, learning as well as dialog and explanation technologies seem to have the potential of offering solutions that better capture user policies while also reducing user burden.

1.4.3 Access Control to Rooms in an Office Building

A distributed, smartphone-based access-control system called *Grey* is deployed in a building on Carnegie Mellon University campus [23,24]. *Grey* can be used to control access to physical resources such as office doors, as well as electronic resources such as computer accounts or electronic files. *Grey*-enabled resources allow access when an individual's smartphone presents a proof that access is permitted. Proofs are assembled from a set of credentials that express authority. The credentials are created and managed by end-users on their *Grey* phones. Instead of relying on a central access-control list, in *Grey* end-users are empowered to create flexible access-control policies for the resources they manage.

Grey users can delegate their authority *proactively* by manually creating credentials that let a user or a group of users access a specified resource during a specified time period. *Grey* users can also create credentials *reactively*, when another user asks for access. In this case, the user who may have the needed credentials is prompted to help the user who is trying to gain access. If she decides to help, *Grey* will forward the relevant credentials from her phone to the user trying to gain access or, if such credentials do not yet exist, intelligently prompt her to first create such credentials, for example, by adding the requestor to a group that already has access to the resource. Over three dozen doors were outfitted in the building with *Grey*-enabled Bluetooth door locks and given smartphones with *Grey* software to 19 users. *Grey* is also used by nine members of the *Grey* project team. *Grey* usage is monitored by collecting log files from phones and doors and by interviewing Grey users every four to eight weeks over a period of several months.

Office building includes a shared workspace with open cubicles, as well as conference rooms, labs, storage closets, and offices. Locked perimeter doors secure the entire workspace in the evening and on weekends. Conference rooms, labs, storage closets, and offices can be individually locked. All *Grey* users were given credentials to unlock the perimeter doors, and users with offices were given credentials to unlock their own office doors. Some *Grey* users were also given additional credentials, for example, to unlock a lab or a storage closet. A user accesses a resource (e.g., a door or a computer login) by selecting its name from the phone's menu, after which the phone and the resource communicate via Bluetooth. The resource grants access (e.g., the door unlocks) when it has verified the credentials and proof submitted by the phone. If a user does not have credentials to access the resource, her phone prompts her to ask another *Grey* user to delegate the necessary authority.

Among the following lessons of the initial deployment of *Grey*, many of which may be broadly applicable to other mobile-device applications and access-control technologies.

- A variety of obstacles were found to acceptance of *Grey*, including user perception that *Grey* was slow (even when it was not) and system failures that caused users to get locked out. While security usually focuses on keeping unauthorized users out, the users were more concerned with how easy it was for them to get in, and in interviews never mentioned security concerns.
- It was being hopped to observe frequent delegation, but since *Grey* relies on network effects, the small number of users and resources limited opportunities for

delegation was being investigated in better ways to bootstrap so that *Grey* will be more useful, even for a small population.

- One of the objectives of this trial deployment was to study the types of access-control policies users would create when no longer constrained by the limitations imposed by difficult-to-obtain physical keys. It is observed that the users creating policies that did not mirror the policies they had with physical keys, and it is found that the low overhead for creating and changing policies with *Grey* encourages policy change and the creation of policies that better fit the users' needs.

- Finally, it is surprised for some of the unanticipated uses that the users made of the *Grey* system. For example, some of the users routinely use *Grey* to unlock doors without having to get out of their chairs. It would not have probably discovered without a field study.

1.5 Conclusion

In this chapter, we have discussed the state of the art of privacy and security for pervasive communication. To determine the security and privacy for pervasive communication, related parameters are determined. Finally, two different existing solutions are discussed targeting different pervasive environments.

References

1. SPPC: Workshop on Security and Privacy in Pervasive Computing, 2004.
2. http://www.atp.nist.gov/iteo/pervasive.htm.
3. http://www.computer.org/portal/web/csdl/doi/10.1109/MPRV.2007.86.
4. http://portal.acm.org/citation.cfm?id=1335023.
5. Rahman, Sk. Md. M., Inomata, A., Mambo, M., and Okamoto, E., Anonymous On-Demand Position-based Routing in Mobile *Ad-hoc* Networks, *IPSJ (Information Processing Society of Japan) Journal*, 2006; 47(8): 2396–2408.
6. Hu, Y. C., Perrig, A., Johnson, D. B., Packet leashes, "A defense against wormhole attacks in wireless *ad hoc* networks," in *Proceedings of the 22nd Annual Joint Conference of the IEEE Computer and Communications Societies (INFOCOM 2003)*, San Francisco, March 30–April 3, 2003.
7. Rahman, Sk. Md. M., Mambo, M., Inomata, A., and Okamoto, E., An anonymous on-demand position-based routing in mobile *ad hoc* networks, in *Proceedings of The 2006 Symposium on Applications & the Internet (SAINT-2006)*, Mesa/Phoenix, AZ, IEEE Computer Society Order Number P2508, Library of Congress Number 2005937742, ISBN 0-7695-2508-3, pp. 300–306, January 23–27, 2006.
8. Wu, X, DISPOSER: distributed secure position service in mobile ad hoc networks. Technical Report CSD TR #04-027, Dept. Computer Sciences, 2004.
9. Xue, Y., Li, B., and Nahrstedt, K., A scalable location management scheme in mobile *ad-hoc* networks, in *Proceedings of the 26th IEEE Annual Conference on Local Computer Networks (LCN 2001)*, Tampa, FL, November 2001, pp. 102–111.

10. Wu, X., VDPS: virtual home region based distributed position service in mobile ad hoc networks, in *Proceedings of ICDCS*, Columbus, OH, 2005.

11. Wu, X. and Bhargava B., AO2P: ad hoc on-demand position-based private routing protocol. *IEEE Trans Mobile Comput* 2005; 4(4): 335–348.

12. R. L. Rivest: The MD5 Message Digest Algorithm, Internet RFC 1321, April 1992.

13. NIST: FIPS 180-1, Secure hash standard, US Department of Commerce, Washington, D.C., April, 1995.

14. Technical Report on the IEEE 802.11 Protocol.

15. Zhang, Y., Security in mobile *Ad-hoc* networks. *Ad hoc Networks Technologies and Protocols*, edited by Mohapatra, P., and Krishnamurthy, S., Springer, ISBN 0-387-22689-3, Springer Science+Business Media, Inc, Boston, pp. 249–268, 2005.

16. Carter, S. and Yasinasc, A., Secure position aided ad hoc routing protocol, in *Proceedings of the IASTED International Conference on Communications and Computer Networks (CCN02)*, November 4–6, 2002, 329–334.

17. Raymond, J-F., Traffic analysis: protocols, attacks, design issues and open problems, in *Proceedings of PET 01*, Vol. 2009, LNCS, Springer, 2001, pp. 10–29.

18. Cornwell, J., Fette, I., Hsieh, G., Prabaker, M., Rao, J., Tang, K., Vaniea, K. et al., User-controllable security and privacy for pervasive computing, in *8th IEEE Workshop on Mobile Computing Systems and Applications*, Tucson, AZ, February 26–27, 2007.

19. Patil, S., and Lai, J., Who gets to know what when: configuring privacy permissions in an awareness application, in *Proceedings of CHI 2005*, pp. 101–110, Portland, OR, April 02–07, 2005.

20. Fogarty, J. and Hudson, S. E. Toolkit support for developing and deploying sensor-based statistical models of human situations, in *Proceedings of CHI 2007*, San Jose, CA, April 28–May 3, 2007.

21. LaMarca, A. et al., Place lab: device positioning using radio beacons in the wild, in *Proceedings of Pervasive 2005*, Munich, Germany, May 8–13, 2005.

22. Sadeh, N., Gandon, F., and Kwon, O. B., Ambient intelligence: the MyCampus experience. *Ambient Intelligence, Wireless Network and Ubiquitous Computing*, edited by Vasilakos T. and Pedrycz W., ArTech House, Inc, Norwood, MA, 2006.

23. Bauer, L., Cranor, L. F., Reiter, M. K., and Vaniea, K., Lessons learned from the deployment of a smartphone-based access-control system, Technical Report CMUCyLab-06-016, CyLab, Carnegie Mellon University, October 2006.

24. Bauer, L., Garriss, S., McCune, J. M., Reiter, M. K., Rouse, J., and Rutenbar, P., Device-enabled authorization in the Grey system, in *Proceedings of the 8th Information Security Conference*, Singapore, September 2005.

2

Challenges in Testing Context-Aware Applications

Rana Ejaz Ahmed
*American University
of Sharjah*

2.1 Introduction

Recent advances in sensors, wireless, and mobile communications technologies and ever-growing popularity of mobile devices have given a new dimension to pervasive computing. The mobile devices are now capable of sensing the surrounding environment in order to offer users a wide selection of services. Context-aware computing is one of the important enabling technologies for pervasive computing. Computing entities in a software application are *context-aware* if they can *sense* and *adapt* their behavior in response to changes in their surrounding physical and logical environment attributes (also known as *contexts*). Context-awareness allows computing entities to intelligently choose resources and provide customized services to the end-users. For example, a context-aware application running on a mobile phone *senses* that it is the meeting time for the user, the user has entered the meeting room, and the meeting has started. The application can then conclude that the user is busy in the meeting and it rejects all incoming calls while the meeting is in progress. Context-awareness is increasingly featured in several application domains such as e-commerce, e-learning, e-healthcare, and so on. Some common factors related to categories of context are: time, location, identity, surrounding infrastructure (including network, IT), and activity of the user. Location-based (LB) context aware systems that deal with the location information of the mobile devices are widespread and, perhaps, well-studied systems. Other context-aware applications such as smart homes, context-aware healthcare systems, and smart sensor networks are also becoming more common now.

Context sensing is typically accomplished via some sort of *sensor*. Earlier, these sensors were usually infrared (IR) or ultrasonic badges for indoor location detection, or global positioning system (GPS) receivers and cellular phones for outdoor location-aware services. The Federal Communications Commission (FCC) in the United States has mandated that the telecom operators must be able to locate the position of a mobile phone making an emergency (911) call within an accuracy of 125 m. This mandate requires that each mobile phone in the United States should have the ability to be location-aware. The technologies in mobile computing are now extended to address more types of context information rather than just the location. Today, the concept of context may denote a wide scope of information ranging from physical environments to computing resources and social situations. In principle, context may refer to any environmental attribute that the applications are aware of and behave in response to the attribute accordingly. Non-physical sensors capture contexts about user profiles and activities and collect more abstract contexts obtained from the aggregation of lower level or raw context data [1].

Traditionally, a non-context-aware application only uses explicit user inputs to provide the output (or services), while a context-aware application uses additional sensed context information gathered from the user's physical environment or derived from the user's logical environment (e.g., IT infrastructure). A general structure of a context-aware application is shown in Figure 2.1.

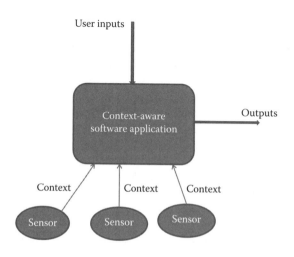

FIGURE 2.1   General structure of a context-aware software application.

Several context-aware systems have recently been developed for pervasive computing. Example systems include *Cabot, Gaia, Carisma, RCSM* [1], and so on. Loke [2] and Poslad [3] provide good introduction to several types of context-aware systems.

Context-aware systems can be designed and implemented in several ways. The approach depends on the requirements and conditions, such as the locations of sensors, number of users, resources on mobile devices, and so on. The method of context-data acquisition is very crucial because it predefines the architectural style of the system. Developing context-aware applications requires an architecture that can evolve in the presence of a large number of platforms, increasing number of sensors, and frequent network and drivers updates. Such requirements usually lead to the use of a layered architecture. Some of these layers are often aggregated into a *context-aware middleware*, which processes the context values on behalf of the application and triggers adaptive behavior in the application. A context-aware middleware collects the context information from the sensing devices and applications, processes them, and delivers to the application. With the aid of context-aware middleware, context-aware applications only need to subscribe the contexts of their interests from the middleware, and adapt their behaviors based on these contexts and the triggering rules. Most existing systems incorporate a middleware-based software architecture [4–6].

Figure 2.2 shows a layered architecture of context-aware application that incorporates a context-aware middleware to support the processing of context data. The middleware typically consists of two parts: *Context Manager* and *Adaptation Manager*. The *Context Manager* collects and maintains low-level context information, whereas the rule-based *Adaptation Manager* queries and processes the current context values on behalf of the application and triggers the adaptive behavior by the application. Adaptation rules define, in parts, the application behavior and are typically specified in terms of predicates over the variables representing the context readings [5].

2.1.1 Overview of Software Testing for Context-Aware Applications

Software testing is the well-known approach for assuring high-quality software products, with the aim of detecting as many faults as possible before shipping the product to the

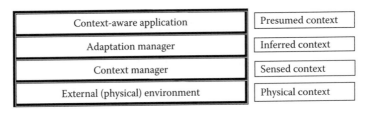

FIGURE 2.2 Architecture of context-aware software applications The corresponding context category is shown on the right. (Adapted from Sama, M. et al., Multi-layer faults in the architecture of mobile, context-aware adaptive applications: a position paper, in *Proceeding of 1st International Workshop on Software Architectures and Mobility (SAM'08)*, May 10, Leipzig, Germany, 2008.)

customers. The software testing process involves generating test cases that are applied to the software implementation under test to obtain the test results, and evaluating the test results against the known specifications. A discrepancy between the test result and the corresponding specification points to a software fault. Test case selection (also known as, *test adequacy problem*) and test result evaluation (also known as *test oracle problem*) are the two most important problems in software testing. *Test adequacy* refers to methods of how to select test cases from a very large input domain for the software unit under test. The evaluation of test results is done by comparing the test execution results with the related specifications or supposedly correct version of the implementation when the same test case is applied. The testing techniques that use some sort of coverage criteria are referred to as *coverage-based testing*. A *coverage criterion* serves both as a stopping rule of the test case selection process and as measurement for the effectiveness of a test suite. This coverage is calculated by applying all the selected test cases and computing the proportion of the test requirements that has been exercised. The test case selection process stops when the coverage of the test requirement meets some pre-defined satisfactory level. Coverage-based testing can be further classified into *structural testing* and *functional testing*. *Structural testing* (also known as *white-box testing* or *code-based testing*) refers to the class of criteria on the coverage of different types of program elements when the implementation (i.e., the source code) is available to the testers. The basic structural criteria include statement coverage, branch coverage, and path coverage. *Functional testing* (also known as *black-box testing* or *specification-based testing*) refers to the testing approaches where the functional specifications are used to guide the test case selection. Different functional testing strategies depending upon the nature of the specification are used.

Software testing for context-aware applications suffers from both the test adequacy problem and the test oracle problem in pervasive environment. We need to test context-aware applications for several types of mobile devices, different platforms, over different types of carrier networks under different contexts changes. Unlike conventional programming paradigms with standardized formats and features, context-aware applications do not have any uniform architecture model so far. There are several challenges to the testing of context-aware applications. Context-aware applications may not only evolve with the changes in features, but may also evolve with environmental changes such as addition, removal, and modification of contexts. When the applications evolve, a major challenge is how to efficiently perform testing (especially, regression testing) on the new evolved system. Furthermore, unlike traditional software applications, software failures in context-aware applications may emerge not only due to program faults but also from unreliable and/or inconsistent context sources. These factors make the debugging process more difficult [1].

This chapter presents the challenges in testing context-aware applications, and surveys the recently reported testing techniques for such applications. A discussion is also be made on the available tools and support for testing context-aware applications.

2.2 Modeling Context-Aware Systems

Models for context-aware systems define and describe the contexts, and how contexts are created and used for adaptation in an application. The models basically define how to represent contexts in a computation form [3]. Several ways to classify contexts have been proposed in the literature. One popular view refers to *external* and *internal* types of context which are similar to the *physical* and *user* contexts, respectively. A classical classification, proposed by Schilit et al. [7], divides contexts into three categories: *where you are* (location context), *who you are with* (social context), and *what computing and networking resources are nearby* (IT Infrastructure context). Some researcher proposed the ideas of *passive* and *active* contexts. A *passive* (or *static*) context describes the aspects of pervasive system, that is, invariant (with respect to time and space, for example), such as a person's identity and date of birth. An *active* (or *dynamic*) context refers to a user or environment, and the context can be highly variable over space and time (e.g., temperature).

New contexts can be created in real-time using, for example, temperature sensors. Lower level raw contexts output from sensors may often need processing into high-level contexts that are relevant to the users and applications. The raw sensed context values may need to be scaled and/or changed into different value ranges, or formats. Some contexts, such as location and time, can act as sources of contextual information from which the other contexts, known as *derived contexts* or *context reuse*, can be derived. In many scenarios, combining several individual context values may generate a more accurate understanding of the current situation than taking into account any individual context. Similarly, a low-level context can be derived from a higher level one. For example, a GPS position can be determined from annotated positions, such as street names and landmarks, and so on. Context-aware systems need to support all the book-keeping and maintenance activities, including the creation, modification, deletion, and interlinking heterogeneous contexts [3].

Context representation can be done in many ways. These are classified by the scheme of data structures which are used to exchange contextual information in the system [1,3,8]. The *Key-value model* is the most simple data structure for modeling contextual information. The data structure is in the form of an ordered pair (c, v), where c is the context variable representing the environment and v is the context value captured from the environment. For example, the pair (user, "John") represents the context that the user is currently "John." Other example could be for the context variable temperature, and the value taken by the variable could be, for example, from the set {"Very Hot," "Hot," "Moderate," "Cold," "Very Cold"}. Several early location-aware applications represented the location using key-value model. Problems in using key-value model include the usage of exact matches, lack of expressive structuring, and the lack of efficient context retrieval algorithms.

The *Markup scheme* model uses a hierarchical data structures (e.g., XML) consisting of user-defined markup tags with attributes that can be arbitrarily nested. It enables hierarchical structuring of context information with persistent and serialized representation as well as lightweight storage, and it favors the usage in context-aware middleware systems. Some example systems that use markup scheme models include *CSCP*, *CARMEN*, *MobiPADS*, *Solar*, *Cabot* [1].

The *Graphical model* uses graph data structures and richer data types, such as Unified Modeling Language (UML) and Entity Relationship (ER) diagrams. This model is more expressive than key-value and hierarchical models. Other context representation models include *Object-oriented* (*OO*), *Logic-based*, and *Strong Ontology* models.

Some of the key challenges in modeling context are summarized as follows [3,8]:

1. User (internal) contexts may be incorrectly, incompletely, imprecisely determined, or predicted. This may be due to the fact that the user may have provided faulty information when explicitly asked, or the user contexts are modeled from too little data over too small time period.
2. Environmental (external) contexts may also be incorrectly, incompletely, imprecisely determined, or predicted. This could be due to delays that can occur in exchanging dynamic information, or path between external context producer and the consumer is disconnected temporarily or permanently.
3. Some contexts may exhibit a range of spatial and/or temporal characteristics; that is, the information generated may change quickly over time and distance.
4. Some contexts may be using different format and may have alternative representations.
5. Some contexts may be distributed and composed of multiple parts that are highly interrelated. They may be related by rules that make a context dependent on other context. These composite contexts may need to be partially validated as all their parts cannot be always accessed.
6. In general, context-awareness generates a huge volume of data due to large state-space of environment to be studied, and many sensors are used.
7. Context use can reduce the security and privacy of users.
8. The awareness, availability, and change of context signals may overload users and distract them from performing their on-going interaction with the application.

A robust context-aware application must ensure that there are reasonable solutions to above-mentioned problems. For example, in order to solve a huge-volume data-generation problem (as mentioned in item 6), we can filter raw context information before storing data. Moreover, some data mining schemes can be used to analyze and filter the raw data.

2.3 Testing Challenges

Context-aware pervasive systems raise several software testing challenges. The middleware architecture of context-aware application, as shown in Figure 2.2, gives rise to four different views of the context: *physical, sensed, inferred,* and *presumed* context [5]. It is quite possible that all four views may differ from each other at any given point in time during the execution of application. It is possible that faults may exist in either context sensors, context manager, the adaptation rule, or in the application logic. However, complex faults may arise due to inconsistencies among various views of the context.

Testing context-aware applications, especially testing for adaptation faults in the applications, are quite challenging due to factors mentioned in the previous section about context modeling and the followings [5,9]:

1. There are several representations of contexts along with different formats used.
2. The context variables are updated asynchronously at different rates by the middleware, causing transient inconsistencies between external physical context value and

its internal representation within the application. A context variable can contain static information about local configuration of the mobile device (such as language preference) or dynamic information (such as GPS latitude and longitude). The dynamic context variables need to be refreshed (periodically or asynchronously) at different rates, which leads to synchronization problems when the adaptation manager tries to relate several context variables for rule triggering.

3. The user may configure its own behavior and, hence, some of the context variables. This may lead to context inconsistencies and/or failure due to buggy user-defined configuration.

4. The space of rules for adaptive actions become very complex to analyze due to several shared context variables, concurrent triggering of rules, and so on. Multiple rule predicates may be satisfied simultaneously and some predicates may be satisfied transiently.

5. It is very difficult to define precise test oracle as the execution differs under various vectors of context input.

6. The real environment to run application may not be available. There may not be, for example, enough sensors, or different types of networks available at the time of testing.

7. As context may include sensitive information about people and their activities, some applications give opportunity to the user to protect their security and privacy. For example, the *Context Toolkit* [6] introduces the concept of context ownership. Under such case, it becomes even more difficult to test such context-aware application.

2.3.1 Example Application

An example application, adapted from [5,9], is now presented to outline some of the challenges mentioned above. The example context-aware application, *PhoneAdapter*, adapts a mobile phone's *profile* according to context variable information. Phone profiles are a set of parameters that determine the behaviors of a phone, such as ring tone volume, screen display intensity, and vibration. The application uses a set of adaptation rules to trigger automatic selection of a profile. The selected profile exists in the system until a more suitable one is chosen through triggering of some other rules. The rule predicates are expressed using context readings from Bluetooth and GPS sensors on the mobile phone, and the phone's internal clock and appointments calendar. Some of the *PhoneAdapter*'s profiles used are: *General, Home, Office, Meeting*. The *General* profile is applied by default when the phone sensors are unable to detect any activity related to one of the other profiles. The *Home* profile increases the ring tone volume and removes the vibration; the *Office* profile mutes the ring tone and activates vibration; while the *Meeting* profile mutes the ring tone and disables the vibration. One can consider a scenario when the application uses GPS to infer that the user is at home, and it uses Bluetooth to discover the user's office laptop PC, from which it infers that the user is at work. In this scenario, the true physical context of the user is his home location, but the context manager senses both home and office locations. This leads the adaptation manager to infer simultaneously the existence of two different (and inconsistent) contexts. Another example of context inconsistency arises when the user's mobile device internal calendar and time indicate that the user is in a meeting in the office; while the GPS indicates that the user has not yet reached the office (perhaps, stuck in a traffic jam!). In this example, the internal (or user) and external (or physical) views of the context are different for the application. Such faults are very difficult to be detected by the conventional software testing approaches.

The timing of context updates may affect the triggering of rules. As context updates asynchronously, the internal view of context can become inconsistent temporarily. This can cause selection of rules to produce incorrect results. The faults can propagate multiple layers in the application architecture, leading to complex inconsistencies in the different views of the context present in the architecture.

2.3.2 Issues in Test Planning

The following issues need to be considered while planning testing for a context-aware application: *Functionality, Performance, Usability, Interoperability,* and *Security and Privacy*. Appropriate test cases to address those issues must be part of any test suite designed to test the application.

Functionality testing verifies whether the application meets the intended specifications and functional requirements. The functionality testing for context-aware applications becomes more difficult due to the above-mentioned itemized factors. Test cases should take into account several different network/device technologies, and operational environment along with the contexts to verify the functionality.

Performance testing determines quantitatively how different components of the context-aware application perform under various well-defined workloads. This type of testing verifies whether the application meets the well-established performance criteria. Performance is one of the critical elements for testing context-aware applications, and it is affected by several sources, including wireless network, sensors, and the middleware. One can also use volume and stress testing to find out the bottlenecks and the level of robustness offered by the application.

Usability testing checks whether the user interfaces (e.g., GUI) are easy to use, navigate, and understand. This type of testing is needed for context-aware applications as those applications mainly run on mobile devices that usually have poor and tedious user interfaces.

Interoperability testing checks whether two (software) entities following same standards and specifications can work together. More specifically, interoperability testing in a context-aware application checks whether the two distributed systems can exchange data and commands in real-time to provide services to the end-users. As there are several entities involved in providing services in a context-aware application, this type of testing verifies high degree of integration capabilities of services offered by different systems.

Security and Privacy testing verifies that the user's personal and sensitive information and the activities log are protected from the accesses from unauthorized entities.

2.4 Recent Trends in Testing Context-Aware Applications

The fundamental aspect for the validation of context-aware applications is that *changes in context can occur and affect the application behavior at any time during its execution.* Although this may happen with other types of inputs, it is particularly prevalent with contextual inputs since they are the continuously streaming drivers of the applications. Test engineers must identify not only *what* context values to provide, but also *when* the stream of variations in context values can impact the behavior of the application. This is an essential difference from the testing of conventional software systems, where the selection of input values can mostly be performed *a priori* [10].

Compared to the research work done in the design and development of context-aware pervasive applications, little work has been done on the testing aspects of such applications. Considering different settings underlying mobile devices, some work has been done on testing platforms and tools for testing pervasive applications [1].

Developing context sources remains a major challenge, as testing context-aware applications with physical context sources in a controllable and reproducible manner is quite difficult. In Broens and Haltesen [11], the *SimuContext* framework is presented where the simulated context sources emulate the real-life context sources.

Satoh [12] developed a software testing framework that can emulate the physical mobility of devices by logical mobility of applications designed to run on them. A mobile agent-based emulator was designed for mobile device and the emulator could perform an application-level emulation of its target devices.

In Delamaro et al. [13], proposed a method for the coverage testing of applications on target mobile devices and device emulators. Bo et al. [14] presented a tool that conducts black-box testing for mobile applications. Calegari et al. [15] proposed performance testing strategies for mobile applications running on *ad hoc* mobile networks.

The usability testing for context-aware services with simulated context data on the top of a game engine was proposed by Bylund and Espinoza [16]. They proposed a tool (called *QuakeSim*) that interactively simulates context information in real-time.

Regehr [17] proposed a technique for testing interrupt-driven software applications that are widely used in the implementation of embedded systems and wireless sensor networks. The technique uses the random criterion for test cases selection.

Sama et al. [9] proposed a new model of adaptive behavior, called an *Adaptation Finite-State Machine* (A-FSM) that enables the detection of faults caused by both erroneous adaptation logic and asynchronous updating of context information. They evaluated their approach on a set of synthetically generated context-aware adaptive applications and on a simple application in which the cell phone's configuration profile changes automatically as a result of changes in the user's location, speed, and surrounding environment.

In Wang et al. [10], an approach that improves the context-awareness of an existing test suite is presented. The technique first identifies key program points where context information can effectively affect the application's behavior. It generates potential variants for each existing test case that explore the execution of different context sequences, and it then attempts to dynamically direct the application execution toward the generated context sequences. The supporting infrastructure for their approach consists of the following components:

- *Context-Aware Program Point Identifier:* This component identifies program points where context changes may affect the application's behavior.
- *Context Driver Generator:* Once context-aware program points have been identified, one would like to explore the context scenarios that are likely to generate different program behaviors. This component forms potential context interleaving that may be of value to fulfill a context-coverage criterion.
- *Program Instrumentor:* This component incorporates a scheduler and context-aware program point identifier controllers into the application to enable direct context manipulation.
- *Context Manipulator:* This component attempts to expose the application to the enumerated context interleavings through the manipulation of the scheduler.

Lu et al. [18] proposed a set of three-test criteria using data flow testing. The control flow graph of the source code is first built and then the life-cycle of data variables (their definitions and the usages) is tracked. The approach leads to test cases that focus on improper use of data due to coding errors. Their approach uses these criteria to create test cases that detect faults in the context-aware interface that are otherwise difficult to be discovered through conventional testing methods.

Taranu and Tiemann [19] proposed a testing approach by applying and extending the classical approach of testing communication systems. In their approach, they isolated the core of a context-aware system, which is the control or algorithmic part for adapting the system to the current scenario and requested service. Their approach uses a context management system that takes internal and external information into account for local decisions. The isolated part (i.e., the decision algorithm or control part) is directly stimulated with context information via the context management. The appropriate context is directly generated, and the context represents the environment or situation of the *System under Test* (SUT). Their work is a unique effort where both foreground and background testers are explicitly considered. The testers represent one or more instances in the communication and contain an implementation of protocol. A foreground tester interacts with the SUT, that is, the generated traffic is also influenced by the SUT. A background tester is an implementation without feedback from the tested system, and it is used to generate the background load efficiently. The context or situation is generated by context generators, categorized as foreground and background generators. The situation generator is implemented in Java programming language and it generates different *situations* in radio network. The situations refer not only to traffic and load generation, but also to contain information generated out of location, interference, or mobility models.

Another issue relevant to testing context-aware application is the context inconsistency, which is becoming increasingly important in the presence of more and smarter sensors. Some researchers have recently proposed inconsistency detection in the applications where patterns identify conflicts among context inputs at run-time before the contexts are fed into an application [20–22]. The patterns are designed beforehand based on the understanding of relevant physical and mathematical laws under which the application is supposed to run. In Lu et al. [23], a framework is proposed where the middleware supports Context Inconsistency Resolution (CIR) services and a dataflow test criteria to test context-aware application is then designed. The criteria focus on the propagation of context variables in the applications, which are potentially affected by CIR services.

2.5 Summary and Conclusions

Context-aware applications running on mobile devices are growing at an enormous rate. While there are lots of new opportunities in developing new applications, testing such applications remains a tedious task. Testing context-aware applications is quite challenging due to the following main factors: there are considerable context variations is a short period over short distances; the context variables are updated asynchronously at different rates by different components of the middleware, causing transient inconsistencies between external physical context value and its internal representation within the application; and it is very difficult to define precise test oracle as the execution differs under various vectors of context input. Several tools and testing methodologies have been proposed, but their application domain is relatively limited. The field of context-aware application testing is still in its infancy state, and a large amount of research work is expected during the next few years.

References

1. Lu, H. A software testing framework for context-aware applications in pervasive computing, PhD Thesis, University of Hong Kong, 2009.
2. Loke, S. *Context-aware Pervasive Systems*, Boston, MA: Auber Publications, 2006.
3. Poslad, S. *Ubiquitous Computing: Smart Devices, Environments and Interactions*, Chichester, West Sussex, UK: John Wiley & Sons, 2009.
4. Ye, C., Cheung, S., Wei, J., Zhong, H., and Huang, T. A study of replaceability of context-aware middleware, in *Internetware'09*, October 17–18, Beijing, China, 2009.
5. Sama, M., Rosenblum, D., Wang, Z., and Elbaum, S. Multi-layer faults in the architecture of mobile, context-aware adaptive applications: a position paper, in *Proceeding of 1st International Workshop on Software Architectures and Mobility (SAM'08)*, May 10, Leipzig, Germany, 2008.
6. Baldauf, M., Dustdar, S., and Rosenberg, F. A survey on context-aware systems. *Int J Ad Hoc Ubiquitous Comput*, 2007; 2(4): 263–277.
7. Schilit, B., Adams, N., and Want, R. Context-aware computing applications, in *Proceedings of First International Workshop on Mobile Computing Systems and Applications*, Santa Cruz, CA, pp. 85–90, 1994.
8. Strang, T. and Linnhoff-Popien, C. A context modeling survey, in *1st International Workshop on Advanced Context Modeling, Reasoning and Management*, UbiCom 2004, Nottingham, UK, pp. 34–41.
9. Sama, M., Elbaum, S., Raimondi, F., Rosenblum, D., and Wang, Z. Context-aware adaptive applications: fault patterns and their automated identification. *IEEE Trans Software Eng*, 2010; 36: 644–661.

10. Wang, Z., Elbaum, S., and Rosenblum, D. Automated generation of context-aware tests, in *29th International Conference on Software Engineering*, 2007, Minneapolis, MN, pp. 406–415.

11. Broens, T. and Halteren, A. SimuContext: simply simulate context, in *Proceedings of International Conference on Autonomic and Autonomous Systems (ICAS'06)*, July 2006.

12. Satoh, I. A testing framework for mobile computing software. *IEEE Trans Softw Eng*, 2003; 29(12): 1112–1121.

13. Delamaro, M., Vincenzi, A., and Maldonado, J. A strategy to perform coverage testing of mobile applications, in *Proceedings of the 2006 International Workshop on Automation of Software Test (AST 2006)*, pp. 118–124, 2006.

14. Bo, J., Xiang, L., and Xiaopeng, G. MobileTest: a tool supporting automatic black box test for software on smart mobile devices, in *Proceedings of the 2nd International Workshop on Automation of Software Test (AST 2007)*, p. 8, 2007.

15. Calegari, R., Musolesi, M., Raimondi, F., and Mascolo, C. CTG: a connectivity trace generator for testing the performance of opportunistic mobile systems, in *Proceedings of the 6th European Software Engineering Conference (held jointly with 2007 ACM SIGSOFT International Symposium on Foundations of Software Engineering (ESEC/FSE-2007))*, pp. 415–424, 2007.

16. Bylund, M. and Espinoza, F. Testing and demonstrating context-aware services with Quake III Arena. *Commun ACM*, 2002; 45(1): 46–48.

17. Regehr, J. Random testing of interrupt-driven software, in *Proceedings of the 5th ACM International Conference on Embedded Software (EMSOFT 2005)*, pp. 290–298, 2005.

18. Lu, H., Chan, W., and Tse, T. Testing context-aware middleware-centric programs: a data flow approach and an RFID-based experimentation, in *14th ACM SIGSOFT International Symposium on Foundations of Software Engineering*, 2006, pp. 242–252.

19. Taranu, S. and Tiemann, J. General method for testing context aware application, in *MUCS2009*, June 2009, pp. 3–7.

20. Xu, C. and Cheung, S. Inconsistency detection and resolution for context-aware middleware support, in *Joint 10th European Software Engineering Conference and 13th ACM SIGSOFT Symposium on the Foundations of Software Engineering*, September 2005, pp. 336–345.

21. Xu, C., Cheung, S., and Chan, W. Incremental consistency checking for pervasive context, in *International Conference on Software Engineering*, May 2006, pp. 292–301.

22. Xu, C., Cheung, S., Chan, W., and Ye, C. Partial constraint checking for context consistency in pervasive computing. *ACM Trans Softw Eng Method*, 2010; 19(3): 9:1–9:61.

23. Lu, H., Chan, W., and Tse, T. Testing pervasive software in the presence of context inconsistency resolution services, in *International Conference on Software Engineering (ICSE'08)*, pp. 61–70, May 2008.

3

Medium Access Control Protocols for Wireless Sensor Networks in a Pervasive Computing Paradigm

Muhammad
K. Dhodhi
Ross Video Limited

Syed Ijlal Shah
Freescale Semiconductor

Marwan Fayed
University of Stirling

3.1 Introduction

In his seminal work [1], Mark Weiser introduced the vision of the twenty-first century's ubiquitous and pervasive computing. In it he wrote, "The most profound technologies are those that disappear. They weave themselves into the fabric of everyday life until they are indistinguishable from it." This invisible computing is accomplished by means of "embodied virtuality," the process of drawing computers into the physical world [2]. In context-aware pervasive computing, devices must be ever-present. They are critical enough to our activities; yet, they must be taken for granted and effectively disappear into the background.

The technological advances needed to create a pervasive computing environment include four broad areas: such as hardware devices, networking, middleware, and applications [3]. Mobility, constant availability, invisibility, adaptability, and privacy are salient characteristics of pervasive computing environments that are intended to free the user from the distractions of having to interact with computers and machines.

Recent advances in VLSI circuit and fabrication technology have produced small and sophisticated devices, such as micro-sensors, to provide the necessary hardware infrastructure for creating pervasive computing environments. Nanotechnology is further helping the proliferation of tiny devices in our daily lives and systems. Not only is the size of these devices shrinking, but also the processing capacity follows the trend of their larger counterparts and doubles every 2 years. This increased processing capacity is in turn enabling the tiny devices to run complex applications and protocols.

Wireless sensor networks (WSNs), whether they are employed in the home, office, industrial plants, healthcare, or the habitat, form one of the building blocks for constructing a truly pervasive environment. The sensors in a WSN, given the processing capabilities that can be embedded into them, will have the ability to act on their own given high-level guidance from their users. They will also be able to detect and adapt to changes in their environment.

WSNs are an emergent multi-disciplinary field that impacts all the layers of the OSI network model. WSNs are characterized mainly by limited availability of energy to power these devices, lossy or challenging environments, and loosely connected networks. Adding to these challenges, sensors have limited radio range and limited storage capability [4], although, with advancement of technology, storage and processing capacities are increasing. Nevertheless, these devices fall well short of the capabilities of their larger counterparts that are available in large wired, power grid connected devices such as radar, or closed circuit TV cameras. WSNs applications, both continuous monitoring and event-driven, are impacted by the resource-constrained nature of the sensor nodes.

Medium access control, or MAC, protocols play a crucial role in the dissemination and transfer of the collected data in an efficient manner. This is because it is the MAC protocol that specifies how nodes share the wireless channel, which in turn dictates the active/sleep modes of the radio transceiver in a sensor node.

In the sections to follow, we will describe major research issues and challenges as well as proposed solutions for the MAC layer in sensor networks.

3.2 MAC Protocols

Nodes in a sensor network collect data and communicate the observed data with one another using short haul radio links with a limited range and bandwidth. Due to limited energy supply, one of the major design goals in sensor networking is energy conservation. MAC protocols play a crucial role in the dissemination and transfer of data in an efficient manner. The MAC protocol is responsible for negotiating and obtaining access to the communication channel, over which a single node at a time may transmit. Effectively, it is the MAC that specifies how the wired or wireless channel is shared. Because of this, MAC protocols are tightly coupled to the active/sleep modes of the radio transceiver in sensor nodes. This coupling is important as it is the radio transceiver that dominates with respect to energy consumption. For this reason, it is advantageous to power down the transceiver whenever possible. This is often referred to as sleep-mode. On the other hand, the receiver of the sensor nodes must be turned-on (active mode) so that neighboring nodes can communicate with each other. As a result the desire to power-down a transceiver conflicts with the need to listen for transmissions. Therefore, nodes must coordinate or compete to establish and maintain active/sleep schedules with their neighbors in order to communicate messages.

Beside energy-efficiency, MAC layer protocols need to meet several other important design goals for their applicability in the pervasive computing paradigm. The additional design goals include mobility, adaptability to changes, network size, network topology, reliability, fault-tolerance, re-configurability, flexibility, and context-awareness. Unlike wired networks such as Ethernet, increased latency, throughput, and bandwidth utilization may be secondary design goals for the MAC protocol in the WSNs [5–9]. Finally, because sensor nodes in a WSN environment collaborate with each other to perform a common task, fairness among nodes is a design criterion that requires little attention [7,10].

3.2.1 Energy Conservation

There are a number of transceiver activities that may lead to excessive or unneeded energy consumption. These include collisions of transmissions in multiple sender environments, control packet overhead, adaptation to network changes when nodes join, leave, or move [6]. Each of these activities demand idle listening, where a node waits and listens for activity on the communication channel. Therefore, the main sources of energy wastage by radio transceiver are due to idle listening, collisions, overhearing, and control packet overhead [11]. A collision occurs when two nodes try to transmit packets at the same time or a transmitted packet is corrupted due to interference. The packet must be discarded and retransmitted. The retransmissions increase energy consumption. Overhearing occurs when a node receives packets that are destined to other nodes. Control packet overhead includes sync packet, RTS-CTS. Idle listening is waiting to receive possible traffic that is not sent. This is especially true in many sensor network applications. If nothing is sensed, the sensor node will be in idle state for most of the time. The study by Reason and Rabaey [12] shows that idle listening accounts for approximately 90% of the energy wasted in sensor networks. Duty-cycling is largely seen as the means by which idle listening periods may be reduced, and this is the basis of many sensor network MAC protocols. Among the earliest examples designed to address these challenges are fixed duty cycling in sensor MAC (S-MAC) [6,11], and wake-up slots in time-division multiple access-wakeup (TDMA-W) [13]), discussed later in the chapter.

3.2.2 Mobility and Adaptability to Network Changes

Most of the MAC protocols are designed for relatively static sensor nodes. Mobility and network changes are ignored. Mobility is a natural requirement in the devices used in pervasive environments. The architecture of the WSNs used in pervasive computing applications must handle the physical mobility of sensor nodes for better coverage. Adaptability to network topology changes, node join (add), and node failures (drop) must be handled gracefully.

3.2.3 Scalability

Pervasive computing environments will likely face a proliferation of users, applications, networked devices, and their interactions on a scale never experienced before [3]. Handling large-scale deployment of sensor nodes over a wide area is crucial.

3.2.4 Reliability/Fault-Tolerance

A number of WSN applications require reliable data delivery while delays can be tolerated. Wireless sensor nodes often operate in inhospitable environments are prone to failures. The ability of a sensor network to operate successfully in the wake of failures (i.e., fault tolerance) will make the goal of data reliability achievable.

3.2.5 Reconfigurability/Flexibility

Versatile and flexible protocols such as Berkeley MAC (B-MAC) [14], where by tuning different parameters can provide application-specific trade-offs. A reusable suite of tools that offer compatibility with a wide variety of sensor and system technologies is needed.

3.2.6 Context Awareness

Context-aware applications require MAC protocols that can dynamically adapt to context and should be able to translate context information into suitable MAC parameters. For example, context-aware body sensor networks [15] are critical for the development of pervasive healthcare monitoring applications.

3.2.7 Categories of MAC Protocols

Despite the differences between MAC protocols for WSNs, they can nevertheless be grouped into four different categories as shown in Figure 3.1. It must be noted that, historically, MAC protocols were grouped into two broad categories: contention-based protocols and schedule-based protocols. We have extended this categorization to include new families of hybrid and consensus-based protocols. This categorization helps in identifying the key attributes of each group and provides a critical review of their usefulness in addressing the design constraints described above.

3.2.7.1 Contention-Based MAC Protocols

Most contention-based protocols are randomized techniques that allow asynchronous access to the wireless medium via collision avoidance mechanisms. Collisions (sometimes described as contentions) cause corruption of data frames. In order to recover from corrupted frames, nodes must retransmit. Carrier Sense Multiple Access with Detection (CSMA/CD) is the classical contention-based MAC protocol in wired networks generally referred to as Ethernet. In CSMA, a node, upon detecting a collision, backs off for a random period of time before retransmission. Other contention-based protocols substitute for the detect-and-retransmit approach. CSMA suffers from the

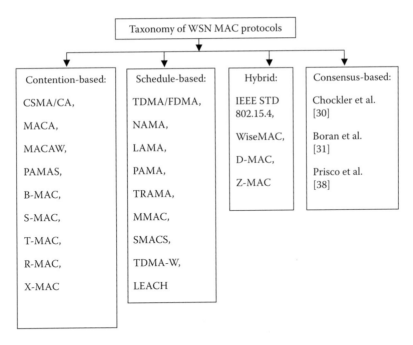

FIGURE 3.1 Taxonomy of MAC layer protocols.

hidden terminal problem [5], and the CSMA/Collision Avoidance (CSMA/CA) was developed to solve the hidden terminal problem. For example, CSMA/CA prevents collisions before they happen by listening before talking. Before transmission of the data packets, CSMA/CA first senses the channel by exchanging control packets' request to send (RTS) and clear to send (CTS). In case of collision, the CSMA/CA protocol requires the sender to back-off, where the back-off period is determined by a uniform distribution. The process is repeated once the back-off period expires. Once obtained, a node occupies the channel until it completes its transmission. Contention-based MAC can be further sub-divided into two groups: contention-based synchronous MAC protocols, such as S-MAC [11] and timeout MAC (T-MAC) [16], and asynchronous MAC protocols, such as B-MAC [17], X-MAC [18], and receiver-initiated MAC (RI-MAC) [19].

These protocols can easily adapt to network topology changes that occur when nodes join or leave. This family of protocols is simple to implement and requires less coordination between nodes. In addition, there is no need for a central coordinating node to exist in the network. However, this simplicity comes at a cost: contention-based protocols suffer from lower throughput and higher delay when compared to other protocols. They may also use more energy as compared to other protocols due to longer idle listening periods, overhearing, and collision recovery.

We now discuss the more popular contention-based protocols used in WSNs (such as the S-MAC and the T-MAC). These protocols regulate idle listening, over-hearing, and data transmission phases with sleep and wake up periods in such a way so as to reduce the overall energy consumption and improve throughput and delay.

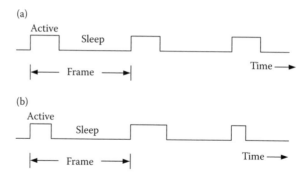

FIGURE 3.2 S-MAC vs. T-MAC frame. (a) S-MAC frame with constant active (sleep) time. (b) T-MAC frame with dynamic active time.

3.2.7.1.1 Sensor MAC

Sensor MAC (S-MAC) is one of the first contention-based MAC protocol [6,7,11] designed for multi-hop WSNs. In order to improve energy efficiency, S-MAC implements a periodic listen (or active) and sleep cycle. This scheme divides time frame into a short constant listen period, followed by a long sleep period as shown in Figure 3.2a. Typically the duty cycle (i.e., the ratio between the listen time and the full listen/sleep cycle time) consists of only 1–10% [11]. All sensor nodes choose an active/sleep schedule for themselves such that nodes listen, exchange data, and control packets (i.e., SYNC, RTS/CTS, ACK) during active periods, and turn off their radios during the sleep periods. Nodes share their schedules with neighbors so that they may communicate. Despite this sharing, it is not always possible to synchronize schedules between neighbors. One reason for this is clock shift. Therefore, to provide clock synchronization, each node periodically broadcast its schedule in a SYNC packet.

In order to avoid collisions, S-MAC uses RTS/CTS handshake procedure similar to the IEEE 802.11 distributed coordination function operation [20]. Another important feature of the S-MAC is the use of a message-passing mechanism to reduce the contention latency of long messages. Long messages are fragmented and transmitted in bursts. Only one RTS/CTS pair is used to reserve the medium for all the fragments, while a separate acknowledgement (ACK) is used per fragment. This scheme will allow only a particular fragment to be retransmitted if required, thus avoiding the cost of retransmission of the full message. S-MAC also allows the formation of virtual clusters of nodes that have common sleep schedules.

We may summarize the S-MAC family of protocols according to the following benefits and drawbacks.

Pros:

- S-MAC tries to reduce energy wastage due to collision by using RTS/CTS.
- The energy waste caused by idle listening is reduced by sleep schedules.
- Its implementation is simple relative to other protocols.

Cons:

- Broadcast data packets do not use RTS/CTS which increases collision probability.
- Due to predefined fixed duty ratios, the efficiency of the algorithm under variable traffic load degrades.
- The fixed duty cycle is the biggest drawback of S-MAC. If the listening period is long, too much energy would be wasted by idle listening as the sensor will be awake even if there is no traffic. On the other hand, if the listening period is short, contention probability is high and energy would be wasted by retransmission efforts.
- With increase in the WSN density, maintaining neighbor's schedules is an additional overhead.

3.2.7.1.2 Timeout MAC

T-MAC protocol proposed by [16] improves the S-MAC protocol by using a novel adaptive listen/sleep duty cycle, demonstrated using Figure 3.2b. T-MAC also uses fixed frames similar to S-MAC; however, T-MAC allows for a variable duration of the active listening period. In T-MAC, nodes dynamically time-out and end the active period, thereby turning off their radio if no activation event has occurred. The duration, T_a, of the active period is given by Equation 3.1. An event may be reception of data packet or start of listen/sleep frame time. For variable traffic, the energy consumption in T-MAC is less than the S-MAC because of the duty cycle adjustment through fine-grained time-outs. It is also better in handling traffic fluctuations. However, the trade-off for increased energy conservation in T-MAC when compared to S-MAC is latency.

$$T_a = 1.5(C + R + T) \tag{3.1}$$

where C is the length of the contention interval, R is the length of the RTS packet transmission time, T is the turn-around time (i.e., the time between the end of the RTS packet and the beginning of the CTS packet).

3.2.7.1.3 Berkeley MAC

Polastre et al. [14] proposed a CSMA-based versatile MAC protocol, called B-MAC, to meet the challenges of a wide range of WSN applications. Instead of a single monolithic protocol for general purpose workloads, B-MAC is an adaptive, reconfigurable light weight protocol that implements only the core functionality of an MAC layer. It employs low power listening (LPL) [21] to reduce duty cycle and to minimize idle listening. It provides well-defined flexible interfaces that allow different applications to implement their own MAC. B-MAC employs clear channel assessment technique and back-offs for channel arbitration for improving the channel utilization, and link layer ACKs for reliability. Another important feature of the B-MAC is that it uses asynchronous duty-cycle approach. That is, B-MAC does require neighboring nodes to synchronize their active-sleep schedules. Each node can independently operate on its own duty-cycle schedule.

LPL scheme introduced by Hill and Culler [21] is an asynchronous wake-up of sleeping radios for CSMA-based protocols. LPL is similar to preamble sampling

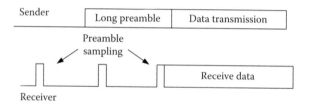

FIGURE 3.3   Low power listening in B-MAC.

scheme presented by El-Hoiydi [22] for ALOHA. In both of these schemes, prior to data transmission, a sender node transmits a long preamble (i.e., a wake-up signal) that lasts longer than the sleep period of receiver nodes and the receiver periodically wakes up according to its schedule and senses the channel as shown in Figure 3.3. If a preamble is detected, the node will remain awake to receive data, otherwise the node will return to sleep mode. The extended preamble of the LPL scheme ensures that the receiver will wake-up at least once during the preamble. The preamble length is provided as a parameter to the upper layer and it depends on the application developer.

B-MAC's flexibility results in better packet delivery rates, throughput, latency, and energy consumption as compared to S-MAC. That is, the application can adjust the sleep schedule to adapt to changing traffic loads. This results in higher performance from B-MAC because with the application-specific knowledge one can turn on and off these low-level features to minimize the overhead of the MAC protocol.

The authors in [14] also present an analytical model for monitoring application in order to calculate and set B-MAC parameters that optimize power consumption. Experiments have shown that B-MAC has better performance in terms of latency, throughput, and energy consumption as compared to S-MAC.

Pros:

- B-MAC provides on the fly reconfiguration and provides bidirectional interfaces by which the user can choose optimal parameters for a given application.
- The sender and receiver can be completely decoupled in their duty cycles. Avoids synchronization overhead.
- No need to share schedules with neighbors.
- B-MAC is currently the default MAC protocol in TinyOS.

Cons:

- Long preamble consumes a significant amount of channel time. Therefore, it is not efficient in case of contending traffic flows [18].
- Lop power listening approach suffer from overhearing at receiver nodes that are not target of a given data packet.
- As a consequence, B-MAC does not suit for highly varying traffic rates.
- The flexibility of B-MAC means the application programmer needs to know how to optimize the parameter. Therefore, it makes the application to be more complicated.

3.2.7.1.4 X-MAC

X-MAC proposed by Buettner et al. [17] is an asynchronous duty-cycled energy-efficient MAC protocol for WSNs that employs a series of short preamble prior to data transmission. X-MAC embeds the ID of target nodes into preambles and adds gaps between preambles to wait for ACK from the target node. On the one hand, by using a series of short preamble instead of a single long preamble X-MAC avoids the overhearing problem and on the other hand, it retains the advantages of LPL scheme. Similarly, by embedding the target address in each short preamble, X-MAC allows irrelevant nodes to go to sleep immediately and if the node is an intended recipient, it sends early ACK *to* the sender and remains awake for the subsequent data. When the sender gets the ACK, it stops preamble transmission and starts transmitting the DATA frame. In this way, X-MAC saves energy by avoiding overhearing while reducing latency almost by half on average [18].

Authors in [17] have shown through implementation that X-MAC. results in significant savings in terms of both energy and latency. X-MAC also uses traffic-specific adaptive duty cycle algorithm to accommodate varying traffic loads in the network.

Pros:

- Because of asynchronous duty-cycled approach, it avoids synchronization overhead.
- No need to share schedule information with neighbors.
- Active period can be significantly shorter than that of synchronized methods.
- Avoids overhearing due to short preamble. Energy efficient for under light traffic loads.

Cons:

- Series of short preamble before the data packet consume a significant amount of channel time. Therefore, it is not efficient in the case of contending traffic flows.

3.2.7.1.5 Receiver-Initiated MAC

RI-MAC proposed by Sun et al. [18] is an energy-efficient and high-performance asynchronous duty cycle MAC protocol for a wide range of traffic loads. RI-MAC separates itself from other contention-based schemes by implementing receiver-initiated data transmission.

In RI-MAC, each node wakes up based on its own schedule and broadcasts a beacon to notify neighbors it is awake. That is, the receiver sends the invitation beacon and stays awake for some time to wait for packets from a sender before going back to sleep if a sender wants to send a data frame, it jumps to the listening state and waits for receiver's beacon. Upon receiving the beacon, it begins data transmission immediately.

In LPL-based asynchronous protocols such as B-MAC and X-MAC, the preamble transmission by a sender may occupy the wireless channel for much longer than the data transmission. In the worst case, this could prevent all neighboring nodes with pending data from transmitting their data. In WSN paradigms, where multiple sensor nodes detect the same event and send their data to the sink node, all nodes could experience significant delay. In contrast, in RI-MAC, instead of a long preamble only a short beacon is transmitted before the data transmission. This minimizes unnecessary channel occupancy. In RI-MAC, collisions are detected by the receiver, if a collision occurs; the receiver sends a new beacon specifying a contention window. In the case of consecutive collisions, the length of contention window is increased.

Sun et al. [18] have demonstrated the performance evaluation of the RI-MAC protocol through *ns-2* simulation and through measurements of an RI-MAC implementation in TinyOS. RI-MAC achieves higher throughput, packet delivery ratio, and power efficiency under a wide range of traffic loads as compared to the prior asynchronous duty cycling approaches such as X-MAC.

Pros:

- High-power efficiency under a wide range of traffic loads.
- Asynchronous duty-cycled means sender and receiver schedules are decoupled.
- More suitable in pull-mode. That is, where the need to receive data dominates over the need to transmit.

Cons:

- Idle listening may be encountered because the sender usually has to be awake for a long time before the receiver sends beacon.

3.2.7.2 Schedule-Based MAC Protocols

Schedule-based MAC protocols such as TDMA, frequency division multiple access, and code-division multiple access are widely used in traditional wireless networks. Schedule-based protocols avoid collisions by dividing the wireless channel into sub-channels (slots) based on time, frequency, or orthogonal codes. The slots are further organized into time frames. In every frame, each sensor node is allocated a slot to occupy the medium and conduct its operation. This provides a mechanism for two nodes to communicate with each other in a uniquely selected slot. These protocols can use either centralized or distributed coordination schemes so that the access to a particular channel is limited to only one sender node at a given slot. Thus, scheduled protocols tend to avoid collisions and idle listening by scheduling the data transmission in advance, but waste energy in other ways. These protocols require strict time synchronizations and also need to exchange their schedules with neighbors periodically [10,13,22].

While these protocols avoid collision, are more energy efficient than contention-based protocols, and offer higher throughputs, they tend to be poorly suited in environments that are un-predictable and dynamic.

3.2.7.2.1 Time Division Multiple Access

In TDMA protocols, time is divided into slotted frames and each node is assigned a particular time slot per frame for transmission. These protocols are collision and re-transmission free, because each node has a dedicated time slot in which only that particular node transmits. These protocols save energy by keeping their radio transceivers off most of the time and turning them on only for short period at their scheduled time slot when they are either transmitting or receiving packets. Hence, in TDMA, sensor nodes are also able to eliminate the idle listening problem which is a significant source of energy wastage in the case of contention-based MAC protocols.

TDMA requires centralized or cluster-based approach. Many variations of the TDMA protocol have been reported [13,23]. TDMA protocols often require 2-hop neighbor information to establish schedules. For example, low-energy adaptive clustering hierarchy (LEACH) proposed by Heinzelman et al. [24] is a well-known routing protocol for static WSNs. LEACH organizes nodes into cluster hierarchies, and applies TDMA within each cluster. The major drawbacks to naïve TDMA approaches such as LEACH are a fixed throughput that is impossible to increase beyond utilization of all available slots. Network topology is also relatively fixed since a given network topology is used to establish a collision-free arrangement and tight synchronization among nodes. Both knowledge of topology and strict synchronization require large overheads and/or expensive hardware and hence render TDMA solutions less attractive in large-scale sensor deployment. It means scalability is a big issue.

3.2.7.2.2 *TDMA-Wakeup*

TDMA-W proposed by Chen and Khokhar [13] is an energy-efficient distributed schedule-based MAC protocol for WSNs environment which introduced the concept of a wakeup slot. The wireless channel is organized as a set of TDMA-W frames of T_{frame} seconds. Each node is assigned two slots per T_{frame}; one slot for data transmission called a send slot (*s-slot*) and another slot for wakeup session (*w-slot*). A node always listens during its *w-slot* and transmits data, if any, in the s-slot. A node utilizes its assigned slot only when it is sending or receiving information; otherwise, its receiver and transmitter are turned off. The wake-up packets are used to activate a sleeping node in order to accelerate their response.

The TDMA-W protocol operates in two phases; a self-organization (setup) phase and a channel access phase. In the network setup (self-organization) phase as described in [13], each node repeatedly executes a 7-step self-organization scheme that will uniquely determine *s-slot* number for all the nodes and their corresponding neighbors. During the setup phase, nodes broadcast their node-ids, *s-lot* number one-hop neighbors'-ids and neighbor s-lot numbers to each other. At the end of the setup phase, each node also selects any unused *s-slot* beyond its two-hop neighbors as its *w-slots* and broadcast it. The *w-slot* can be shared with neighbors. The *w-slot* may not be a unique, whereas *s-lots* are unique in two-hop range.

In the channel access phase, each sensor node maintains a pair of counters per neighbor (an outgoing packet counter and an incoming packet counter). These counters are used to determine the status of a particular neighbor, whether the corresponding neighbor is in active or sleep mode. Whenever a source node has a data packet to be delivered to a given destination node, it determines the status of the destination node and if it is active, the data packet will be sent out during the *s-slot*. If the destination node is in sleep mode, the source node sends a control packet (including the source ID) in the wake-up slot of the particular destination node prior to the data packet. The intended destination node will turn on its receiver in the *s-slot* corresponding to the source node ID. This effectively creates a pairing between source and destination within a slot.

Nodes in TDMA-W send control packets during scheduled data slots, and that is how the neighbors' allocation information is disseminated. Based on its neighbors' allocation information, a node is able to select a collision-free transmission slot which will be used for subsequent data and control packet transmissions. It handles both the synchronized and non-synchronized networks.

Pros:

- Energy is saved in TDMA-W using transmission slots and avoiding contention.
- Special wake-up packets may be to expedite a sleeping receiver's response.

Cons:

- Absolute slot identifications must be exchanged via control packets.
- TDMA-W is dependent on strict global framing and time synchronization.

It should be noted that the benefits in energy consumption involve predictability rather than savings. This is because in each frame a node must remain active during its wake-up slot even if it is not transmitting or receiving during that frame. This implies that a node must expend one full data slot worth of energy in each frame, irrespective of its communication activities. This drawback is addressed by the traffic adaptive medium access (TRAMA) protocol, discussed in the following section.

3.2.7.2.3 Traffic Adaptive Medium Access

TRAMA proposed by Rajendran et al. [23] is an energy-efficient, collision-free traffic-adaptive medium access protocol for static multi-hop WSNs. It is designed to be a dynamic time-division scheme in which the size of both the frames and the allocation of slots within the frames are determined dynamically based on the traffic flow.

A distributed election scheme uses information about the traffic at each node to determine which node can transmit in each time slot. TRAMA does not assign time slots to those nodes that have no traffic to send and, therefore, nodes can keep on sleeping instead of waking-up. This reduces idle listening and overhearing to zero. Energy consumption is reduced by ensuring collision-free transmissions and by providing facilities that allow nodes to remain in a low-power mode whenever they are not transmitting or receiving.

TRAMA consists of three major components: (a) the neighbor protocol that collects information about neighboring nodes, (b) the schedule exchange protocol that is used to exchange two-hop neighbor information and their schedules, and (c) the adaptive election algorithm (AEA) that determines the sender and receiver nodes for the current time slot using the neighborhood as well as schedule information.

In TRAMA, time frame is divided into a contention-free *scheduled access* period and a contention-based *random access* period, as shown in Figure 3.4a adopted from [10]. The overall frame time is fixed. Data transmission is performed in the scheduled *access* slot and the neighbor information is collected in the *random access* slot. Each node uses an AEA to determine the slot which can be used to transmit packets. The schedule access period is then used to announce the schedule and perform the actual data transmission.

The main advantage of TRAMA over S-MAC [6,11] is improvement in channel utilization. The tradeoff is longer delay and higher energy consumption in comparison with S-MAC protocol.

The main issue with TRAMA is its complexity and the assumption that nodes are synchronized network-wide.

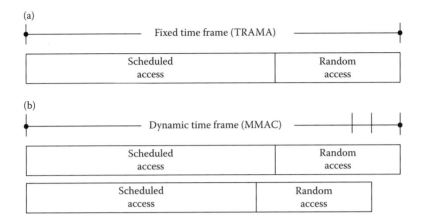

FIGURE 3.4   Fixed vs. dynamic frame. (a) Fixed time frame as used in TRAMA. (b) Dynamic Time frame used in MMAC. (Adapted from Ali, M. and Uzmi, Z., *Int J Sensor Netw* 2006; 1(3/4): 134–142.)

3.2.7.2.4 Mobility Adaptive MAC

MMAC proposed by Ali et al. [10] is a collision-free distributed MAC protocol for mobile WSNs that are a variant of TRAMA [23]. MMAC has two different access periods similar to TRAMA, a random access period for schedule reservation and a scheduled access period for transmitting data packets, as shown in Figure 3.4b.

MMAC is motivated by the idea that, in mobile environments, the use of a fixed-length static frame causes performance degradation because mobile nodes may be forced to wait for unacceptable lengths of times before being able to communicate, thereby rendering schedules obsolete. To alleviate this problem, MMAC uses a dynamic frame time that is inversely proportional to the level of mobility. This is possible in MMAC only if nodes are aware of their locations so that mobility patterns of nodes may be predicted. The time frame length is dynamically adjusted according to the mobility patterns, thus making it suitable for handling both the low mobility (e.g., topology changes, node joins, and node failures) and the high mobility (e.g., concurrent node joins and failures, and physical mobility of nodes).

MMAC uses a probabilistic autoregressive model to predict the mobility of two-hop neighbors. It then adjusts the time frame and random access time according to the mobility of those nodes. The result is a collision-free schedule based on estimates of traffic flow, mobility patterns.

Pros:

- A time-division scheme that allows for node mobility.

Cons:

- Location information is assumed to exist.
- Node localization may consume a lot of energy.
- MMAC is the highly complex and computation intensive scheduling algorithm for determining the transmitter for each slot in a frame.

- In case of high mobility, packet schedule causes a higher control packet overhead, and results in a higher duty cycle.

3.2.7.3 Hybrid MAC Protocol

We have seen that both collision detection and time-division schemes have their drawbacks. Hybrids approaches incorporate the advantages of both contention-based and schedule-based schemes, but tend to avoid their disadvantages. IEEE 802.15.4 [25], wireless sensor MAC (WiseMAC) [26], and Z-MAC [27] are examples of hybrid MAC protocols.

3.2.7.3.1 Wireless Sensor MAC

The WiseMAC protocol [26,28] builds on spatial TDMA and CSMA first proposed in [29], using a preamble sampling protocol. In the original work, nodes have two communication channels such as TDMA, which is used for accessing data channel, and CSMA, which is used for accessing control channel. WiseMAC reduces the number of data channels back down to one.

To reduce the idle listening, it uses non-persistent CSMA with preamble sampling technique.

In WiseMAC, the sampling schedules of the direct neighbors are shared by piggybacking them into ACK frames. A node receiving a data frame includes in the following ACK frame the remaining time until its next sampling time. The length of the wakeup preamble is minimized by exploiting the knowledge of sampling of the schedules of its direct neighbors.

With this knowledge, the sender of the next data frame estimates when the target receiver will wake up next, and starts transmitting its preamble just before then. All nodes in the network perform channel sampling periodically with independent sampling offsets.

This scheme allows the reduction in the energy consumption caused by unnecessarily long preambles by the sender as well as the receive power consumption. It also brings a drastic reduction in the energy wasted due to overhearing.

WiseMAC requires no set-up signaling, no network-wide time synchronization, and is adaptive to the traffic load. It provides ultra-low average power consumption in low traffic conditions and high energy efficiency in high traffic conditions.

While WiseMAC solves many of the problems associated with low power communications, it does not provide a mechanism by which nodes can adapt to changing traffic patterns.

3.2.7.3.2 Zebra-MAC

Z-MAC [27] is a hybrid CSMA/TDMA MAC protocol for WSNs.

As we have learned, CSMA is ideal for low traffic loads while it suffers from traffic contention under high traffic. On the other hand, TDMA avoids contentions and provides high channel utilizations under high traffic. In TDMA, nodes can only transmit during scheduled slots and under low contentions slots are wasted and channel utilization is reduced. The Z-MAC protocol combines the advantages of both TDMA and CSMA while offsetting their weaknesses. This is achieved by utilizing CSMA as a baseline protocol while switching to TDMA under high traffic contentions. Therefore, Z-MAC achieves high channel utilization and low-latency under low contention as in CSMA and it achieves high channel utilization under high contention and reduces collision among two-hop neighbors at a low cost similar to TDMA. Therefore, Z-MAC can dynamically adjust the behavior of MAC between CSMA and TDMA depending on the traffic contention and provide high performance at both ends of the utilization spectrum.

Authors [27] have demonstrated that Z-MAC performs better than B-MAC under medium to high contentions. However, its performance is inferior to B-MAC under low contention.

3.2.7.4 Consensus-Based MAC Protocol

A relatively unexplored approach rests in the family of consensus-based MAC protocols. These protocols are primarily concerned with achieving agreement on a particular value or state within the wireless nodes. This feature also provides a level of fault-tolerance absent in the other MAC approaches. This may be highly desirable, given that some WSNs can be categorized by rugged terrain, high node failure rates, battery depletion, or hibernation where an agreed upon state by the nodes in the network may get lost. A loss of state can result in temporary disruption of connectivity in the network. Hence, fault tolerance is necessary to maintain a consistent state that all active nodes in the network can act on. Studies in [30,31] show that even sophisticated MAC protocols can result in excessive message loss of up to 50%.

There has been extensive research done on fault tolerance in wired networks with reliable or eventually reliable communications channels. The challenges in wireless networks are more complex as messages can be lost due to collision, node failures, and signal fading. This is only accentuated by environments which are more dynamic, allowing nodes to enter and exit from the sensor community. Developing a consensus algorithm that addresses all of these concerns is non-trivial. In recent years, several papers have been published to address the issue of consensus in WSNs. We will briefly discuss a couple of algorithms on fault tolerance and consensus in WSNs.

Most algorithms [32–34] that work on consensus define the agreement over a set of processes, where each process may propose a value. A leader then decides which value to accept and asks for approval of that value from the processes. Once the leader receives the acceptance from the processes, it then declares the value to be the final value for that round.

Different algorithms stipulate slightly different conditions, but all of them in general follow the process defined above. In Chockler et al. [32], for example, the number of nodes in the system is not known *a priori*; however, they do assume a synchronous system with bounded delays and a message loss detection scheme. In Borran [33], the consensus algorithm based on the Paxos algorithm [35,36] requires that the number of nodes in the system are known *a priori*, so that when the leader sends out the proposed value, and receives responses back from the other sensors in the network, it knows when to stop and move on to the next phase. The leader moves on to the next stage when it receives responses from a majority of the active nodes. The algorithm described in [37], besides using the three basic steps described above, also requires a failure detection scheme.

In WSN, because of the conditions that these sensors operate in, fault tolerance and consensus in the system are vital to improve the efficiency of the system.

3.3 Conclusions and Future Directions

In this chapter, we have discussed the issues and challenges associated with MAC protocols for WSNs.

Sensors have always been a part of our lives. However, with advancements in process technology, it is possible to miniaturize sensor nodes to the point where they can be embedded inconspicuously in most things we use. The inevitable proliferation of sensors is enabling pervasive communication and computing, where sensor nodes collect data and process it before relaying it to the host node where decisions can be made on how and whether to act on the information received.

Sensor-based pervasive computing also requires sensor networks to be energy-efficient, scalable, fault tolerant, and cost-effective. WSNs are composed of tiny sensors which are connected to one another, where each node is constraint by energy, storage, and processing capacity, despite the advances made in process technology and VLSI design. WSNs impact all layers of the OSI model and in particular the MAC layer. Several different categories of MAC protocols have been suggested in literature addressing different problems and applications. These MAC protocols can be categorized into four different families. These are: Contention; Scheduled; Hybrid; and Consensus based protocols.

Contention-based algorithms are simple to implement; they require very little coordination between different nodes in the network. The existence of a central coordinating node is not a requirement. This family of protocols is also robust to failures and changes in the network topology. But this flexibility comes at a cost: these protocols may use higher energy; have lower throughput; and have higher transmission delay. Contention-based protocols are well suited to environments which are dynamic; that is, they are constantly changing due to either node failures or mobility and are characterized by low message activity.

Schedule-based protocols, on the other hand, offer higher network efficiency, and lower delay. However, these protocols require a central node to arbitrate the schedule and assign time slots. Most of the protocols in this category are based on the TDMA protocol. The schedule-based protocols are well suited for sensor environments that require lower message delay, have high message traffic, and have a stable topology. They are also much more complicated to use, in part because of the added complexity of time-synchronization.

Hybrid protocols, as the name suggests, are a mixture of both schedule- and contention-based protocols. Protocols like the one suggested in [26] use the contention-based protocol for downstream information transfer and the schedule-based protocols for upstream communication. Having two separate protocols may be reasonable when they are implemented in software, but implementing two different MACs in hardware adds to the cost of the MAC.

Finally, the fourth category that we discussed belongs to consensus-based protocols. Consensus-based protocols are becoming important as pervasiveness of sensors in our environment and daily lives increases. Wireless sensor nodes often operate in inhospitable environments, are prone to failure, and therefore need a protocol that either implements in the MAC layer or utilizes the functions of the underlying MAC protocol to achieve consensus. Consensus is a growing requirement in WSN networks when all nodes in the network need to have a consistent view of the state of the network.

As WSN evolve, MAC protocols will also evolve to accommodate the needs of the applications and the environment. Since the applications and the environment that sensors support and operate in are varied, it is difficult to foresee a single dominant category. Consensus-based protocols will gain more traction as fault tolerance requirements increase.

A simple and lightweight modular MAC that can dynamically adaptable to the context, handles mobility, fault-tolerant, flexible, and easily extensible to accommodate new services will be a step forward toward achieving the vision of ubiquitous pervasive computing.

References

1. Weiser, M. The Computer for the 21st Century, *Scientific American*, pp. 94–104, September 1991.
2. Soylu, A., De Causmaecker, P., and Desmet, P. Context and adaptivity in context-aware pervasive computing environments: links with software engineering and ontological engineering. *J Softw*, 2009; 4(9): 992–1013.
3. Saha, D. and Mukherjee, A. Pervasive Computing—a paradigm for the 21st century. *IEEE Comput* 2003; 36(3): 25–31.
4. Akyildiz, I.F., Su, W., Sankarasubramaniam, Y., and Cayirci, E. A survey on sensor networks. *IEEE Commun Mag* 2002; 40(8): 102–114.
5. Ye, W. and Heidemann, J. Medium Access Control in Wireless Sensor Networks, USC/ISI Technical Report, Tech. Rep. ISI-TR-580, 2003.
6. Ye, W., Heidemann, J., and Estrin, D. Medium access control with coordinated adaptive sleeping for wireless sensor networks. *IEEE/ACM Trans Netw* 2004; 12(3): 493–506.
7. Demirkol, I., Ersoy, C., and Alagoz, F. MAC protocols for wireless sensor networks: a survey. *IEEE Commun Mag* 2006; 44(4): 115–121.
8. Bachir, A., Dohler, M., Watteyne, T., and Leung, K. MAC essentials for wireless sensor networks. *IEEE Commun Surveys Tutorials* 2010; 12(2): 222–248.
9. Ali, M., Saif, U., Dunkels, A., Voigt, T., Römer, K., Langendoen, K., Polastre, J., and Zartash, U. Medium access control issues in sensor networks. *ACM SIGCOMM Comput Commun Rev* 2006; 36(2): 33–36.
10. Ali, M. and Uzmi, Z. Medium access control with mobility-adaptive mechanisms for wireless sensor networks Source. *Int J Sensor Netw* 2006; 1(3/4): 134–142.
11. Ye, W., Heidemann, J., and Estrin, D. An energy-efficient MAC protocol for wireless sensor networks. *IEEE Infocom*, 2002; 2: 1567–1576.
12. Reason, J. and Rabaey, J.M. A study of energy consumption and reliability in a multi-hop sensor network. *ACM SIGMOBILE Mobile Comput Commun Rev* 2004; 8(1): 84–97.

13. Chen, Z. and Khokhar, A. Self-organization and energy-efficient TDMA MAC protocol by wake up for wireless sensor networks, in *Proceedings of the first IEEE Conferences Sensor and Ad Hoc Communication and Networks (SECON '04)*, pp. 335–341, October 2004.

14. Polastre, J., Hill, J., and Culler, D. Versatile low power media access for wireless sensor networks, in *Proceedings of the 2nd international Conference on Embedded Networked Sensor Systems (SenSys'04)*, Baltimore, MD, USA, 3–5 November 2004, pp. 95–107.

15. Chiti, F., Fantacci, R., Archetti, F., Messina, E., and Toscani, D. An integrated communications framework for context aware continuous monitoring with body sensor networks. *IEEE J Select Areas Commun* 2009; 27(4): 379–386.

16. van Dam, T., Langendoen, K. An adaptive energy efficient MAC protocol for wireless networks, in *Proceedings of the First ACM Conference on Embedded Networked Sensor Systems,* Los Angeles, CA, November 2003.

17. Buettner, M., Yee, G., Anderson, E., and Han, R. X-MAC: a short preamble MAC protocol for duty cycled wireless sensor networks, in *Proceedings of the 4th International Conference on Embedded Networked Sensor Systems (Senses 2006)*, pp. 307–320, 2006.

18. Sun, Y., Gurewitz, O., and Johnson, D.B. RI-MAC: a receiver initiated asynchronous duty cycle MAC protocol for dynamic traffic loads in wireless sensor networks, in *Proceedings of the 6th ACM Conference on Embedded Networked Sensor Systems,* November, 2008.

19. IEEE 802.11, Wireless LAN Medium Access Control (MAC) and Physical Layer (PHY) Specifications, 1999.

20. Hill, J.L. and Culler, D.E. Mica: a wireless platform for deeply embedded networks, *IEEE Micro* 2002; 22: 12–24.

21. El-Hoiydi, A. Aloha with preamble sampling for sporadic traffic in ad hoc wireless sensor networks, in *Proceedings of IEEE International Conference on Communications*, New York, NY, 2002, pp. 3418–3423.

22. Yadav, R. et al. A survey of MAC protocols for wireless sensor networks. *Ubiquit Comput Comput J* 2009; 4(3): 827–833.

23. Rajendran, V., Obraczka, K., and Garcia-Luna-Aceves, J.J. Energy efficient, collision-free medium access control for wireless sensor networks, in *International Conference on Embedded Networked Sensor Systems*, November 2003, pp. 181–192.

24. Heinzelman, W., Chandrakasan, A., and Balakrishnan, H. Energy-efficient communication protocol for wireless microsensor networks, in *Proceedings of the 33rd Hawaii International Conference on System Sciences*, January 2000, pp. 1–10.

25. IEEE 802.15.4: Wireless Medium Access Control (MAC) and Physical Layer (PHY) Specifications for Low-Rate Wireless Personal Area Networks (LR-WPANs), 2006.

26. El-Hoiydi, A. and Decotignie, J-D. WiseMAC: an ultra low power MAC protocol for multi-hop wireless sensor networks, in *Proceedings of the First International Workshop on Algorithmic Aspects of Wireless Sensor Networks (ALGOSENSORS 2004), Lecture Notes in Computer Science, LNCS 3121*, pp. 18–31, July 2004.

27. Rhee, I., Warrier, A., Aia, M., and Min, J. Z-MAC: a hybrid MAC for wireless sensor networks, in *Proceedings of the 3rd international Conference on Embedded Networked Sensor Systems (SenSys'05)*, San Diego, CA, USA, November, 2005.

28. Enz, C.C., El-Hoiydi, A., Decotignie, J-D., and Peiris, V. WiseNET: an ultralow-power wireless sensor network solution. *IEEE Comput* 2004; 37(8): 62–70.

29. El-Hoiydi, A. Spatial TDMA and CSMA with preamble sampling for low power ad-hoc wireless sensor network, in *Proceedings of ISCC'02, Seventh International Symposium on Computers and Communications*, pp. 685–692, July 2002.

30. Woo, A., Tong, T., and Culler, D. Taming the underlying challenges of multihop routing in sensor networks, in *The First ACM Conference on Embedded Networked Sensor Systems (SENSYS)*, pp. 14–27, 2003.

31. Zhao, J. and Govindan, R. Understanding packet delivery performance in dense wireless sensor networks, in *The First ACM Conference on Embedded Networked Sensor Systems (SENSYS)*, pp. 1–13, 2003.

32. Chockler, G., Demirbas, M., Gilbert, S., Newport, C., and Nolte, T. Consensus and collision detectors in wireless ad hoc networks, in *Proceedings of ACM Symposium on Principles of Distributed Computing (PODC'05)*, Las Vegas, NV, USA, 17–20 July 2005.

33. Borran, F. et al. Extending paxos/last voting with an adequate communication layer for wireless ad hoc networks, in *Symposium on Reliable Distributed Systems*, 2008.

34. Chockler, G., Demirbas, M., Gilbert, S., Lynch, N., Newport, C., and Nolte, T. Consensus and collision detectors in radio networks. *Distrib Comput* 2008; 21(1): 55–84.

35. Lamport, L. The part-time parliament. *ACM Trans Comput Syst* 1998; 16(2): 133–169.

36. Lamport, L. Paxos made simple. *ACM SIGACT News* 2001; 32(4): 18–25.

37. Chandra, T.D., Hadzilacos, V., and Toueg, S. The weakest failure detector for solving consensus. *J ACM* 1996; 43(4): 685–722.

38. Prisco, R.D. et al. Revisiting the PAXOS algorithm. *Theor Comput Sci* 2000; 243(2): 35–91.

4

On the Quality of Service Routing in Mobile *Ad Hoc* Networks

Jamal N. Al-Karaki
The Hashemite University

Ibrahim Al-Oqily
The Hashemite University

4.1 Introduction

Ad hoc networks (AHNs) is a new type of wireless networks that can be established on the fly and without the need for infrastructure or centralized control. Mobile AHNs (MANETs), wireless sensor networks (WSNs), and wireless mesh networks (WMNs) are three prominent classes of AHNs. MANETs are described as multi-hop wireless networks that exhibit dynamic topology, temporary link associations, and scarce resources.

WSNs consist of a large number of low-power, low-cost, and possibly mobile but more likely at fixed locations sensor nodes that collaborate to gather data and disseminate it to some external points of centralized control called base stations [1]. On the other hand, WMNs consist of mesh routers and mesh clients, where the mesh routers form the backbone of WMNs with minimal mobility. WMNs have many characteristics such as multi-hop wireless network, support for *ad hoc* networking, and have the capability of self-forming, self-healing, and self-organization.

In the absence of a fixed infrastructure in AHNs, mobile nodes need to cooperate to provide routing services, relying on each other to forward packets to their destination. Hence, the path between any pair of users can traverse multiple wireless links resulting in multi-hop routing in general (Figure 4.1 shows an example of a MANET where a multi-hop route between a pair of mobile nodes is shown). Mobile nodes can also be heterogeneous, thus enabling an assortment of different types of links to be part of the same AHN [2,3]. In general, MANETs are characterized by energy-constrained nodes, bandwidth-constrained and variable capacity wireless links, and dynamically changing topology. Such characteristics lead to frequent and unpredictable topological and connectivity changes. Nevertheless, MANETs have tremendous number of military applications as well as civil applications [1].

While many routing protocols were proposed to manage the node movement and topology changes in MANETs (see [4] for a survey), most of these protocols were not designed with the necessity to guarantee data delivery in mind. Due to the profileration of multimedia applications on mobile devices, the necessity to support quality of service

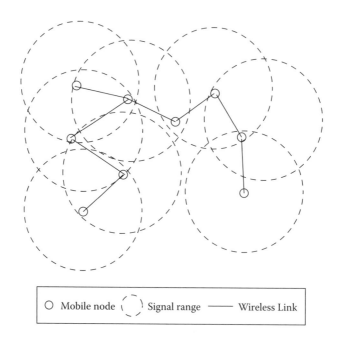

FIGURE 4.1 Mobile *ad hoc* networks: an example.

(QoS) in MANETs becomes important. In fact, the issue of QoS support is receiving attention in all other networking platforms. However, the support of QoS and QoS routing in MANETs is a challenging task due to the following reasons. First, the absence of a fixed infrastructure coupled with the ability of nodes to move freely cause frequent route breakage and unpredictable topology changes. Second, the overhead of QoS routing is too high for bandwidth limited MANETS because the mobile node should have some mechanisms to store, update, and maintain the precise link state information. Third, the limited bandwidth resource is usually shared among adjacent nodes due to the wireless medium. Fourth, the nodes themselves can be heterogeneous, thus enabling an assortment of different types of links to be part of the same network. Finally, the traditional meaning that the required QoS should be ensured once a feasible path is established is no longer true because of the mobility-caused path breakage or power depletion of the mobile nodes. QoS routing should rapidly find a feasible alternate route that will guarantee continuous service.

Moreover, QoS routing protocols that were developed for wired and other types of wireless networks do not apply directly to AHNs. To provide a complete QoS solution for AHNs, the interaction and cooperation of several components (e.g., QoS routing and resource reservation, QoS capable medium access control (MAC) layer, and physical layer) must be studied and quantified. This is due to the strong interdependencies between the different layers in the protocol stack of wireless networks. In most of the proposed QoS architectures, the emphasis was on the QoS routing component. Before a connection can be admitted or any resources can be reserved, a feasible path between a source-destination pair with sufficient available resources to satisfy QoS requirements must be found [4]. Indeed, the routing problem is exacerbated when the dynamically changing route need to satisfy certain QoS guarantees.

Although QoS resource reservation and QoS routing problems are closely related and may be coupled or decoupled in many QoS architectures, the two mechanisms have distinct responsibilities. QoS routing is defined as the process of choosing the routes to be used by the connections in attaining the associated QoS guarantees and are determined based on the knowledge of resource availability in the network [5]. On the one hand, QoS resource reservation is used to reserve resources such as buffer space and bandwidth, setup and maintain virtual connections, release resources and tear down connections in the network. On the other hand, the QoS routing algorithm must first select a feasible path that has a good chance of meeting the QoS requirements [5]. This paper focuses on the critical issue of QoS routing in AHNs.

In order to provide QoS in AHNs, it is very important to discover and react to future effects that may cause service disruption or QoS violation, for example, route breakage or topology changes. Designers of QoS routing algorithms for MANETs must consider several design issues such as: (i) QoS metric selection (e.g., bandwidth, delay, etc.), (ii) QoS path computation methods (e.g., reactive, proactive, etc.), (iii) QoS state propagation and maintenance, (iv) distributed versus centralized algorithmic design, (v) the routing architecture (e.g., flat or hierarchical), and (vi) scalability for large networks. The QoS routing protocol must also deal with imprecise state information due to node (router) movement and topology changes. Furthermore, a QoS routing scheme for AHNs must balance efficiency and adaptivity, while maintaining low control overhead.

Therefore, a MANET QoS routing protocol should have the following properties [6]. First, a routing protocol should be distributed in order to increase reliability. Where all nodes are mobile, it is unacceptable to have a centralized routing protocol. Hence, a distributed operation is a basic requirement in the design of any routing protocols in MANETs. Second, a routing protocol should assume routes as unidirectional or directed links in a heterogeneous environment. Wireless medium may cause a wireless link to be available in one direction only due to physical factors, and it may not be possible to communicate bidirectionally. Third, a routing protocol should be power efficient due to energy limitations. Finally, a protocol should be more reactive than proactive in order to avoid high protocol overhead. As this paper will show, many of the proposed QoS routing protocols for AHNS differ with regard to these design choices.

QoS routing in MANETs has recently received great attention from a growing number of researchers [7–12]. The purpose of this paper is to provide a comprehensive survey on the recent developments in the area of QoS routing in AHNs. In particular, we survey the state-of-the-art QoS routing techniques in the three classes of AHNs, namely MANETs, WSNs, and WMNs along the design issues discussed earlier. We start by presenting the problems and challenges of QoS routing in AHNs. We then categorize the QoS routing strategies for each class of AHNs. In each category, the different approaches are discussed and contrasted. Our strategy is to outline for each routing protocol, the strengths and limitations. In general, the operation of the routing protocol is affected by the network structure, that is, flat and hierarchical. Also, the methods used to search for a route distinguishes the proposed routing protocols. Since network topology changes more frequently in MANETs, many routing protocols adopt different techniques to find, reserve, or even predict the existence of a route between nodes. Furthermore, comparisons along multiple design metrics are presented toward the end of the protocols description. Routing performance metrics also include route optimality, route latency, and route diversity. Route optimality (e.g., shortest-path routes) is a secondary objective of some protocols. Because routes are often short-lived, it may be preferable to use a suboptimal route while it is available than to spend time and network resources finding optimal routes. Route latency refers to the time it takes for a source to obtain initial routing information for a new destination. Also, route diversity is desirable due to both bandwidth and energy constraints. If the protocol tends to focus on using the same path, those nodes may suffer excessive battery consumption. Spatial diversity of routes optimizes the use of available bandwidth. The ongoing research on this topic as well as open future issues related to the QoS routing problem in MANETs are outlined toward the end of this paper. In particular, suggestions as to what are the most promising directions in QoS routing are analyzed.

The rest of the paper is organized as follows. In Section 2, we present the challenges and difficulties associated with the problem of QoS routing in AHNs. A comprehensive and detailed literature review of the QoS routing schemes in the three classes of AHNs is then presented in Section 3. Different approaches are grouped according to many parameters with emphasis on similarities and differences among these approaches. Section 4 presents a perspective of current and future directions of QoS routing in MANETs. The paper concludes with a summary of the outcomes of the paper in Section 5.

4.2 Challenges of QoS Support in MANETs

QoS support in MANETs is different from QoS provisioning in other wire line or infra-structured wireless networks in the sense that free node mobility may easily result in service disruption or cause QoS guarantees violations. In fact, the characteristics of MANETs preclude any tight bounds on QoS performance measures. For example, it would be useless to reserve sufficient resources via a resource reservation protocol to guarantee a worst case delay for a high priority flow, if we cannot guarantee the delay on wireless links, especially that those links are subject to outage. This motivated applications to require at least statistical QoS guarantees, and not deterministic guarantees. QoS routing is the facilitator for QoS provisioning in MANETs, and the solution of the issues related to QoS routing is fundamental for enabling QoS in MANETs.

QoS is usually specified as a set of service requirements or constraints that need to be met by the network. These service requirements are in terms of end-to-end performance, such as delay, bandwidth, probability of packet loss, delay jitter, and so on. Power con-sumption, security, and service coverage area are other QoS metrics that are specific to MANETs. Although loss probability, cost, and delay jitter are very useful QoS metrics, delay and bandwidth are the two most popular QoS metrics. In general, the QoS metric could be concave, multiplicative, or additive. For example, the bandwidth metric is con-cave, that is, a certain amount of bandwidth must be available on each link along the path. Delay, delay jitter, and cost are additive, while the probability of packet loss can be expressed using a multiplicative relation. In some recent work, researchers argue that network security should be regarded as a QoS metric [13]. We will not consider security issues in this paper, since it is an independent and a broad topic by itself. However, it is worth noting that most security protocols increase the overhead in terms of extra mes-sages and increased data, and therefore the required security level may also be subject to a number of trade-offs applied by the QoS scheme. On the other hand, a QoS scheme may opt to satisfy a set of objectives such as maximizing throughput, call admission ratio, or packet delivery ratio.

Several technical challenges face the design of efficient QoS routing protocols in MANETs. Some of these challenges are introduced by the nature of the participating nodes, the node mobility, the nature of wireless links, and the employed-network-management strategy. We elaborate on these challenges in detail as follows:

- *Challenges due to the Dynamic Topology:* The issue of mobility does not exist in fixed networks. Even in infrastructure wireless networks, the mobile nodes move from the domain of one access point (AP) to the domain of another AP. In MANETs, there is a high possibility that the topology may vary at a fast rate. The complications imposed by mobility in MANETs may severely degrade the net-work quality. The frequent route breakage is a natural consequence of mobility, which complicates routing. This problem is even more exacerbated when paths need to satisfy certain QoS guarantees during the connection lifetime. When the network topology changes too frequently, it would be a difficult task for any pro-tocol with reasonable overhead to discover the paths, and establish the connec-tions that provide QoS guarantees. As a result, design of QoS routing protocols is challenged by frequent topological changes in MANETs. Such topological dynamics are further complicated by the natural grouping behavior in mobile user's movement, which leads to frequent network partitioning. Network partitioning poses significant challenges not only to the provisioning of QoS in AHNs, but also to connection establishment at large. This is because the partitioning disconnects many mobile users from the rest of the network.

- *Challenges due to the Scarce Resources:* The wireless spectrum is a limited resource which must be utilized efficiently. In addition, the wireless medium is a shared medium where signal attenuation, interference, multipath propagation effects, such as fading, and the unguided nature of the transmitted wave all contribute to wasting the bandwidth resource. Moreover, some overhead is often required to support reliable data transmission. Since bandwidth availability has direct effect on the QoS routing, effective management of this resource is a key factor in QoS routing.

- *Challenges due to the Absence of Communication Infrastructure:* Standard networks use an infrastructure. In MANETs, there is no pre-existing infrastructure and there is no default router, and every mobile node should be able to act as a router and be able to forward packets to other nodes. Therefore, a QoS routing protocol must consider the self-creating and self-organizing features of MANETs.

- *Challenges due to Power Limitations:* Power is a limited resource in MANETs. Solutions that reduce power consumption will often be favored; all other factors being equal. Mobile nodes need to use their battery limited power supply in a manner that prolongs the lifetime of the battery. If the battery power is used blindly, mobile nodes will fail quickly and this affects the network availability and functionality. Service disruption due to power failure is therefore a problem that needs to be avoided. Power-aware routing schemes, therefore, are designed to provide solutions for this problem. Power-aware routing routes are selected such that nodes with high remaining power are selected. Overall, power consumption can be considered as a quality metric for many protocols.

- *Challenges due to Heterogeneous Nodes and Networks:* Mobile nodes can be heterogeneous, thus enabling an assortment of different types of links to be part of the same AHN. MANETs are typically heterogeneous networks with various types of mobile nodes. In a military application, different military units ranging from soldiers to tanks can come together, hence forming an AHN. Nodes differ in their power capacities and computational powers. Thus, mobile nodes will have different packet generation rates, routing responsibilities, network activities, and power draining rates. Dealing with node heterogeneity is a key factor for the successful operation of QoS routing in heterogeneous MANETs.

- *Challenges due to Link Quality:* The problem of link quality is particularly significant in MANETs. The essential effect on MANETs is that the link quality can become extremely variable, often in a random manner. Although some parts of this effect can be predicted since variations in link quality impact packet delivery, and triggers error recovery procedures, the main QoS parameters such as bandwidth availability, latency, reliability, and jitter are all affected. This effect can happen either during or between connections. In the former case, a link quality might become too bad while a connection is in place. In the latter, a new connection with the same requirements as a previously established one is rejected because the link status variability increases and, as a result, the link becomes unreliable.

- *Challenges due to the Maintenance of State Information:* QoS routing consists of two basic tasks. The first task is to collect the state information and keep it up-to-date. The second task is to find a feasible path for a new connection based on the collected information. The performance of any routing algorithm directly depends on how well the first task is solved. State information can be local or global. In local state, each node is assumed to maintain its up-to-date local state, including the queuing and propagation delay, the residual bandwidth of the outgoing links, and the availability of other resources. The combination of the local states of all nodes is called a global state, and every node is able to maintain the global state by using a link-state-based routing protocol, which exchanges the local states among the nodes periodically. The global state kept by a node is always an approximation of the current network state due to the constantly varying network topology, and consequently link states which encounter a non-negligible delay for propagating between nodes. In general, as the network size grows, the imprecision increases.

- *Challenges due to Interactions with Other Layers:* The choice of the medium access scheme in MANETs is difficult due to the time-varying network topology and the lack of centralized control. Time division multiple access (TDMA) or dynamic time assignment of frequency bands is complex since there is no centralized control. FDMA is inefficient in dense networks, and CDMA is difficult to implement due to node mobility and the subsequent need to keep track of the frequency-hopping patterns and/or spreading codes for nodes in a time-varying neighborhood. At the MAC layer, we also have link layer reliability problems which are related to the high bit error rate, in addition to the possible packet collision problems. QoS at the MAC layer needs further research (QoS aware MAC) which can serve as an infrastructure for facilitating the QoS routing.

- *Challenges due to Lack of Centralized Control:* The QoS path discovery process may be centralized or distributed. In MANETs, distributed algorithms are more preferable due to the lack of a central entity. Although distributed applications are better suited for MANETs due to its peer-to-peer architecture, important network applications and services such as web servers, location information databases, and network services (DHCP, SNMP) are inherently centralized. These services are often critical to the mobile node's operation such that every node requires constant and guaranteed access to them. Therefore, designing protocols for MANETs requires paying attention to both issues.

Despite these challenges, many QoS routing approaches were proposed in the literature in an attempt to overcome or tackle some of these difficulties. In the next section, we review the efforts that have been exerted in the area of QoS routing in MANETs.

4.3 QoS Routing Protocols in MANETs

In this section, we present an extensive survey of the QoS routing protocols in MANETs. Compared to the abundant work on QoS routing for fixed wireline networks, results for QoS routing in MANETs are relatively scarce due to the difficulties mentioned earlier, as well as the relative recency of MANETs. However, many promising studies on QoS routing in MANETs have emerged recently. These studies may be classified according to several criteria that reflect different choices in design and implementation. In order to streamline this survey, we use a classification according to the following criteria: route scheduling (i.e., reactive, proactive, or position), network structure (i.e., flat or hierarchical), route cardinality (single versus multi-path), state information maintenance (i.e., local or global), QoS constraints (i.e., bandwidth, delay, energy, etc.), and design space (i.e., single layer versus cross-layer). The classification is summarized in Figure 4.2.

Some of the QoS routing protocols in MANETs also present variations of an original idea, for example, protocols that build on predictive or ticket-based probing [14]. Others are dependent on local computations of bandwidth. In the hierarchical networks, some of the protocols form a multi-tier network while others sustained two-tier network in its simple form. When node positions were used in the routing decisions, nodes locations were either determined using an external location determination server (e.g., global positioning system (GPS)), or using localized algorithms or GPS-free schemes by using methods like triangulation or trilateration [15]. For each routing protocol, we will mention if the protocol uses single or multi-channel for communication, single or multi-path for routing, maintains local or global states, and also distinguish the QoS metrics used in each protocol. A new trend in QoS routing is to utilize the interactions among various layers in the protocol stack to perform efficient QoS routing in MANETs, hence called cross-layer QoS routing in comparison to the traditional trend of supporting routing at single layer, that is, network layer.

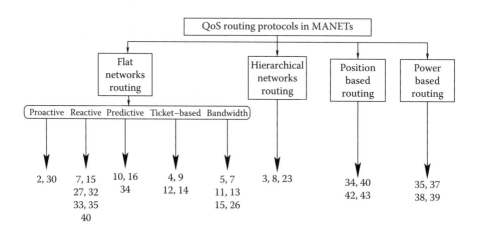

FIGURE 4.2 QoS routing protocols in MANETs: classification map.

4.3.1 QoS Routing Based on Network Structure

In this section, we will describe QoS routing protocols that are based on the network structure, namely flat or hierarchical networks.

4.3.2 QoS Routing in Flat Networks

The protocols we review here fall into three categories: (a) predictive routing, (b) ticket-based routing, and (c) bandwidth-based routing. Predictive routing detects changes in link status and network topology *a priori* and uses this information to build stable routes that have low probability of failing. The basic idea of ticket-based probing is described as follows. A ticket is the permission to search one path. The source node issues a number of tickets based on the available state information. Probes (routing messages) are sent from the source toward the destination to search for a low-cost path that satisfies the QoS requirement. Each probe is required to carry at least one ticket. At an intermediate node, a probe with more than one ticket is allowed to be split into multiple ones, each searching a different downstream subpath. The maximum number of probes at any time is bounded by the total number of tickets. Since each probe searches a path, the maximum number of paths searched is also bounded by the number of tickets. If a QoS routing protocol supports QoS via separate end-to-end bandwidth calculation and allocation mechanisms, it is called bandwidth-based routing. The bandwidth-based routing scheme depends on the use of TDMA medium access scheme in which the wireless channel is time slotted and the transmission scale is organized as frames (each containing a fixed number of time slots). A global clock or time-synchronization mechanism is utilized such that the entire network is synchronized on a frame and slot basis.

4.3.2.1 Predictive QoS Routing Protocols

In this kind of QoS routing, the protocol attempts to predict the link state change or topology change beforehand, and tries to avoid using unstable links in calculating a path. This scheme is hence probabilistic in nature and its success is highly dependent on the ability to predict well with high probability.

In [16], Shah and Nahrstedt discussed the use of the updates of geographic location to develop a *predictive* location-based QoS routing scheme based on a location resource update protocol, which assists a QoS routing protocol. The motivation is that state information in a dynamic environment like MANETs does not remain current for very long. The predicted locations are used to build future routes before existing routes break, and thus avoid route re-computation delay. The proposed protocol is heavily dependent on the prediction of nodes locations. The direction of motion of a mobile node is taken into account when attempting to predict its future location. The approach involves an update protocol, a location-delay prediction scheme, and the QoS routing protocol to route multimedia data to the destination. The location-delay prediction scheme is based on a location-resource update protocol, which assists the QoS routing protocol. The update protocol is used to distribute nodes geographical location and resource information (e.g., battery power, queuing space, processor speed, transmission range, etc.). The update packets used by the update protocol contain timestamps, current geometric coordinates, direction of motion, velocity, and also resource information pertaining to the node that is used in the QoS routing protocol. The location-delay prediction scheme is used to predict the new location and the end-to-end delay of a node based on the last update messages received from that node. The scheme assumes that the end-to-end delay for a data packet from, say *a* to *b*, is equivalent to the delay experienced by the latest update from *b* to *a*. It has been found that the end-to-end delays for packets traveling between *a* and *b* usually remain more or less similar for no more than 0.5 seconds. This means that the end-to-end delay to node *b* will be predicted to be equal to the delay experienced by the most recent update that has arrived from *b*. The delay prediction for each node is based on the end-to-end propagation delay for that node from a certain node, say *a*.

In QoS routing, each node maintains two tables, the update table and routing table, in order to compute paths at the source node. Although this protocol seems promising, the overhead in maintaining and updating tables is high. Moreover, the prediction accuracy is highly dependent on the previous states of mobile nodes which might be misleading. A QoS-aware admission control is also proposed in [16]. A small mathematical model is used to predict the new location of a mobile node. As shown in Figure 4.3, a mobile node uses the following two equations to determine its new position (x_p, y_p) at a future time t_p:

$$x_p = x_2 + \frac{v(t_p - t2)(x_2 - x_1)}{[(x_2 - x_1)^2 + (y_2 - y_1)^2]^{1/2}}, \quad y_p = y_1 + \frac{(y_2 - y_1)(x_p - x_2)}{x_2 - x_1}$$

Using the proposed scheme and in order to be able to do the above calculations, each mobile node has information about the whole topology of the network. It can thus compute a source route from itself to any other node, using the information it has, and can include this source route in the packet header when it is routed. The prediction scheme proves a reduction in the overhead involved in the updates and the QoS routing scheme. However, the prediction scheme keeps track of many information (e.g., tables) and must update these tables frequently when topology or link states change, which results in large computation and communication costs.

In [17], the authors propose a QoS routing scheme which is an extension of the pre-computation-based selective probing (PCSP) scheme. PCSP pre-computes QoS variations as well as the cost and QoS metrics of the least-cost path (LC path) and the best-quality path (BQ path) at each node, taking into account the imprecision of state information. The information is then used with two selective probing methods where one uses the QoS-satisfying LC paths and the other using the least cost of found feasible paths, thus excluding the paths that can never be optimal. In order to be able to execute the algorithm, the set of links are divided into two different sets—stationary link set and transient link set. A routing path is supposed to use stationary links whenever possible in order to reduce the path failure probability. A newly formed link is regarded as a transient link but becomes a stationary link if it survives for a certain duration of time. The proposed PCSP scheme can achieve low routing overhead and small route setup time while guaranteeing high success ratio. However, the measurements of the lifetime of the links are based on predictions of the link states, which might change more often as nodes are free to move. The proposed scheme cannot guarantee that by trying to capture network state in advance, that the paths found are optimal.

In [18], a QoS routing protocol with mobility prediction was proposed. This protocol searches for paths that consist of low mobility nodes. Therefore, the lifetime of this path will be longer than other paths. The protocol uses a metric of path expiration time which is based on the node's mobility speed and location with respect to other nodes. Although this protocol shows a good performance, the nodes' mobility patterns and speeds in MANETs cannot be enforced or assumed before hand. Hence, the accuracy of the calculated path or the expected lifetime of that path cannot really be measured.

An adaptive QoS routing scheme based on the prediction of the local performance in AHNs is proposed in [19]. The scheme is implemented by a link performance prediction strategy. In each local area, the QoS is estimated based on translating the effects of the lower layer parameters into the link state information. In the prediction approach, several mechanisms are built to complete the location information management process (i.e., information monitoring, collecting, and updating functions). The node movement is characterized by the probabilities of the link state and the prediction of the local QoS performance. In this sense, the proposed QoS routing is adaptive to node's mobility. To summarize, the local QoS performance information in this scheme is built based on node's mobility, but does not depend on node's location information where the QoS routing is adaptive to the changes in the mobile network due to the special local exchange mechanism. Local exchange of information would slightly increase network overhead due to the distributed structure of the routing scheme. Finally, the scheme only provides statistical QoS guarantee.

4.3.2.2 Ticket-Based Probing Routing

In core extraction distributed *ad hoc* routing algorithm (CEDAR), the bandwidth is used as the only QoS parameter for routing. Also, in ticket-based routing method, the delay and bandwidth are used for QoS routing but not together. They are implemented as different algorithms.

The basic idea of the ticket-based probing scheme [14] is to utilize tickets to limit the number of paths searched during route discovery. A ticket is the permission to search a single path. When a source wishes to discover an admissible route to a destination, it issues a probe (routing message) to the destination. A probe is required to carry at least one ticket, but may carry more tickets (i.e., connection requests with tighter requirements are issued more tickets). At an intermediate node, a probe with more than one ticket is allowed to split into multiple ones, each searching a different downstream subpath. Hence, when an intermediate node receives a probe, it decides, on the basis of its available state information, whether the received probe should be split and to which neighbors the probe(s) should be forwarded. In the case of route failures, ticket-based probing utilizes three mechanisms: path re-routing, three-level path redundancy, and path repairing. Re-routing requires that the source node be informed of a path failure. After which, the source initiates the ticket-based algorithm to locate another admissible route. The path redundancy scheme establishes multiple routes for the same connection. For the highest level of redundancy, resources are reserved along multiple paths and every packet is routed along each path. In the second level of redundancy, resources are reserved along multiple paths; however, only one is used as the primary path while others serve as backup paths. In the third level of redundancy, multiple paths are selected, but resources are only reserved on the primary path. The path-repairing mechanism tries to avoid the cost of re-routing by attempting to repair the route at the point of failure. Ticket-based probing is proposed as a general QoS routing approach for MANETs and can handle different QoS constraints (i.e., bandwidth, delay, packet loss, and jitter). For example, ticket-based QoS routing solutions for the bandwidth and delay-constrained routing problems were presented in [14].

In [14], backup paths for maintenance of the routing paths when nodes move, join, or leave the network are also used. The proposed algorithms require that each node knows (with some imprecision) the delay and bandwidth on the least delay and largest band-width paths to any possible destination. Thus, in effect, each node should maintain the states of all links in the network, which requires a quadratic communication overhead for the state updates. Figure 4.3 shows an example of how source uses three probes p_1, p_2, and p_3 to find a path to the target D. When one or more probe(s) arrive(s) at the destination node, the hop-by-hop path is known and delay/bandwidth information can be used to perform resource reservation for the QoS satisfying path.

In wireline networks, a probability distribution can be calculated for a path, based on the delay and bandwidth information. In MANETs, however, building such probability distribution is not difficult and not feasible. This is because of the fact that wireless links are subject to breakage at any time and state information of link and network topology is imprecise in nature. Hence, a simple model that provides information about the state of the network, although still imprecise, is needed. Such a model was proposed in the

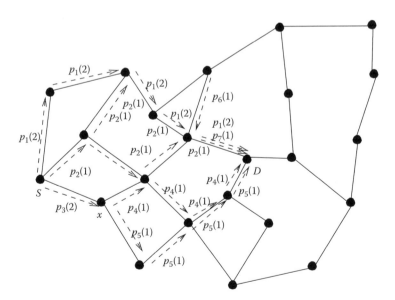

FIGURE 4.3   Source S uses three probes p_1, p_2, p_3 with the number of tickets each probe has been given between parentheses. p_3 is further split to two probes p_4 and p_5 at node x.

ticket-based probing algorithm in [14]. It uses recent history and current estimated delay variations and smoothing formula to find the actual minimum end-to-end delay. Moreover, to adapt to the dynamic topology of MANETs, this algorithm allows a different level of route redundancy mentioned earlier. A route maintenance technique that provides re-routing in case the original path breaks was also proposed. Two routing techniques are considered in [14], both limited to low mobility networks. The first technique is based on the availability of only local state information and the other assumes possibly inaccurate knowledge of global states. For QoS routing using only the local state information, [14] introduces two different distributed routing algorithms, such as source-initiated routing algorithm and destination-initiated routing algorithm. Both rely on the use of probe packets (ticket-based probing) for identifying a feasible route where the information for multiple feasible routes is stored in the probes instead of the intermediate routers. A broken route in this scheme is detected by using the beaconing protocol that detects adjacent neighbors. Another modified scheme for finding a QoS route based on the ticket-based probing is presented in [20] where the authors proposed an algorithm with two modified terminating strategies for the original method proposed in [14]. The objective of the modification is to reduce the message overhead per connection. The terminating strategies are designed for delay-constrained routing where the routing process ends at the destination nodes either when all tickets reach destination or if a probe timeout occurs. The timeout is needed for invalidated probes which traverses the network till it reaches the destination. The invalid probe traversal increases the message overhead. Hence, it was proposed to discard all invalid probes instead of sending them to their destinations. To do that, the authors proposed to modify the data structure of the ticket distribution process where a new field called hotness is added. Hence, the termination will depend both on the total number of tickets and the hotness field value. The hotness field represents the degree of importance of the connection, the higher the value the more important is the connection. The first termination strategy deals with the hot value probes while the second termination strategy deals with cold probes. When the probe is hot, destination responds quickly while if the probe is cold, the destination will wait for more probes to arrive before the timeout. At the end of timeout, destination selects the path that suits the connection.

Liao et al. have proposed a multi-path QoS routing protocol in [21]. The protocol also searches for multiple QoS paths at the same time. The QoS metric used is the available bandwidth. The multiple paths found collectively satisfy the required QoS. In general, the multi-path QoS routing algorithm is suitable for MANETs with very limited bandwidth. For the duration of the connection and due to mobility, a single path satisfying the QoS requirements is unlikely to exist all the time. Hence, multiple paths can be used in parallel to deliver packets to destination given that all these paths satisfy the required QoS guarantees. The difference between this protocol and that proposed in [8] is that multiple paths are used at the same time to deliver packets while in [8], different paths are tested and the most suitable path is selected by the destination to deliver the data packets.

In [22], the authors developed an algorithm that is based on the ticket probing and they used fuzzy logic to model imprecise network information. The proposed algorithm is used to generate a maximum number of probe messages that searches for a path. A hop-by-hop path selection criterion is made to select the best path among the candidate paths. The difference between [14] and [22] is the way in dealing with imprecise state information for the QoS metrics considered. In [22], the number of generated probe messages is based on an inference rule derived by a fuzzy logic system.

4.3.2.3 Bandwidth Calculation-based Routing

Using a TDMA scheme in wireless MANETs allows each node to know about the free slots between itself and its neighbors through a simple broadcasting scheme. Based on this information, bandwidth calculation and assignment can be performed distributively. However, determining slot assignment while searching for the available bandwidth along the path is an non-deterministic polynomial time (NP)-complete problem. Bandwidth calculation requires the knowledge of the available bandwidth on each link along the path as well as resolving the scheduling of free slots.

Note that in wireline networks, the path bandwidth is the minimum available bandwidth of the links along the path. In time-slotted MANETs, however, bandwidth calculation is much harder. In general, we not only need to know the free slots on the links along the path, but also need to determine how to assign the free slots at each hop.

A QoS-aware routing protocol that incorporates an admission control scheme and a feedback scheme to meet the QoS requirements of real-time applications is presented in [23]. The QoS-aware routing protocol is built off the well-known ad-hoc on demand distance vector (AODV) routing protocol. The protocol only considers the bandwidth constraint when supporting real-time video or audio transmission. The emphasis is on soft QoS or better than best-effort service rather than guaranteed hard QoS. The soft QoS means that there may exist transient time periods when the required QoS is not guaranteed due path breaking or network partition. The protocol uses approximate bandwidth estimation during route set-up to react to network traffic by using two bandwidth estimation methods to find the residual bandwidth available at each node in order to support new QoS streams. The QoS-aware routing protocol either provides feedback about the available bandwidth to the application (feedback scheme), or admits a flow with the requested bandwidth (admission scheme). Both the feedback scheme and the admission scheme require the knowledge of the end-to-end bandwidth available along the route from the source to the destination. Simulation results show that the packet delivery ratio increases and packet delay and energy dissipation decrease while the overall end-to-end throughput is not effected, compared with routing protocols that do not provide QoS support.

In [24], a destination sequenced distance vector (DSDV)-based QoS routing scheme was proposed. The routing protocol provides QoS support via separate end-to-end bandwidth calculation and allocation mechanisms, thus called bandwidth-based routing. The proposed bandwidth routing scheme depends on the use of a CDMA over TDMA [25] medium access scheme in which the wireless channel is time slotted, the transmission scale is organized as frames (each containing a fixed number of time slots) and a global clock or time-synchronization mechanism is utilized. That is, the entire network is synchronized on a frame and slot basis. The path bandwidth between a source and destination is defined as the number of free or available time slots between them. Bandwidth calculation requires the knowledge of the available bandwidth on each link along the path as well as resolving the scheduling of free slots. This problem is an NP-complete and thus requires a heuristic approach. To support fast rerouting during path failures (e.g., a topological change), the bandwidth routing protocol maintains secondary paths. When the primary path fails, the secondary route is used (i.e., becomes the primary route) and another secondary is discovered. A similar algorithm has been proposed by Lin and Liu [24] where an end-to-end bandwidth calculation and bandwidth allocation scheme is also used to assign link bandwidth. The algorithm in [24] also maintains secondary and ternary paths between source and destination for the use in re-routing when the primary path fails.

In [26], an on-demand QoS routing protocol based on AODV for TDMA-based MANETs was proposed. A QoS-aware route reserves bandwidth from source to destination by reserving free time slots. In the route discovery process of AODV, a distributed algorithm is used to calculate the available bandwidth on a hop-by-hop basis and using the RREQ/RREP query cycle as shown in Figure 4.4a and b where node n_4 initiates the route discovery process. When a route fails, a route error message (RERR) is sent back to the source as illustrated in Figure 4.4c. Only the destination node can reply to a route request (RREQ) message that has come along the path with sufficient bandwidth. If an intermediate node learns of such a path, RREQ messages received by this intermediate node can also be sent back by this intermediate node, hence quicker responses can be obtained. This is explained in Figure 4.4d,e, and f where node n_1 replied to the RREQ and data start flowing on the returned path. If the RREQ is received by a node with inadequate bandwidth, it will be dropped by the intermediate nodes. The protocol can handle limited mobility by restoring broken paths and it is suited for small MANETs with short routes.

In [2], a bandwidth reservation scheme is also used to provide QoS guarantees in MANETs. A route discovery protocol is proposed, which is able to find a route with a given bandwidth (represented by number of slots). The proposed protocol can reserve routes while addressing both the hidden-terminal and exposed-terminal problems. Unlike previous protocols, this protocol addresses the previous two problems by not assuming that the bandwidth of a link can be determined independently of its neighboring links. Recall that a QoS request is represented in terms of number of slots. Figure 4.5 shows an example of how bandwidth reservations are applied in this protocol. The QoS route discovery is an on-demand source routing basis, and works similar to the dynamic source routing (DSR) protocol on disseminating route-searching packets. Take the path from A to C going through node B. In Figure 4.5a, the white slots associated with each node are free and the gray slots are busy. Matching the free slots between nodes, we obtain five common free time slots, that is, 1, 2, 3, 4, 5 between A and B and four common free time slots, that is, 3, 4, 5, 6 between B and C. In Figure 4.5b, an attempt is made to make a reservation from A to C with three slots. Unfortunately, this is not achievable for obvious reasons. Hence, the possible amount of reservation from A to C is only two slots as shown in Figure 4.5c and the situation cannot be improved, unless we change the assignment for A.

In [27], an AODV variant protocol, called QoS–AODV, is proposed. Similar to [2], QoS–AODV incorporates slot scheduling to find QoS routes over the network. The differences between [27] and [2] are that the protocol in [27] uses an integrated route discovery and bandwidth reservation protocol. Unlike other path finding and route

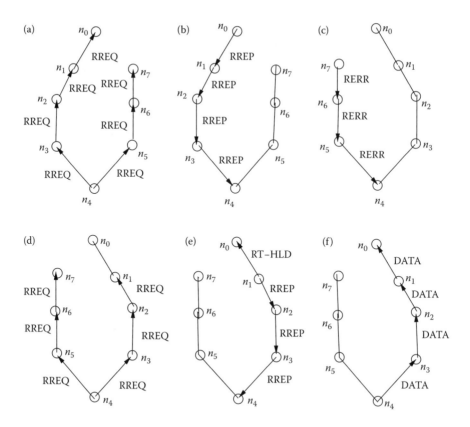

FIGURE 4.4 An example of route setup and route repair with the QoS routing protocol in [11]. (a) Route discovery starts by transmitting a RREQ packet; (b) n_1 sends a RREP to n_4 in the opposite direction; (c) the node upstream of the broken link (n_2) detects its next hop node (n_1) is gone and sends a RERR packet back to the source; (d) the source node n_4 receives the RERR packet, it sends out a new RREQ and starts a new round of route discovery; (e) n_1 generates a local reply and sends out the RREP back to the source in the reverse direction; (f) as data packets flow through the route, the RESV state at every node on the route is being refreshed periodically. (Adapted from Zhu, C. and Corson, M. S., QoS routing for mobile ad hoc networks, in INFOCOM 2002. vol. 2, 2002, pp. 958–967.)

discovery protocols that ignore the impact of the data link layer, QoS–AODV incorporates slot scheduling information to ensure that end-to-end bandwidth is actually reserved. QoS–AODV performs path search simultaneously with time slot scheduling by using simple heuristic algorithms. The protocol creates virtual connections by reserving MAC TDMA time slots along with one of the discovered paths. Another protocol proposed in [28], which is a DSDV variant protocol, also performs bandwidth calculation and reservation using TDMA. The CDMA-over-TDMA channel model is used in this protocol, where the use of a time slot on a link is only dependent on the status of its one-hop neighboring links. Multi-path routes are searched to the destination which satisfies the bandwidth requirement. The destination determines a QoS

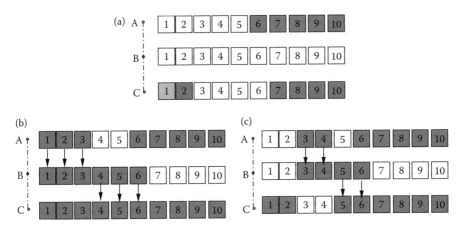

FIGURE 4.5   TDMA-based bandwidth reservation protocol. (a) Matching the free slots between nodes; (b) an attempt is made to make a reservation from A to C with three slots; (c) the possible amount of reservation from A to C is only two slots.

multi-path routing and replies to the source node. This protocol collects link-state information from source to destination in a reactive manner. A mobile node knows the available bandwidth to each of its neighbors. When a source node S needs a route to a destination D of bandwidth B, it will send out some RREQ packets, each of which carries the path history and link-state information. Each RREQ packet records all link-state information from source to destination. The destination collects all possible link-state information from different RREQ packets sent from the source. A partial network, which is a flow-network, is constructed in the destination node after receiving multiple information packets. An algorithm is applied at the destination to determine a better result for QoS multi-path routing. After determining a multi-path route, a reply packet is sent from the destination to the source. On the reply's way back to the source, the bandwidths are confirmed and reserved. The protocol works in a number of phases outlined in detail in the paper. Multi-constrained QoS routing problems have been proven to be NP-complete and cannot be solved by a simple and efficient algorithm. In [29], an on-demand routing protocol called FLMQRP that incorporates service quality and energy constraints has been proposed to solve this problem. It is based on a flooding-limited concept with a built-in quick-pruning mechanism, unlike other QoS routing schemes. The established connections are guaranteed and all QoS metrics are specified (bandwidth, end-to-end delay, and battery energy). Multi-constrained QoS routing finds a path that satisfies multiple independent path constraints. One example is delay-cost-constrained routing, that is, finding a route in the network with a bounded end-to-end delay and bounded end-to-end cost. The FLMQRP is a reactive protocol and its operation is like the DSR protocol. the bandwidth-constraint routing problem (BCRP) were studied in [30], BCRP is polynomial-time solvable for wired networks, because each link is physically independent from the other links, even if they are attached to the same node. However, BCRP becomes more difficult in wireless AHNs due to spatial contention in a shared wireless medium. Traffics carried by wireless links may be interfered with one another. The available bandwidth of a wireless link depends not only on the traffics carried by its neighboring links, but also on how well transmissions on its neighboring links are scheduled at the MAC layer. Therefore, BCRP should be considered with the MAC and network layers. In their work and without considering the optimization of scheduling, they showed that BCRP is an NP-complete under contention-based MAC protocols. However, both the collision-free channel model and the First In-First Out (FIFO) scheduling policy are considered to relieve the link interference and coordinate neighboring flows, respectively. Therefore, BCRP can be polynomial-time solvable. In [31], the authors proposes a QoS preemptive routing protocol with bandwidth estimation (QPRB) that computes the available bandwidth in the route and then sets up the route based on the network traffic and maintains the route using preemptive routing procedure. This protocol is intended to provide QoS support to real-time applications by providing a feedback about the network status. The protocol executes in three main phases: (a) bandwidth estimation, where the sender's current bandwidth usage and sender's one hop neighbors current bandwidth usage are disseminated via hello messages, (b) route discovery, where the source broadcasts RREQ messages. Each neighbor which receives the message checks if it is the destination. If it is not the destination, then the request is rebroadcast until the requested destination is found, (c) and route maintenance to avoid link failures that are mainly due to mobility using preemptive method. The authors in [32] proposed multiple access collision avoidance with piggyback reservations (MACA/PR), a network architecture which is based on CSMA/CA and combines the asynchronous operation of WLANs and the QoS support of traditional TDMA-based network architecture. MACA/PR can be viewed as an extension of IEEE 802.11 and Floor Acquisition Multiple Access (FAMA) [33] which provides guaranteed bandwidth support (via reservation) to real-time traffic. In MACA/PR, the network layer is equipped with a bandwidth reservation mechanism and a QoS routing protocol which are inspired to our earlier work on cluster TDMA and adaptive clustering. The main contribution to QoS routing is the introduction of the bandwidth constraint. The bandwidth is a slot within the cycle frame. In its most complete formulation, the bandwidth routing algorithm keeps track of shortest paths for all bandwidth values. To compute these paths, each node periodically broadcasts to its neighbors the bandwidth and hop distance pairs for the preferred paths (one per bandwidth value) to each destination. In our case, the number of preferred path is the maximum number of slots (or packets) in a cycle. As such, MACA/PR allows the establishment of real-time connections over a single hop only. However, by complementing MACA/PR with a QoS routing algorithm and a fast connection setup mechanism, routing with end-to-end multimedia connectivity in a mobile and multihop network can be supported. In evaluating the protocol, two bandwidth values were kept: the bandwidth on the shortest path and the maximum bandwidth (over all possible paths). Performance results show that MACA/PR can provide a good compromise between low latency, low packet loss, good acceptance control, fair bandwidth sharing among connections, good mobility handling, and good scaling properties.

A secure routing protocol with QoS support has been proposed in [34]. The protocol includes secure route discovery, secure route setup, and trustworthiness-based QoS routing metrics. The routing control messages are secured by using both public and shared keys, which can be generated on-demand and maintained dynamically. The proposed protocol were designed to detect the difficult internal attacks, including Byzantine attacks, to build a trustworthiness repository by the node on its neighboring nodes that deliver the message, and to guarantee QoS requirement on the links along a route, such as packet delay and link quality. Therefore, the protocol uses both route and message redundancies during topology discovery. The attacks can be detected by verifying various copies of a received message, which reaches a node via different paths at different times.

4.3.3 Hierarchical QoS Routing Protocols

Hierarchical or cluster-based routing, originally proposed in wireline networks, is a well-known technique with advantages related to scalability and efficient communication. As such, the concept of hierarchical routing is also utilized to perform QoS routing in MANETs.

Using concepts from multi-layer adaptive control, Chen et al. proposed in [35] an approach for controlling QoS in large AHNs by using hierarchically structured multi-clustered organizations. In the proposed approach to QoS routing in MANETs, the network can be stand alone or connected to wireline networks. The proposed scheme uses bandwidth as the QoS metric for nodes grouped in clusters and using a time slotted scheme (TDMA) inside each cluster. The role of cluster dynamics and mobility management, as well as resource reservation and route repair and router movement on QoS are addressed in detail. First, the network is partitioned into clusters where neighboring clusters use different spreading codes in order to reduce interference and to enhance spatial reuse of channels. Within each cluster, the MAC is implemented using time-slotted frame which is divided into a control phase and data phase as described earlier. The QoS metric considered is the bandwidth represented in terms of time slots. The available bandwidth computation is carried out independently at each node. A loop-free DSDV scheme is used for routing in this architecture. To avoid loops in DSDV, modifications to the well-known Bellman–Ford algorithm, which is the basis for DSDV, is applied. Although the proposed scheme makes what is called *fast reservation* in order to adapt to network mobility, it is still hard to guarantee QoS requirement based on reservations only. The protocol is an extension to the DSDV routing algorithm. Figure 4.6 shows an example of clustering used in this protocol where cluster heads maintain the TDMA-based schedules in each cluster.

Two new QoS routing schemes, both based on link state protocols as the underlying mechanism, appear in [22]. Both protocols attempt to reduce routing update overhead, one by selectively adjusting the frequencies of routing table updates and the other by reducing the size of the table. The QoS path is computed locally at each source node based on the routing table. Although the states can never be guaranteed to be accurate, the update selection strategy managed to avail helpful network state that improves the network performance.

In [4], an extension to the Fisheye state routing routing protocol, and hierarchical state routing protocols with QoS guarantees were proposed. QoS support is offered by adding an entry to the link bandwidth and channel quality information in the routing tables.

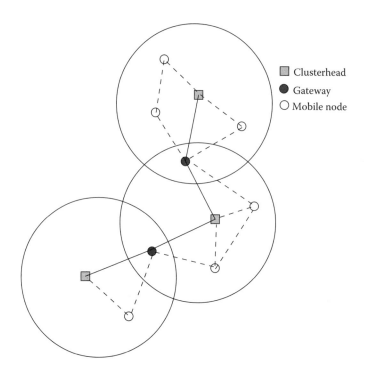

FIGURE 4.6   Clustering in MANETs: an example.

However, no specific QoS routing algorithm was discussed in [4]. In [36], a CEDAR has been proposed as a QoS routing algorithm for small-to-medium size mobile AHNs, consisting of tens to hundreds of nodes. The core of the network is a subset of nodes selected to perform network management and routing function in the network. CEDAR is an on-demand source routing algorithm, which includes three key components: (i) core extraction, (ii) link state propagation, and (iii) route computation. In core extraction phase, the core of the network is extracted by approximating a minimum dominating set (MDS) of the AHN using only local computation and local state information. The MDS is the minimum subset of nodes, such that every node in the network is in the MDS (i.e., is a core node) or is a neighbor of a node in the MDS. Each node in the core then establishes a unicast virtual link with other core nodes over a distance of three hops or less away in the AHN. Each node that is not in the core chooses a core neighbor as its dominator. The core nodes are responsible for collecting local topology information and performing routing on behalf of the nodes in their respective domain (or immediate neighborhood). Figure 4.7 shows a MANET with three nodes selected to act as core nodes, that is, virtual topology. The core nodes are linked by virtual links in the virtual topology. A virtual link represents one or more links in the physical topology.

A protocol that uses the core (i.e., MDS) of the network is called core-based routing protocol. CEDAR uses core-based routing mechanisms for two primary reasons. First,

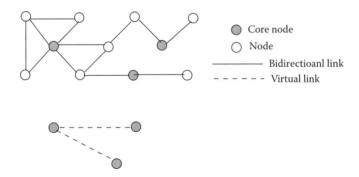

FIGURE 4.7 An example showing a network with a possible set of core nodes and the corresponding core graph in CEDAR.

because of the bandwidth and power constraints, reducing the number of nodes participating in route maintenance (i.e., state propagation and path restoration) is expected to increase network performance and increase network scalability. Second, because of the hidden terminal and exposed terminal problems, local broadcast may be highly unreliable in mobile AHNs. Using only a subset of nodes should reduce the negative effects of local broadcast. QoS routing in CEDAR is achieved by propagating the bandwidth availability information of stable links in the core subgraph. When a link (a,b) experiences a significant change (i.e., changes by some threshold value) in available bandwidth, a and b must inform their respective dominators. The dominators are responsible for propagating state information via slow-moving increase waves and fast-moving decrease waves (i.e., messages) to all other core nodes via the core broadcast mechanism.

In [37], a QoS routing protocol is proposed for AHN with mobile backbones (MBN). In such a network, nodes are grouped into multi-hop local group (LG), and each group elects a group-head to be a backbone node (BN). An example of the two-level MBN is shown in Figures 4.8 with six MBN, six LGs (represented by dash circles), six corresponding BNs (represented by solid circle), and the backbone network connected by wireless links (represented by thick solid lines). The protocol is based on a proactive intra-group link state updating protocol to avoid the expensive global link state update. It reactively discovers multiple segmented QoS paths sequentially from the source to

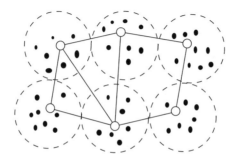

FIGURE 4.8 A two-level *ad hoc* network with MBNs.

the destination. The protocol also uses multiple paths to account for the imprecision of the link state and to increase the probability to find a feasible path. The protocol can handle various QoS routing problems in a unified framework. It can be applied to various QoS routing problems including the minimal delay routing, least cost routing, bandwidth-constrained, or delay-constrained least cost routing. The route search algorithm is modified from the k-shortest path algorithm and the limited path Dijkstra's algorithm. The protocol uses segmented reactive path discovery to find feasible QoS path from group to group. The link state is updated periodically only within each LG or backbone group (LG or BG) in an attempt to reduce the control overhead. Each backbone has the membership information of its LG and broadcasts this information periodically within the BG, and hence each backbone has the BG and all LG membership information. The reactive routing protocol is composed of the two main mechanisms of "Path Discovery" and "Path Maintenance." Each node performs the neighbor discovery and the topology broadcast tasks. Neighbor discovery is implemented by exchanging "hello" message periodically. It is used for a node to identify its neighbors and update its local link state of all its outgoing links. Topology broadcast of a node is used to advertise its local link state to its group. Each node has the link state information of its LG and a backbone has both the link state information of its LG and the BG. Although the results of this protocol seem promising, the clustering algorithm used to construct MBN is not efficient. Moreover, the link state can be easily outdated or become imprecise and the protocol will suffer from high overhead in this case.

In [38], a simple end-to-end QoS routing mechanism in a physical hierarchical environment is proposed. The algorithm operates an end-to-end signaling at each level of the hierarchy (assumed two levels in this paper) for efficient and scalable QoS route construction and maintenance (see Figure 4.9). Nodes are organized into different interconnected domains, called mobile groups, and the nodes have reference point group mobility (RPGM). In the RPGM, each mobile group has a logical mobility center, which is a reference point (or leader node) that follows the group movement. The movement of leader node represents the motion of the group, including location, speed, direction and acceleration, and so on. The other nodes (normal nodes) are uniformly positioned within the group. All the nodes

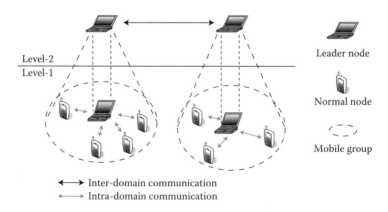

Level-2
Level-1

Leader node

Normal node

Mobile group

⟷ Inter-domain communication
⟵ Intra-domain communication

FIGURE 4.9 A physically hierarchical *ad hoc* network.

within same mobile group have the same group mobility pattern, but have different individual mobility patterns. It is assumed that the leader node has stronger radio transmission capability and better computing power compared to the other nodes in the group. It is also assumed that a normal node in level-1 is controlled by at least one leader node. Also neighboring leader nodes in lever-2 are able to communicate directly with each other through separated channel, called hyper-channel. The algorithm is divided into the following three phases: RREQ phase, QoS path computation phase, and QoS path maintenance phase.

A priority-based QoS routing scheme is proposed in [39] that uses the notion of zone-disjoint routes with the objective of providing an interference-free communication to high-priority traffic. The contention between high- and low-priority routes is avoided by reserving the high-priority zone of communication. Low-priority flows will try to avoid this zone by selecting routes that is maximally zone disjoint with respect to the high-priority reserved zone and will consequently allow a contention-free transmission of high-priority traffic in that reserved zone. The protocol assigns a path to a high-priority flow that is shortest as well as maximally zone-disjoint with respect to other high-priority communications. Each low-priority flow will try to take an adaptive zone-disjoint path avoiding all high-priority zones. If such a path is not available, it will block the flow adaptively to protect high-priority flows. However, unless the hop-count or path length of low-priority flows, packets belonging to low-priority flows may get diverted toward longer path unnecessarily?, increasing the end-to-end delay. Moreover, there is no assurance of convergence, that is, the packets may move around the network in search of zone-disjoint paths but may not reach the destination at all. Adaptive call blocking of low-priority flows were also used to improve the high-priority throughput. Furthermore, narrow-beam directional antennas are used to accommodate multiple numbers of non-overlapping high-priority zones. However, the paths would become less stable with narrow-beam directional antenna especially when the nodes are mobile. It is not obvious how the use of directional antenna will improve the throughput of prioritized flow without degrading the low-priority flows in the network.

4.3.4 QoS Routing Based on Route Scheduling

Based on route scheduling, there are four categories, namely reactive, proactive, location, and hybrid protocols. Proactive table-based routing schemes require each of the nodes in the network to maintain and update routing information, which is used to determine the next hop for a packet transmission in order to reach destination. On-demand routing is an emerging routing philosophy intended to overcome the scalability and routing overhead problems of proactive protocols in MANETs. Routes are created when necessary based on a query-reply approach. It differs from the conventional proactive routing protocols in that no permanent routing information are maintained at network nodes, thus providing a scalable routing solution in large MANETs. Examples of reactive protocols are DSDV [40] and DSR [41] that splits routing into discovering a path and maintaining a path. Maintaining a path happens only while the path is in use in order to make sure that it can still be used. Thus, no periodic updates are needed. The key distinguishing feature of DSR is the use of source routing. That is, the sender knows the complete hop-by-hop route to the destination.

4.3.4.1 Reactive QoS Routing Protocols

The AODV routing protocol [42] has been proposed for best effort routing in mobile AHNs. When a route to a new destination is needed, the node broadcasts an RREQ packet to find a route to the destination. Each node that participates in the route acquisition process places in its routing table the reverse route to the source-node. A route reply (RREP) packet, which contains the number of hops required to reach the destination node, D, and the most recently seen sequence number for the node D, can be created whenever the RREQ reaches either the destination node, or an intermediate mode with a valid route to the destination. To provide QoS support using AODV, a minimal set of QoS extensions has been specified for the RREQ and RREP messages [43]. Specifically, a mobile host may specify one of the two services: a maximum delay and minimum bandwidth. Before a node can rebroadcast an RREQ or unicast an RREP to the source, it must be capable of meeting the QoS constraints. Upon detecting that the requested QoS can no longer be maintained, a node must send an Internet control message protocol QoS LOST message back to the source. The specific extensions for the routing table and control packets (e.g., RREQ and RREP messages) are shown in [43].

In [26], a QoS routing protocol that can establish QoS routes with reserved bandwidth in a network employing TDMA is presented. A small network is assumed, where the topology changes at a relatively slow rate, and sessions transmit at constant bit rates. Assuming TDMA is used at the MAC layer, the QoS measure is the amount of bandwidth (given in number of time slots). The authors showed that the path bandwidth calculation problem is an NP-complete. Hence, an algorithm for calculating the end-to-end bandwidth on a path is developed and used together with the route discovery mechanism of AODV to set up QoS routes. A session specifies its QoS requirement as the number of transmission time slots it needs on its route from a source to a destination. For each session (flow), the QoS routing protocol will both find the route and the slots for each link on the route. To provide a bandwidth of R slots on a given path P, it is necessary that every node along the path find at least R slots to transmit to its downstream neighbor, and these slots do not interfere with other transmissions. Because of these constraints, the end-to-end bandwidth on the path is not simply the bandwidth on the bottleneck link. In addition to building QoS routes, the protocol also builds a best-effort route when it learns of such a route. The best-effort route is used when a QoS route is not available.

A shortcoming of the QoS routing protocol in [26] is that it is designed without considering the situation when multiple QoS routes are being setup simultaneously. An RREQ is processed under the assumption that it is the only one in the network at the moment. When multiple routes are being setup simultaneously, they reserve their own transmission time slots independently, and may interfere with one another. It is possible that two QoS routes will block each other when they are trying to reserve the same time slots simultaneously which may lead to a deadlock situation. This protocol works well in small networks (or over short routes) and under low network mobility. For a large or highly mobile network, it lacks the scalability and the flexibility to deal with frequent route failures.

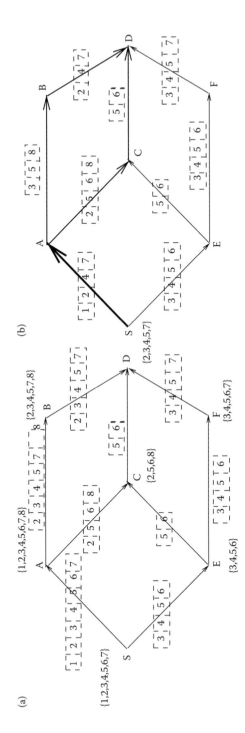

FIGURE 4.10   Example of multi-path in CDMA-over-TDMA channel model. (a) Collecting link state information; (b) the destination selected two paths to receive data from the source node.

In [28], Chen et al. presented an on-demand, link state-based routing protocol. The proposed protocol can find multiple paths between a source-destination pair. CDMA-over-TDMA channel techniques were used to calculate the end-to-end path bandwidth of a QoS multi-path routing (Figure 4.10 that describes a reservation example where in (a) the link state information are collected and in (b) the destination selected two paths to receive data from the source node). The basic idea of this protocol is to reactively collect link-state information from source to destination. The information will be used to construct a flow network, which is a network topology sketched from source to destination. If there are multiple paths between the source–destination pair, destination node will select the path that satisfies its bandwidth requirement and replies back to the source node. The protocol is able to produce high success rate for the RREQs. However, the method used to calculate or reserve bandwidth is not clear. The scalability of the proposed protocol for large networks is questionable. In fact, all protocols that use TDMA to calculate bandwidth are only efficient for small and low mobile networks.

In [44], Lin proposed an on-demand QoS routing protocol that also performs admission control. The QoS metric is bandwidth. Multiple paths are searched at the same time between a source–destination pair. If multiple QoS paths are found, the shortest path will be selected. Since the scheme is pure on-demand, there is no maintenance of any routing tables and there is no exchange of routing information. The path is only discovered upon request. Therefore, in large networks, this path discovery process may incur significant setup times. The network is using a TDMA slot via CDMA channel allocation among different nodes, that is, CDMA is overlaid on top of the TDMA infrastructure. Hence, multiple sessions can share a common TDMA slot via CDMA. The process is carried through two phases: control phase and data phase, as shown in Figure 4.11. In the control phase, all control functions, such as slot and frame synchronization, power measurement, code assignment, slots request, and so on, are performed. The amount of data slots per frame assigned to a virtual circuit (VC) is determined according to bandwidth requirement. In the control phase, each node uses pure TDMA with full power transmission to broadcast its information to all of its neighbors in a predefined slot, such that the network control functions can be performed in a distributed fashion. All nodes take turn in this process. It is assumed that the information can be heard by all of a node's adjacent nodes.

By the end of the control phase, each node should know the channel reservation status and can decide which free slots to request, if any. Therefore, the available path

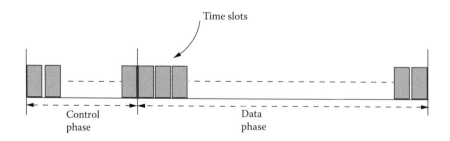

FIGURE 4.11   TDMA time frame divided into two phases.

between two nodes is the set of free slots between them. When the destination node receives an RREQ packet from the source node, it returns an RREP packet by unicasting back to the source. If the reservation process fails, then the scheme gives up and sends an appropriate message back to the destination. The control messages are also used to calculate bandwidth hop-by-hop. The success of this scheme is heavily dependent on the reservation procedure and it may incur large setup times in large networks. It also does not provide fairness among different connections as slots are not distributed among nodes in a fair fashion.

In [45], a QoS routing framework that consists of admission control, resource reservation, and QoS routing were proposed. The shortest path was always used for routing packets. The framework deals with finding shortest paths that satisfy a minimum bandwidth through on-demand channel assignment methods. The framework is heavily dependent on the resource reservation and admission control components where resources are tentatively reserved when a new connection with certain QoS bounds is required. Since nodes are mobile, resource reservation is performed at places where mobile node is expected to visit. Two on-demand channel assignments proposed by the authors in a different work are employed in the proposed framework. The authors claim that their proposed framework is capable of performing well consistently in all performed experiments. The disadvantage of this approach is that its complexity is high and it is hard to make resource reservation in MANETs as the nodes are highly mobile. Furthermore, the proposed bandwidth assignment does not really guarantee the bandwidth as links are prone to failure. In addition, the use of shortest path does not always guarantee the required quality, hence affecting the percentage of accepted calls and could cause congestion on some paths.

The authors in [46] consider the issue of route maintenance and repair of active routes for QoS-sensitive applications. Through simulations, two on-demand route repair techniques tailored for operation in MANETS were demonstrated. Specifically, an FN/RDM-based ROUTE REPAIR scheme, first presented earlier by the authors, is analyzed in this paper and its performance is compared with that of an FN-based ROUTE REPAIR scheme. Moreover, the effects of relative node velocity, node density, and communication range on the communication cost and scalability of a routing protocol are studied. Results illustrated the importance and significance of the route repair process that decreases the network resources needed for route reestablishment when route failures occur as well as to the perceived quality of the call.

Li et al. [47] proposed an on-demand route discovery algorithm that can find multiple disjoint paths. The motivation is that when using disjoint paths, a single link or node failure will only cause a single path to fail. Therefore, disjoint paths can provide higher fault tolerance. The goal of the routing scheme is to find multiple disjoint paths whose aggregate bandwidth can meet the bandwidth requirement. The routing protocol reactively collects link state information from source to destination. A method for choosing proper paths, namely largest bandwidth-hop-bandwidth first was proposed. Bandwidth is measured in the unit of the free timeslots. The timeslots used by a host should differ from that used by any of its two hop neighbors. The MAC sublayer adopts the CDMA-over-TDMA channel model. The source node initiates an RREQ packet and floods it to its neighbors only when it is necessary. Unlike some protocols that consider resource reservation during the route discovery, the proposed protocol initially collects link bandwidth information from source to destination in order to find multiple disjoint paths. Then, proper paths at the destination node are chosen according to the bandwidth and hop-count of each path, and finally necessary bandwidth resource is reserved over each selected path by sending RREP packets. When a source receives a request with bandwidth requirement B and maximal delay bound D, it initiates a RREQ packet by setting the TTL to D and floods the RREQ packet to its neighbors. The TTL is used to limit hop-count of the path. The time slot assignment algorithm is used to acquire the maximal available bandwidth over a path. The largest hop-bandwidth first considers bandwidth and hop-count as two important factors in selecting multiple paths. However, the bandwidth is the most important factor and the hop-count is the second important factor in selecting multiple disjoint paths in this paper. The protocol was shown to improve the success rate by means of finding multiple disjoint paths that provide more aggregate bandwidth. Moreover, simulation results showed a reduced network cost.

An on-demand delay-constrained unicast routing protocol (ODRP) for wireless AHNs was proposed in [48]. Various strategies are employed in the protocol to reduce the communication overhead in acquiring cost-effective delay-constrained routes. ODRP reduces the communication overhead by acquiring a low-cost delay constrained path by employing the following strategies in its route-searching operations: hybrid routing, directional search, and link-delay-based scheduling of (control) packet forwarding. Thus, the operations of ODRP accordingly can be divided into the following two phases: probe the feasibility of the min-hop path and enforcement to perform destination-initiated route discovery. The hybrid routing strategy checks to see if the minimum hop path will satisfy arriving QoS request. If not, a destination-initiated route-searching process via restricted flooding is enforced. Directional search is employed to restrict the search range of the route-searching process. Link-delay-based scheduling of control packet forwarding is for an intermediate node to decide on when it retransmits an RREQ packet it receives. In ODRP, this retransmission is scheduled at a speed proportional to the delay of the link over which the packet was received. If multiple RREQs are received, the one carrying the least delay value is chosen. Simulations showed that ODRP provide a better success ratio, lower average control message overhead, and lower average path cost.

In [49], the authors build on (GAMAN) by adding an effective topology extraction algorithm, called E-GAMAN, to reduce the search space of GAMAN. The E-GAMAN uses two QoS parameters for routing, namely delay and transmission success rate to decide the QoS path. E-GAMAN has two algorithms: search space reduction algorithm (SSRA) and GAMAN. GAMAN and E-GAMAN support soft QoS without hard guarantees. E-GAMAN is a source-based routing algorithm where a small population size of few nodes is involved in route computation. The genetic algorithm (GA) search different routes and they are sorted such that the first one is the best route, but other ranked routes can be used as backup routes. By using a tree-based GA method, the loops can be avoided. By using SSRA, the algorithm extracts the effective topology of the MANET by avoiding transient links and hidden terminal problems. The performance evaluation via simulations shows that E-GAMAN has a good performance compared to GAMAN. However, the proposed algorithms can only be applied for small and medium scale AHNs.

The suburban ad-hoc network (SAHN) aims to provide suburban area connectivity for local area networks at broadband speed with a low initial cost and zero service charges using wireless technology. One of the important requirements of the SAHN implementation is to employ an appropriate routing solution. A number of existing *ad hoc* routing algorithms can be regarded, for example, DSR, *ad hoc* on demand distance vector (AODV) routing, as feasible candidates to be deployed in an SAHN. A routing protocol for SAHN, however, should fulfill many requirements such as routing with guaranteed QoS and balancing load among feasible routes. In [50], a hybrid routing approach, known as SAHNR and based on the principles of DSR and AODV, was proposed. A node tries to discover routes to a destination if no route is found in its routing table or existing routes are unable to provide the required QoS. The route discovery process is accomplished with the help of RREQ and RREP packets. Node authentication and negotiation of shared key are also integrated in this stage with digital signature mechanism. At each intermediate node, local QoS information (available bandwidth, error rate) are appended in the route information list (RIL) of the RREQ packets. Each intermediate node retrieves new routes and their QoS values from the RIL. An RREP packet is generated whenever an intermediate node has one or more routes to the destination, or is the destination node itself. If the node is not the destination node but has a route to the destination, the RREP is constructed by joining RIL of the RREQ with the RTL of the route to the destination. When an intermediate node receives an RREP, the node updates its routing table with previously unknown routes and QoS values contained in the RIL.

In [51], a QoS routing protocol that is based on AODV (QS–AODV) and that creates routes according to application QoS requirements is proposed. The IEEE 802.11 was used as the MAC protocol. A QoS extension is added to the AODV routing table and the control packets, and routes are sought which have sufficient bandwidth for each application. A local repair mechanism is also used to provide a better packet delivery ratio. The only QoS metric considered here is bandwidth for a QoS flow. The effects of mobility on performance are also presented. The performance of QS–AODV is evaluated by measuring three parameters: data packet delivery ratio, normalized routing overhead, and end-to-end data packet delay. It is shown that QS–AODV provides performance comparable to AODV under light traffic conditions. In heavy traffic, QS–AODV provides higher packet delivery ratios and lower routing overheads, at a cost of slightly longer end-to-end delays.

In [52], a protocol named QoS–adaptive source routing (QoS–ASR) protocol is proposed. It is an adaptive soft-QoS protocol with aggregate flows. It applies the source routing mechanism defined by the DSR unicast protocol to avoid channel overhead and to improve scalability. QoS–ASR handles QoS criteria taking into account application requirements (link transmission delay, available bandwidth, packet loss rate) combined with network state-related constraints (battery life, link stability, node congestion state). Actually, QoS–ASR constitutes an extension of protocol DSR, to which QoS features are embedded in the selection, the maintenance of routes, and the traffic management. Moreover, QoS–ASR applies the principle of flow aggregation introduced in the well-known mechanism Diffserv to reduce overhead. The performance of the proposed scheme is evaluated via simulations and is compared to DSR and it is found to perform better than traditional DSR.

The work in [53] build on previous work that demonstrated a framework for supporting QoS routing in MANETs and where two mechanisms for dynamic channel assignment, called the minimum-blocking and bandwidth-reallocation channel assignment (MBCA/BRCA) algorithms, were proposed. MBCA/BRCA are on-demand channel assignment methods that reactively provide a differentiated service treatment to multimedia traffic flows at the link level using novel techniques for end-to-end path QoS maximization. Efficient QoS routing is then accomplished by giving the routing mechanism access to QoS information, thus coupling the coarse grain (routing) and fine grain (congestion control) resource allocations. In [53], the specifics and individual mechanisms of the MBCA/BRCA algorithms and their interaction affect on the overall protocol performance are examined and documented. To be specific, a framework of bandwidth-based routing for QoS support in MANETs is described. A method for slot reservation and distributed channel assignment in mobile AHNs, called MBCA, is presented. In addition, a mechanism that accounts for the channel usage and seeks alternative channel-usage reconfigurations among the active channels is introduced. The objective is to increase the available link/path bandwidth. This method is referred to as BRCA. The proposed system performance is studied through simulations and it is found that MBCA/BRCA methods are able to increase system's aggregate traffic by 2.8 kb/s, on average, comparing to a non-MBCA/BRCA dynamic channel-allocation scheme.

The AODV routing protocol is taken into consideration and an extension to the protocol has been suggested to support QoS in [54]. The extension will work with AHNs that have stationary links, that is, those in which the nodes are not moving that rapidly so as to make QoS routing meaningless. The authors introduce a stability field into the working of the protocol. The stability of the node indicates if the node in consideration should be used for routing or should be deleted from the route tables, which will lead to predictive routes. Depending upon how mobile the node is, a history table for each node is created and the nodes are classified as stable, partially stable, and unstable. Simulation results of this AODV extension showed a better packet delivery ratio when source nodes use stable nodes as intermediate routers. However, with the increase in the speed of the node movement, the stability factor does not significantly enhance the packet delivery ratio. Also the increase in the number of nodes in the networks reduces the packet delivery ratio.

4.3.4.2 Proactive QoS Routing Protocols

Proactive protocols have many desirable properties, especially for applications with real-time communications. This type of communication requires QoS guarantees, such as low-latency route establishment, alternate QoS path support, and resource state monitoring. Since the satisfaction of such requirement is dependent on the accuracy of the routing information stored in the tables, frequent network topology changes may render this information obsolete. In MANETs, the performance of proactive routing protocols deteriorates with frequent node movement.

In [55], the optimized link state routing (OLSR) protocol was proposed. OLSR is a proactive routing protocol, which exchanges topology information with other nodes in the network regularly. The protocol inherits the stability of a link state algorithm and has the advantage of having routes immediately available when needed due to its proactive nature. OLSR is an optimization over the classical link state protocol, tailored for mobile AHNs. The key concept used in the protocol is that of multi-point relays (MPRs). MPRs are selected nodes which forward broadcast messages during the flooding process. MPR selection is the key point in OLSR. The smaller the MPR set is, the less overhead the protocol introduces. The idea of MPR is to minimize the overhead of flooding messages in the network by reducing duplicate retransmissions in the same region. Each node in the network selects a set of nodes in its neighborhood which may retransmit its messages. Each node selects its MPR set among its one hop neighbors. This set is selected such that it covers (in terms of radio range) all nodes that are two hops away. The nodes which are selected as a MPR by some neighbor nodes announce this information periodically in their control messages. Therefore, a node announces to the network, that it has reachability to the nodes which have selected it as MPR. In route calculation, the MPRs are used to form the route from a given node to any destination in the network. The protocol uses the MPRs to facilitate efficient flooding of control messages in the network. OLSR inherits the concept of forwarding and relaying from HIPERLAN (an MAC layer protocol). The QoS support in the protocol is implemented by extending the routing table to include two parameters: minimum available bandwidth and maximum delay. It is worth mentioning that selecting MPRs is similar to finding the MDS of an arbitrary graph.

The authors proposed the QOLSR protocol, a QoS routing over OLSR protocol in [56]. OLSR is a proactive routing protocol, which inherits the stability of a link state algorithm and has the advantage of having the routes immediately available when needed. The OLSR protocol uses a kind of Dijkstra's shortest path algorithm to provide optimal routes in terms of number of hops. QOLSR is an enhancement of the OLSR routing protocol that supports multiple-metric routing criteria. The protocol uses QoS metrics of bandwidth and delay rather than the hop distance. Paths selected by source nodes satisfy the end-to-end bandwidth and delay requirements. In QOLSR protocol, a source node continuously changes next hop in response to the change in the available bandwidth on the path. The paper describes a way to achieve QoS routing without using explicit reservation mechanisms and gives new distributed solutions to the oscillation and collision of flows. The authors claim that the flows will take the appropriate paths without interferences. The QOLSR protocol is an enhancement of the OLSR routing protocol to support multiple-metric routing criteria. QoS parameter reservation is more difficult in proactive than reactive protocols. However, QOLSR protocol suffers from some defects related to QoS information diffused in exchanged messages. For example, the problems of oscillation and collision, although a solution is proposed in the paper, can still happen in some cases. Moreover, no simulations were performed to compare the performance of the enhanced QOLSR, OLSR, and the standard QOLSR protocols.

Three algorithms were developed in [57] that allow OLSR to find the maximum bandwidth path. Through simulation, these algorithms were shown to improve OLSR in the static network case. Two of the proposed algorithms, called $OLSR_{R2}$ and $OLSR_{R3}$ are discussed below, were found to perform well (i.e., guarantee that the highest bandwidth path between any two nodes is found). In the first algorithm, called $OLSR_{R1}$, MPR selection is almost the same as that of OLSR described earlier. However, when there are more than one 1-hop neighbors covering the same number of uncovered 2-hop neighbors, the one with the largest bandwidth link to the current node is selected as MPR. The idea behind the second algorithm, called $OLSR_{R2}$, is to select the best bandwidth neighbors as MPRs until all the 2-hop neighbors are covered. The third algorithm, called $OSLR_{R3}$, selects the MPRs in a way such that all the 2-hop neighbors have the optimal bandwidth path through the MPRs to the current node. Here, optimal bandwidth path means the bottleneck bandwidth path is the largest among all the possible paths. Note that the overhead when using $OLSR_{R3}$ may increase compared with the original OLSR algorithm because we may increase the number of MPRs in the network. This is because $OLSR_{R3}$ may select a different MPR for each 2-hop neighbor.

The authors in [58] focus on the proactive routing and link bandwidth allocation problems in MANETs. The authors proposed to combine the well-known ant-based routing or swarm intelligence algorithm and multi-path approaches in order to improve the QoS required by multimedia traffic carried over MANETs. The core of this approach is the routing table which assigns probabilities to next hops and the special agents that use these probabilities to choose the next hop. The special agents that explore the routes are called ants. The algorithm uses three kinds of agents: regular forward ants, uniform forward ants, and backward ants. Regular and uniform forward ants explore and reinforce the paths of the network proactively. They create a probability distribution at each node for its neighbors. Backward ants are used to propagate the information collected by forward ants through the network, and to adjust the routing table entries.

A routing scheme, called localized adaptive QOS routing (LAQR), was proposed in [59]. The source node makes its routing decision based on its local information and some infrequently exchanged network information. Instead of exchanging information with other routers to obtain a global view of the network QoS state, a source node tries to infer the network QoS state from locally collected flow statistics such as flow arrival and departure rates and flow blocking probabilities, and performs adaptive proportioning of flows among a set of candidate paths based on this local information. However, this strategy will brought about frequent QOS re-routing and complex QOS maintenance process. In the scheme, it is assumed that the re-route information is always known to the source node. It is also assumed that the longer the route exists, the less likely it will vanish in a short period. In the LAQR scheme that is based on a table-driven routing strategy, all nodes will exchange their route information periodically. So, the approximate existing time of the route can be estimated. Due to the frequently changing network conditions of MANETs, the candidate paths are selected by a path selection procedure in a dynamic fashion. Based on what the source node knows about the network state, paths that satisfy the QOS requirements and stable will be included in the path set. Unstable routes, which are the routes that exist for time t, that is, less than a critical value T, will be excluded. The path selection procedure is executed whenever the network state is updated.

In [60], a methodology to a support QoS routing in terms of numerical probability of successful message delivery in a specified time is proposed. This is an important feature to users who want rapid message delivery, even as network topology changes. The case of multiple message priorities is also included. Routing knowledge is distributed by adding path and error rate information to packets as they traverse the network without adversely protocols that employ selective reject retransmission error control. The protocol and methodologies presented in this paper are designed to provide maximum speed and shortest path delivery of messages in a fixed network with adaptability to the environment when the network's topology changes. Rather than developing such a reactive protocol, a protocol is presented that proactively seeks to rapidly deliver messages with a predictable level of quality in a network which is fixed, variable path, or a combination of both. It is shown how the value added network provider can use this QoS to provide many different types of service trading speed of delivery against priority of delivery and cost of communications capacity.

The authors in [61] propose a proactive routing technique that satisfies multiple QoS constraints. It computes the feasible path by using connectivity index (CI) and delay to provide hop-by-hop QoS routing in AHNs. CI is routing metric which indicates connectivity of each node in AHNs; a large value of CI indicates more branch node which makes a node more robust to link failure. They argue that CI parameters (2-hop CI and 2-hop modified CI) can be used as routing metric to improve the performance in terms of throughput, packet delivery ratio (PDR), and delay. Which they operate well in either light or heavy load situation since it illustrates more actual branch connectivity which is quite appropriate for applying to AHN, since more branch nodes make a node more robust to link failures.

4.3.5 Location-Based QoS Routing Protocol

Little work has been done in position-based QoS routing. In using position-based routing, the research community usually assumed the use of GPS as an external position determination server. GPS provides location information (latitude, longitude, and possibly altitude) to mobile nodes in MANETs that are equipped with GPS cards. The use of GPS is justified by the fact that GPS cards will be, in the near future, very inexpensive and will be deployed in each car, and possibly in every user terminal. Accuracy measurement in GPS can be enhanced by using differential GPS, which offers accuracy up to a few meters. Position-based routing [62] is a new trend in performing routing in MANETs. Routing in this approach requires including the destination's position in the packet, in addition to the position of neighbor nodes that will forward the packet. Therefore, no route establishment and maintenance are needed. As a result, the efficiency of any position-based routing protocol is dependent on both the selection of the position server(s) and the selection of the forwarding strategy. If the use of GPS is not feasible, a GPS-free approach can also be used to build a local coordinate system based on the exchange of node's relative position [15].

In [63], a DSR-based protocol, which uses GPS for location determination, was proposed. By letting each node maintain a table of the position of all other nodes, the mobile node maintains a snapshot for the network connectivity graph and therefore will be able to compute paths locally without the need for route discovery delay. By exploiting location information obtained from GPS, the proposed protocol enhances the end-to-end delay for packet delivery compared to the original DSR. This enhancement is possible because a source node has information about other nodes' location, and hence selects the path, which will satisfy the end-to-end delay with high probability based on its location database. However, the exchange of position information in this protocol may use much of the network resources. Furthermore, if the network topology changes frequently, which is the normal case in MANETs, the positions maintained in the position tables becomes stale quickly and they need to be updated more often. The update process consumes bandwidth in a limited-bandwidth MANETs.

In [18], the authors used GPS to obtain the position and speed of the mobile nodes in the network area. They used this information to build routes that will be valid for a certain period of time and can therefore be used to provide QoS paths which satisfy certain delay guarantees. Therefore, the proposed protocol is also a predictive protocol in the sense that the motion of the nodes is predicted to build routes in the network. The lifetime of any link between two mobile nodes is defined as a function of their current locations, current speed, and moving directions. Therefore, this information is used on every link to find the expected lifetime of the whole path. The authors claimed that by having a stable path, the path setup time will be reduced, as well as the network control overhead. The authors did not describe how the position information is distributed among nodes and how the prediction information is updated. The GPS-based routing algorithm has two drawbacks. One is that GPS cannot provide the nodes much information about the physical environment and the second is the power dissipation overhead of the GPS device itself.

In [15], the authors proposed a distributed mechanism for GPS-free positioning in mobile AHNs. They used the method of triangulation and coordinate translation to form the coordination system. Being independent of the external position server, the coordinate system provides continuous information about positions of different nodes which can be used to perform QoS routing in MANETs. The disadvantage of this solution is that it is computationally intensive and is also expensive in terms of the number of messages to be exchanged before a coordinate system can be established since each node individually re-orients its coordinates to the reference node's coordinates.

In [64], an approach for integrating QoS in the flooding-based route discovery process was proposed. The proposed positional attribute-based next-hop determination approach discriminates the next hop node based on their location or capabilities. When a RREQ is broadcast, the intermediate node has two options. First, either it will randomly rebroadcast it to its neighbors and the delay in this case is also random and called random rebroadcast delay. Second, instead of random rebroadcast, the receivers opt for a delay inversely proportional to their abilities in meeting the QoS requirement of the path. The decision at the receiver side is made on the basis of a predefined set of rules. Thus, the end-to-end path will be able to satisfy the QoS constraints as long as it is intact. A broken path will initiate a QoS-aware route discovery process, which restarts the whole process again.

An integrated framework is proposed for performing QoS routing in MANETs. The GPS information* were utilized to divide the network area into fixed and square zones (clusters) and an optimal election algorithm is executed inside each zone to select one node as a cluster head. Figure 4.12 shows an example of the clustering process in both homogeneous (identical transmission ranges and heterogeneous (variable transmission

---

* A GPS-free approach is also possible.

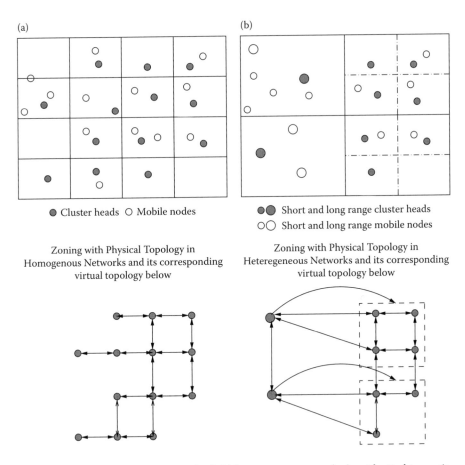

(a)

(b)

● Cluster heads   ○ Mobile nodes

●● Short and long range cluster heads
○○ Short and long range mobile nodes

Zoning with Physical Topology in
Homogenous Networks and its corresponding
virtual topology below

Zoning with Physical Topology in
Heteregeneous Networks and its corresponding
virtual topology below

FIGURE 4.12 The zoning process in both (a) homogeneous networks (i.e., identical transmission range) and (b) heterogeneous networks (i.e., variable transmission range): an example.

ranges)) networks by using a location service. The role of the cluster head is dynamically changed when the network status change. The set of the selected cluster heads form a rectilinear virtual topology. The virtual topology is used as a wireless virtual backbone to perform QoS routing. An extended version of open shortest path first (OSPF) coupled with an extended version of the weighted fair queuing (WFQ) operates on the virtual topology to provide end-to-end statistical guarantees in terms of bandwidth and maximum delay. The motivation of using OSPF is that OSPF is currently used in the Internet, hence integration between MANETs and the Internet is implemented easily. The extended version of OSPF employs an efficient link update mechanism and the extended version of WFQ to obtain link costs. A hybrid criterion is used to compute QoS routes on the virtual topology. The objective is to maximize the call acceptance rate in MANETs.

An energy- and QoS-aware route discovery protocol for a large-scale MANETs was developed in [65]. The objective is to find a route between a source and a destination (whose location is usually unknown to node) satisfying the energy-efficiency or QoS constraints (bandwidth-efficiency), while incurring as small a routing overhead as possible. Each node in the network maintains the location information for nodes within its proximity and for a small subset of the network nodes. The heart of the proposed protocol is the location-update scheme. The routing protocol is a hybrid of existing location-based protocols with support for efficient re-routing. To achieve shortest-path QoS routing, a combinations of the following routing metrics were used: the number of hops, the total transmission power, the residual battery energy, and the end-to-end delay.

An on-demand location-aided, ticket-based QoS routing protocol (LTBR) was proposed in [66]. Two special cases of LTBR, LTBR-1, and LTBR-2, are discussed. LTBR-1 uses a single ticket to find a route satisfying a given QoS constraint. LTBR-2 uses multiple tickets to search valid routes in a limited area. All tickets are guided via both location and QoS information. The proposed approach differs from the original ticket-based probing scheme in using the on-demand feature. The tickets are dynamically generated based on timely position and QoS information and with no routing table maintenance needed. Therefore, LTBR has lower overhead compared with the original ticket-based routing. On the other hand, LTBR can find routes with better QoS qualities than traditional location-based protocols. In LTBR, a ticket is a special control message for searching QoS paths. A ticket contains the information of the QoS routing request and traversed path. In the routing process, one ticket is originally generated at the source and sent to neighbors selected by certain ticket assignment rules. Any intermediate node receiving a ticket either ignores it or sends tickets to successors selected in the same way as the source. A ticket will be forwarded until reaching the destination, or stopped at an intermediate node if it violates the QoS constraints. A ticket arriving at the destination indicates the success in finding a path. Overall, the routing mechanism of LTBR differs from the original ticket-based routing in two aspects: (i) LTBR uses location information to guide ticket forwarding without the need for the routing table in TBP and (ii) LTBR allows issuing multiple tickets in the intermediate nodes to improve the success rate with controllable overhead. Although the protocol seems to enhance on the original TBR, the problems with the frequently changing topology will render the routing algorithm useless. Moreover, the setup time to find a path might be long in many cases.

A scheme that relies on a location-based *ad hoc* on-demand routing protocol (greedy perimeter stateless routing (GPSR)) to discover routes to the destination of a new traffic flow is proposed in [67]. The scheme is a cross-layer design as the resource allocation process takes into account the characteristics of the IEEE 802.11 DCF and the dynamics of its service process by using both traffic and link-layer channel models. The scheme provides probabilistic end-to-end delay guarantees, instead of average delay guarantees without consuming the limited processing power of the AHN nodes or the channel bandwidth in frequent measurements or traffic monitoring. The scheme consists of two phases. The first phase is the discovery part, which is responsible for discovering possible routes to be tested for admission by the resource allocation process. The second phase is the route maintenance, which is invoked either during the resource allocation process or when the route is broken.

Another cross-layer solution was proposed in [68]. Cross-layer optimizations were introduced to improve the performance of the different protocols that conform to it. Their proposal builds on the IEEE 802.11e standard by adding a probe-based admission control system, along with an enhanced version of the DSR protocol to make it efficient also at high degrees of mobility. The proposed architecture is modular in that it allows the plug-in of different protocols, which indeed offers great flexibility. It does not rely on intermediate stations along an end-to-end path for admission control or signaling purposes, avoiding resource consuming tasks such as continuous channel measurements, traffic shaping, and resource reservation. By restricting requirements to a minimum, they are able to use devices with reduced computing power.

4.3.6 QoS Routing Based on Path Cardinality

Papadimitratos et al. [69] proposed a real-time multi-path routing algorithm, called disjoint path set selection protocol (DPSP), based on a heuristic that picks a set of reliable paths. The convergence to a highly reliable path set is fast, and the protocol provides flexibility in path selection and routing algorithm. Furthermore, DPSP is suitable for real-time execution with nearly no message exchange overhead and with minimal additional storage requirements. This paper presents evidence that multipath routing can mask a substantial number of failures in the network compared to single path routing protocols, and that the selection of paths according to DPSP can be beneficial for mobile AHNs, since it dramatically reduces the rate of route discoveries. It is shown that the mean time to failure under a multipath routing algorithm is roughly a factor of three higher than that of a single path routing algorithm. Furthermore, the authors say that their scheme is twice as effective as currently proposed schemes that select a path set by minimizing the number of hops.

4.3.7 QoS Routing Based on Design Space

The paper in [70] studied three MANET routing protocols: OLSR, DSR, and AODV, with an emphasis on the effect they have on various QoS metrics. Specifically, AODV with and without link layer feedback, DSR with link layer feedback, and OLSR with and without link layer feedback were considered. In particular, the study focused on the path selection, broken links detection, and message buffering differences of the various protocols. The effects of these differences are quantified in terms of packet delivery ratio, end-to-end hop count, end-to-end latency, and mechanism overhead. We compare the performance of the protocols on random movement scenarios. The empirical results presented in this paper. It is found that the proactive protocol, OLSR, builds paths with consistently lower hop counts than the reactive protocols, AODV, and DSR, a fact that leads to a reduction in end-to-end latency that assists a QoS model in meeting timing requirements and improves global network performance. On the other hand, a routing protocol that cannot quickly recover from link breakage caused by mobility renders a QoS model incapable of meeting delivery requirements.

4.3.8 QoS Routing Based on QoS Metric

4.3.8.1 Bandwidth-Aware QoS Routing in MANETs

In [71], a GA-based routing method for mobile ad-hoc networks (GAMAN) is proposed. Robustness rather than optimality is the primary concern of GAMAN. The authors argued that it is better to find a route very fast in order to have a good response time to the speed of topology change, than to search for the optimal route in MANETs because the network topology is changed frequently. However, the authors only consider a type of MANET whose topologies are not changing that fast which contradicts their original assumptions. The GAMAN uses two QoS parameters (delay and transmission success rate) to decide the QoS path with soft QoS support. However, the required QoS should be ensured when the established paths remain unbroken. The GAMAN algorithm is a source-based routing algorithm where few nodes are involved in route computation. These few nodes are called population. By taking a subpopulation, the nodes broadcast messages only for the nodes in this subpopulation. The GA search different routes and they are sorted such that the first one is the best route, but other ranked routes can be used as backup routes. By using a tree-based GA method, the loops can be avoided. The performance evaluation via simulations shows that GAMAN is a promising QoS routing algorithm for MANETs.

4.3.9 Power-Aware QoS Routing in MANETs

There is an increasing interest in the power-aware routing in MANETs. *Ad hoc* wireless networks are power constrained since nodes operate with limited battery energy. To maximize the lifetime of these networks (defined by the condition that a fixed percentage of the nodes in the network will "die out" due to lack of energy), network-related transactions through each mobile node must be controlled such that the power dissipation rates of all nodes are nearly the same, that is, avoid overloading a subset of nodes.

There have been some studies on power-aware routing protocols for MANETs. In [72], a source-initiated (on-demand) routing protocol for mobile AHNs that increases the network lifetime was proposed. Simulation results show that the proposed power-aware source routing protocol has a better performance than other source-initiated routing protocols in terms of the network lifetime. A greedy policy was applied to the fetched paths from the cache in order to make sure no path would be overused and also make sure that each selected path has the minimum battery cost among all possible path between two nodes. Power-aware source routing solves the problem of finding a route π at route discovery time t such that the following cost function is minimized:

$$C(\pi, t) = \sum_{i \in \pi} C_i(t)$$

where $C_i(t) = \rho_i (\frac{F_i}{R_i(t)})^\alpha$ and ρ_i is the transmit power of node i, F_i is the full capacity of node i battery, R_i is the remaining battery capacity, and α is a weighting factor. The authors also presented a route discovery and maintenance of the routes based on the

DSR techniques. When an intermediate node receives an RREQ packet, it starts a timer (*T*) and keeps the cost in the header of that packet as Min–Cost. If additional RREQs arrive with same destination and sequence number, the cost of the newly arrived RREQ packet is compared to the Min–Cost. If the new packet has a lower cost, Min–Cost is changed to this new value and the new RREQ packet is forwarded. Otherwise, the new RREQ packet is dropped. The destination node waits the threshold number (*T*) of seconds after the first RREQ packet arrives. During that time, the destination examines the cost of the route of every arrived RREQ packet. When the timer (*T*) expires, the destination node selects the route with the minimum cost and replies. Subsequently, it will drop any received RREQs. Route maintenance is needed when connections between some nodes on the path are lost due to their movement or the energy resources of some nodes may be depleting quickly. A local approach was adopted for route maintenance because it minimizes control traffic. In the local approach, each intermediate node in the path monitors the decrease in its remaining energy level, hence increases its link cost when necessary. When the link cost increase goes beyond a threshold level, the node sends a route error back to the source and the route is invalidated.

In [73], the authors proposed a routing algorithm based on minimizing the amount of power (or energy per bit) required to deliver a packet from source to destination. The algorithm proposes to use a function *f*(*A*) to denote node *A*'s reluctance to forward packets, and to choose a path that minimizes the sum of *f*(*A*) for nodes on the path. This routing protocol addresses the issue of energy critical nodes. As a particular choice for *f*, [73] proposes $f(A) = 1/g(A)$, where $g(A)$ denotes the remaining lifetime, normalized to be in the interval [0,1]. To minimize the total energy consumed per packet, it has been observed in [73] that the routes selected will be identical to routes selected by shortest hop routing, since the energy consumed in transmitting (and receiving) one packet over one hop is considered constant. More precisely, the problem is stated as:

$$\text{Minimize} \sum_{i \in R} P(i, i+1) \qquad (4.1)$$

where $P(i, i + 1)$ denotes the power needed for transmitting (and receiving) between two consecutive nodes, *i* and *i* + 1 (i.e., link cost), in the route *R*. The link cost can be defined for two cases: (i) when the transmission power is fixed and (ii) when the transmission power is varied dynamically as a function of the distance between the transmitter and intended receiver. For the first case, energy for each operation (receive, transmit, broadcast, discard, etc.) on a packet is given in [74] by $E_{packet} = b \times \text{Packet} - \text{size} + c$, where *b* and *c* are the appropriate coefficients for each operation. Coefficient *b* denotes the packet size-dependent energy consumption and *c* is a fixed cost that accounts for acquiring the channel and for MAC layer control negotiation. In the second case, packets are retransmitted via some intermediate nodes that may be available in MANETs. The idea is to vary the transmitted power level (*T*) in accordance with the distance *d* such that the received power follows the relation $R = (T/d^{\alpha})$. Several algorithms were proposed for this second case where the value of the required power transmission levels to achieve communication is calculated by making use of the positions of the intermediate nodes with relative to the sender node. Assuming that we can set additional nodes in arbitrary positions between the source and destination, the power optimal packet transmissions can be estimated as discussed in [75].

In [75], a localized routing algorithm for the second case is proposed. The authors [75] assume that the power needed for transmission and reception is a linear function of d^α where d is the distance between the two neighboring nodes and α is a parameter that depends on the physical environment. They make use of the GPS position information to transmit packets with the minimum required transmit energy. The key requirement of this technique is that the relative positions of nodes are available to all nodes. However, this information may not be easy readily available. The main disadvantage of the problem formulation of the previous approach is that it always selects the least-power cost routes. As a result, nodes along these routes tend to *die* sooner because of the battery energy exhaustion. This is doubly harmful since the nodes that die early are precisely the ones that are needed most to maintain the network connectivity (and hence useful service life). Therefore, it is better to use a higher power cost route if it avoids using nodes that have a small amount of remaining battery energy.

The minimum battery cost routing algorithm, proposed in [76], minimizes the total cost of the route. It minimizes the summation of the inverse of the remaining battery capacity for all nodes on the routing path. A conditional max–min battery capacity routing algorithm is also proposed in [76]. This algorithm chooses the route with minimal total transmission power if all nodes in the route have remaining battery capacities higher than a threshold; otherwise, routes including nodes with the lowest remaining battery capacities are avoided. Minimum battery cost routing showed better performance than min–max routing in terms of expiration time of all nodes. Conditional max–min routing showed different behavior that depended on the value of the chosen threshold. However, since there is no guarantee that minimum total transmission power paths will be selected under all circumstances, it can consume more power to transmit user traffic from a source to a destination, which actually reduces the lifetime of all nodes.

4.4 QoS Routing in WMNs

With the increasing interest and developments in wireless technologies, recently wireless mesh networks (WMNs) started to emerge with the aim of integrating all types of different wireless networks, providing broadband access, availability, and coverage to various wireless clients. The key idea in WMNs is to deploy many mesh routers, which can both route data through multi-hopping and connect to the Internet through a gateway. Any wireless client then needs to communicate with a mesh router in order to be part of the mesh network as shown in Figures 4.13 through 4.16. Note that the wireless clients, called mesh clients can also utilize multi-hop communication in order to relay data among themselves. Thus, the mesh network is a multi-hop topology among both mesh routers and clients which is connected to the Internet. This can be regarded as the combination of AHNs (MANETs) and infrastructure-based wireless LANs which can provide better connectivity and bandwidth to the users. While MANETs increase the possibility of multi-hop communication and thus coverage, wireless LANs can provide high-speed and high-bandwidth communication. Possible applications of WMNs

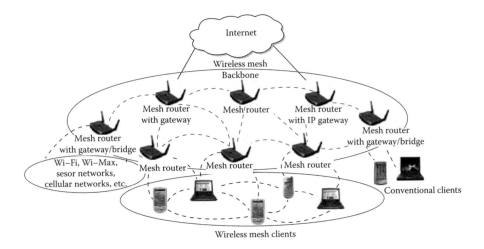

FIGURE 4.13   Wireless mesh network architecture.

include broadband home networking, community networking, metropolitan area networking, building automation, and peer-to-peer networking. WMNs have their own features and characteristics which distinguish them from MANETs, WSNs, and wireless LANs. For instance, the mobility and power consumption are not much of concern in WMNs since the mesh routers usually do not move and they have power supplies. Only the mesh clients can move partially and may have limited power supplies. In addition, WMNs can have clients and routers with multiple radios/channels. Routing with multiple channels can provide better throughput and bandwidth, but it is very challenging due to interference issues among the radios. This is neither the case in MANETs nor in WSNs. They all use single channel communication.

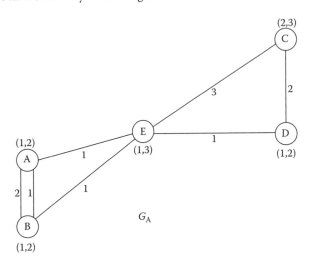

FIGURE 4.14   Example of WMN: two channels, three NIC wireless network.

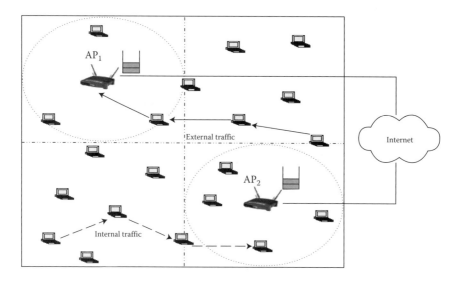

FIGURE 4.15   Internal and external traffic.

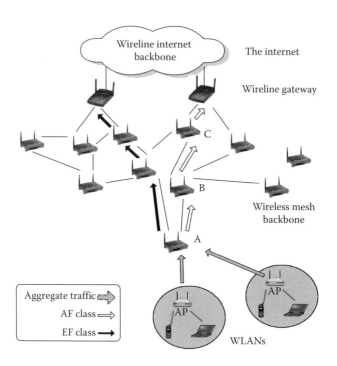

FIGURE 4.16   QoS routing in WMRs. Different classes are routed on different paths.

Integration of different types of networks is another concern. This is not easy since new standards should be set for interoperability and communication. Due to these unique characteristics, new routing protocols for WMNs have started to be developed. Given that one of the aims in mesh networking is to provide broadband networking for applications such as video on demand, VoIP, and so on, a lot of researchers focused on the QoS routing aspects in WMNs. Specifically, they try to apply previous research on QoS routing in MANETs to multi-hop, multi-channel, and resource-rich wireless mesh routers in order to provide desired services to the end users. While there are many technical challenges and open problems in this area, the research is in the very early stages. In this section, we will survey the current routing protocols for WMNs which strive to provide some QoS in terms of various metrics.

4.4.1 Interference-Aware Topology Control and QoS Routing in Multi-Channel Wireless Mesh Networks

A bandwidth aware routing (BAR) protocol for multi-channel, multi-NIC (network interface card), multi-hop wireless networks is presented in [4]. The nodes in this network are assumed to have multiple NICs and can use multiple channels to transmit data. For example, in Figure 4.1 [4] all the nodes have two NICs and there are three channels available to be used: 1, 2, and 3. BAR is one of the first routing protocols which consider QoS support in multi-channel multi-hop WMNs. With the usage of multiple channels, flow-based routing is possible. This means that some data can be fragmented for transmitting on multiple outgoing links (channels). In this case, one of the most important technical challenges is to deal with interferences of channels. There are however new interferences introduced with multi-channel environments. These are inter- and intra-flow interferences. While inter-flow interference occurs among the transmissions of two different routes, intra-flow interference can occur among the transmissions within the same route. The BAR problem particularly deals with these interferences. Given a new connection request with source node s, destination node t, and bandwidth requirement B, BAR finds a route with a required bandwidth on the given topology. The authors reduce this problem to a linear programming (LP) problem by constructing an auxiliary graph from the initial graph. The LP solution is shown to run in polynomial time in the paper. The authors also propose a heuristic for the BAR problem, which computes a single path (rather than a multi-path) for packet routing. This solves the problem of packet fragmentation and reassembly in the case of multi-channel routing. No splitting means that for a specific connection, traffic can only be received and transmitted by a single NIC at any involved node. Note that this NIC can vary for different nodes on the path.

4.4.2 QoS Routing for Mesh-Based Wireless LANs

The current wireless LANs which are based on IEEE 802.11 standard only works for one-hop connection. That is, the user should be within a certain distance of the AP in order to have an Internet connection. Of course, this necessitates employing many numbers of APs for scalability. If any user can act as a router in these networks, then a multi-hop network in the form of a mesh can be formed. While this is not exactly consistent with the broad definition of WMNs, it is still a primitive WMN. Given that the network is multi-hop and the users can be mobile, the idea of QoS routing in WMNs would be very similar to MANETs as shown in [77]. In this work, the authors propose a QoS routing protocol which can provide minimum bandwidth and maximum end-to-end delay guarantees. The protocol is designed to sit between MAC layer and IP layer so that it will work with any of the MAC protocols for wireless networks. The WMN protocol proposed in that work have the following features: (i) virtual destination support, (ii) full on-demand hop-by-hop routing with no route caching, (iii) per-flow granularity, (iv) neighborhood maintenance, (v) temporary bandwidth reservation, and (vi) quick route violation detection and recovery. The virtual destinations are those which are beyond the AP. Thus, the traffic in WMNs is divided into two classes: external traffic and internal traffic. Each node discovers its neighbors through hello messages as in the case of other MANET protocols. The route discovery is also similar which is on demand. It aims to find the shortest path (in terms of hop counts) to the APs. While the bandwidth is reserved for bandwidth constrained routing, end-to-end latency is discovered through round trip delay. The paper also includes mechanisms to maintain QoS routes when the data are flowing. While the paper claims that it merges MANETs and infrastructure-based LANs, the proposed techniques in the paper do not provide general solutions to WMNs where multi-hopping also exists among the APs. The approach can only be applied to multi-hop wireless LANs.

4.4.3 Differentiated Services for Wireless Mesh Backbone

One of the QoS routing solutions for single-channel but multi-hop AP-based WMNs is proposed in [5]. The paper in fact proposes a DiffServ architecture for WMNs which also includes QoS provisioning at both network and MAC layers. DiffServ architecture defines three types of classes: (i) expedited forwarding (e.g., real-time data), (ii) assured forwarding (e.g., reliable delivery), and (iii) best effort (normal delivery). The routers differentiate the packets based on these classes and the packets within the same class are aggregated and forwarded to the upper AP as shown in Figure 4.3. Note that this mechanism is different than providing per-flow guarantees. The QoS routing protocol has two functions in this architecture. It should determine the route as well as the wireline gateway where the data will be forwarded. The paper proposes an on-demand QoS routing protocol which has four components: load classifier, path selector, call admission control (CAC) routine, and route repair routine. Load classifier monitors the traffic load and based on the load the path selector can switch to another path. The path selector mainly determines the gateway and a path to that gateway. The path is found through GPSR [14] protocol. Once it is found, CAC procedure is run to check whether the path satisfies the bandwidth requirements. If not, path selector should find another path with the given requirements. Route repair routine is used whenever a QoS path is broken. An interesting extension to this paper would be to consider multi-hop channels when designing the DiffServ and the corresponding QoS routing protocol.

In addition to the above work, there are other protocols/studies which are worth mentioning here, given that they are related to QoS routing in WMNs. For instance, ROMER

[55] is such a protocol which aims to improve the throughput of the network. It does not provide any guarantees but still can be regarded as a protocol striving to improve the performance in terms of throughput and provide robustness. The idea in this protocol is that for each packet a mesh is created on the fly and each intermediate node opportunistically selects the instantaneous, higher quality wireless links to maximize the delivery throughput. It avoids the performance penalty of repetitive transmissions at the MAC layer by avoiding extremely poor links. With this greedy approach it can provide near-optimal throughput. In fact, the simulations results confirmed that it can increase the throughput up to 195% with respect to conventional single channel routing. The work in [36] studies the performance of multimedia traffic in WMNs. The mesh-based test bed is used to transmit video and voice data with different network configurations and network cards. The authors observed the capacity of the network, the number of flows to be supported, impact of RTS/CTS, and MAC retransmissions on the performance. The conclusions of these experiments stated that particularly packet jitter variations should be improved and an optimal value for the number of maximum retransmissions at the MAC layer should be determined.

4.5 Open Issues in QoS Routing in AHNs

In this section, we discuss some of the most important issues that still need further investigation and more research in the area of QoS provisioning in general, and QoS routing in MANETs in particular. In the following, we list some of those problems pertaining to QoS provisioning and routing in MANETs:

- *Multi-Class Traffic:* The issue of accommodating user traffic with multiple classes is difficult, especially in heavily loaded setups (i.e., guaranteeing QoS for lower level classes may be extremely difficult or impossible). Packet exchanges should not be treated with equal priority in a QoS network. The exchange of control packets should receive higher priority than user data packets in a network designed for QoS. The QoS policy must allow different priorities to exist even among different flows of user packets.
- *Pre-emptive QoS Routing:* Allowing pre-emption in QoS routing is an open area for further research. The development of QoS routing policies, algorithms, and protocols for handling user data with multiple classes coupled with preemption is also an open area for further research. Differentiated services technology proposed for the Internet might help in this situation.
- *Different Operational Conditions:* The issue of providing QoS under different operational conditions (e.g., link and node failure) is worth of further investigation. The type of QoS provided will be dependent on the type of error and its place. How to preserve QoS guarantees under various failure conditions in MANETs is an open issue.
- *Mobile Nodes Position Identification:* The provisioning of QoS is dependent on the position of mobile nodes. Each mobile node must know its adjacent neighbors and other mobile nodes. This is difficult because of dynamic changes in the network topology occur frequently in an AHN. This has a direct impact on the QoS routing protocol. The use of external entities (e.g., GPS) is promising, despite the need to

solve the problems associated with the use of GPS system. A suitable position determination system and a position upgrade mechanism are highly required.

- *Mobility Model:* Guaranteeing QoS in such a network may be impossible if the nodes are too mobile. The challenges increase even more for those AHNs that support both best effort services and those with QoS guarantees and are required to interwork with each other. Most protocols used impractical mobility models for their evaluations. There is a need for a mobility test bed that reflects the actual nodes mobility patterns.

- *Layer Integration:* Recent modeling and simulations of mobile AHNs are incomplete and do not clearly combine the physical layer and the MAC layer with the data link layer and network layer. The impact of layer integration of the QoS provisioning and performance of the network is not quantified yet.

- *Internet–MANET Interaction:* The interaction between mobile AHNs and the existing global information infrastructure, the Internet, is another major area of research. A form of gateway that provides this kind of interaction is still in its early design stages and need further investigation. Being able to connect to wireline network (e.g., Internet) can help in the QoS routing. For example, when MANET becomes partitioned, an alternate routing between the network partitions can be carried out through the wireline network.

- *MANET Security:* Security is an important issue which is a desirable feature in any QoS routing protocol. There should be a form of network-level or link-layer security aspects in the protocol such that malicious retransmissions, manipulation, snooping, and redirection of packets are not allowed.

- *Network Partitioning:* In MANET, natural grouping behavior in mobile user's movement leads to frequent network partitioning. Network partitioning poses significant challenges to the QoS routing protocol since partitioning disconnects many mobile users from each other.

- *Nodes Cooperation:* Some mobile nodes may misbehave by agreeing to forward packets and then failing to do so. Obviously, this has a direct impact on QoS in MANETs and may lead to an effectively disconnected network. A node may misbehave because it is overloaded, selfish, malicious, or broken. Hence, node cooperation is essential for the network to perform its duties. Efficient algorithms that ensure or force nodes cooperation in MANETs are still highly needed.

- *Heterogeneous Networks:* It is generally assumed that all nodes in MANETs are homogenous both in terms of capacity and in terms of functionality. MANETs are typically heterogeneous networks with various types of mobile nodes. In military application, different military units ranging from soldiers to tanks can come together; hence forming an AHN. In conference application, different types of mobile devices such as personal digital assistants, smart badges, and laptops may exist in the AHN at the same time. Nodes differ in their power capacities and computational speeds. Thus, mobile nodes will have different packet generation rates, routing responsibilities, network activities, and power draining rates. QoS issues in heterogeneous networks and especially QoS routing should be investigated in more depth and this issue just started to receive attention in the literature.

- *Adaptive and Context-Based QoS:* Need to write something here as suggested by Dr. Younis.

4.6 Conclusion

This chapter extensively surveyed the QoS routing protocols in AHNs. The issue of QoS routing is important and challenging in AHNs due to the unique characteristics and requirements of these networks. In this chapter, we presented an extensive survey of QoS routing protocols in three classes of AHNs, namely MANETs, WSNs, and WMNs. The design challenges in achieving QoS routing in AHNs were first outlined. In addition, the chapter defined a taxonomy that is suitable for examining a wide variety of protocols in a structured way and has also explored tradeoffs associated with various design choices. For each protocol, the strengths and limitations were also discussed in detail. Some possible future directions of QoS routing in AHNs were also outlined.

References

1. Akyildiz, I. F., Su, W., Sankarasubramaniam, Y., and Cyirci, E., Wireless sensor networks: a survey. *Comput Netw* 2002; 38(4): 393–422.
2. Liao, W., Tseng, Y., and Shih, K., A TDMA-based bandwidth reservation protocol for QoS routing in a wireless mobile ad hoc network, in ICC 2002, Vol. 5, 2002, pp. 3186–3190.
3. Frodigh, M., Parkvall, S., Roobol, C., Johansson, P., and Larsson, P., Future-generation wireless networks. *IEEE Person Commun* 2001; 8(5): 10–17.
4. Iwata, A., Chiang, C., Pei, G., Gerla, M., and Chen, T. W., Scalable routing strategies for ad hoc wireless networks. *IEEE J Select Areas Commun* 1999; 17(8): 1369–1379.
5. Crawley, E., Nair, R., Rajagopalan, B., and Sandrick, H., A framework for QoS based routing in the Internet, in RFC 2386, August 1998.
6. Richard Lin, C., QoS routing in ad hoc wireless networks, in *Proceedings of 23rd Annual Conference on Local Computer Networks (LCN)*, 1998, pp. 31–40.
7. Chen, S., Routing support for providing guaranteed end-to-end quality-of-service, PhD Thesis, University of IL at Urbana-Champaign, http://cairo.cs.uiuc.edu/papers/SCthesis.ps, 1999.
8. Ariza, A., Casilari, E., and Sandoval, F., QoS routing with adaptive updating of link states. *Electron Lett* 2001; 37(9): 604–606.
9. Chen, K., Samarth, S., and Nahrstedt, K., Cross-layer design for data accessibility in mobile ad hoc networks. *Wireless Person Commun* 2002; 21: 49–76.
10. Leung, R., Liu, J., Poon, E., Chan, A., and Li, B., MP-DSR: a QoS-aware multi-path dynamic source routing protocol for wireless ad-hoc networks, in *Proceedings of Local Computer Networks (LCN)*, 2001, pp. 132–141.
11. Al-Karaki, J. N. and Kamal, AE., End-to-end support for statistical quality of service in heterogeneous mobile ad hoc networks. *Comput Commun* 2005; 28(18): 2119–2132.
12. Aggelou, G., On the performance analysis of the minimum-blocking and bandwidth-reallocation channel-assignment (MBCA/BRCA) methods for quality-of-service routing support in mobile multimedia ad hoc networks, in *IEEE Transactions on Vehicular Technology,* Vol. 53, no. 3, May 2004.

13. Holeman, S., Manimaran, G., Davis, J., Differentially secure multicasting and its implementation methods, in *Proceedings of ICCCN, Phoenix, USA*, October 2001, pp. 212–217.

14. Chen, S. and Nahrstedt, K., Distributed quality-of-service routing in ad hoc networks. *IEEE J Select Areas Commun* (Special Issue on Ad hoc Networks) 1999; 17(8): 1488–1505.

15. Capkun, S., Hamdi, M., and Hubaux, J-P., GPS-free positioning in mobile ad-hoc networks, in *Proceedings of Hawaii International Conference on System Sciences*, Maui, HW, January 2001, pp. 3481–3490.

16. Shah, S. and Nahrstedt, K., Predictive location-based QoS routing in mobile ad hoc networks, in *ICC 2002. IEEE International Conference on Communications, 2002*, Vol. 2, 2002, pp. 1022–1027.

17. Byeong, G. and Lee, W., Extended pre-computation based selective probing (PCSP) scheme for QoS routing in ad-hoc networks, in *IEEE 56th Vehicular Technology Conference, 2002*, Vol. 3, 2002, pp. 1342–1346.

18. Chen, J., Wang, J., Deng, S., and Tang, Y., QoS routing with mobility prediction in MANETs, in *IEEE Pacific Rim Conference on Communications, Computers and Signal Processing, 2001*, PACRIM, 2001, Vol. 2, 2001, pp. 357–360.

19. Sun, H. and Hughes, H., Adaptive QoS routing based on prediction of local performance in ad hoc networks, in *IEEE Wireless Communications and Networking (WCNC)*, Vol. 2, 2003, pp. 1191–1195.

20. Hashem, M., Hamdy, M., and Ghoniemy, S., Modified distributed quality-of-service routing in wireless mobile ad-hoc networks, in MELECON 2002, pp. 368–378.

21. Liao, W. H., Tseng, Y. C., Wang, S. L., and Sheu, J. P., A multi-path QoS routing protocol in a wireless mobile ad hoc networks, in *Proceedings of the First International Conference on Networking-Part 2 (ICN '01)*, Edited by Lorenz, P. Springer-Verlag, London, UK, pp. 158–167, 2001.

22. Raju, G., Hernandez, G., and Zou, Q., Quality of service routing in ad hoc networks, in *Wireless Communications and Networking Conference, 2000, WCNC, 2000 IEEE*, Vol. 1, 2000, pp. 263–265.

23. Chen, L. and Heinzelman, W. B., QoS-aware routing based on bandwidth estimation for mobile ad hoc networks. *IEEE J Select Areas Commun* 2005; 23(3): 561–572.

24. Lin, C. R. and Liu, J-S., QoS routing in ad hoc wireless networks. *IEEE J Select Areas Commun* 1999; 17(8): 1426–1438.

25. Kurose, J. and Ross, K., *Computer Networking: A Top-down Approach Featuring the Internet*. Reading, MA: Addison-Wesley, 2001.

26. Zhu, C. and Corson, M. S., QoS routing for mobile ad hoc networks, in INFOCOM 2002. vol. 2, 2002, pp. 958–967.

27. Gerasimov, I. and Simon, R., A bandwidth-reservation mechanism for on-demand ad hoc path finding, in *Simulation Symposium, 2002. 35th Annual Proceedings*, 2002, pp. 20–27.

28. Chen, Y., Tseng, Y., Sheu, J., and Kuo, P., On-demand, link-state, multi-path QoS routing in a wireless mobile ad-hoc network, in *Proceedings of European Wireless Conference 2002 (EW2002)*, Florence, Italy, February, 2002.

29. Yen, Y.-S., Chang, R.-S., and Chao, H.-C., Flooding-limited for multi-constrained quality-of-service routing protocol in mobile ad hoc networks, in *Communications, IET*, Vol. 2, no. 7, August 2008, pp. 972–981.

30. Chiu, C.-Y., Kuo, Y.-L., Wu, E. H-K., and Chen, G-H., Bandwidth-constrained routing problem in wireless ad hoc networks, in *Parallel and Distributed Systems, IEEE Transactions on*, Vol. 19, no. 1, January 2008, pp. 4–14.

31. Thriveni, J., Alekhya, V., Deepa, N., Uma, B., Alice, A., Prakash, G., Venugopal, K., and Patnaik, L., QoS preemptive routing with bandwidth estimation for improved performance in ad hoc networks, in *4th International Conference on Information and Automation for Sustainability*, vol. 1, December 2008, pp .443–448.

32. Lin, C. R. and Gerla, M., Real-time support in multihop wireless networks. *Wireless Netw* 1999; 5(2): 125–135.

33. Fullmer, C. L. and Garcia-Luna-Aceves, J. J. Floor Acquisition Multiple Access (FAMA) for packet-radio networks, *Proceedings of ACM SIGCOMM '95*, 1995.

34. Yu, M. and Leung, K. K., A trustworthiness-based QoS routing protocol for wireless ad hoc networks, in *IEEE Transactions on Wireless Communications*, Vol. 8, no. 4, April 2009, pp. 1888–1898.

35. Chen, T. W., Tsai, J.T., and Gerla, M., QoS routing performance in multihop, multimedia, wireless networks. *IEEE Universal Personal Commun* 1997; 2: 557–561.

36. Sinha, P., Sivakumar, R., and Bharghavan, V., CEDAR: a core-extraction distributed ad hoc routing algorithm, in *Proceedings of IEEE INFOCOM 99*, New York, March 1999.

37. Xiao, W., Soong, B., Law, C., and Guan, Y., QoS routing protocol for ad hoc networks with mobile backbones, in *IEEE International Conference on Networking*, Sensing and Control, Vol. 2, 2004, pp. 1212–1217.

38. Lee, K., Hwang, J., and Ryoo, J., End-to-end QoS routing in physically hierarchical wireless ad-hoc networks, in *IEEE 61st Vehicular Technology Conference*, Vol. 4, 2005, pp. 2488–2492.

39. Ueda, T., Tanaka, S., Roy, S., Saha, D., and Bandyopadhyay, S., A priority-based QoS routing protocol with zone reservation and adaptive call blocking for mobile ad hoc networks with directional antenna, in *IEEE Global Telecommunications Conference Workshops*, 2004, pp. 50–55.

40. Perkins, C. E. and Bhagwat, P., Highly dynamic destination-sequenced distance-vector routing (DSDV) for mobile computers, in Computer Communications Review, October 1994, pp. 234–244.

41. Johnson, D. and Maltz, D., Dynamic source routing in ad hoc wireless networks, *Mobile Computing*, edited by Imielinski, T., Korth, H., Kluwer Academic Publishers, Chapter 5, pp. 153–181, 1996.

42. Perkins, C. and Royer, E., Ad hoc On-Demand Distance Vector (AODV) routing, Internet-Draft, November 1998, http://tools.ietf.org/html/rfc3561.

43. Perkins, C., Royer, E., and Das, SR., Quality of service for Ad Hoc On-Demand Distance Vector (AODV) routing, Internet-Draft, July 2000, http://www.cs.ucsb.edu/~ebelding/txt/aodv/aodvidv3.txt.

44. Lin, C. R., On-demand QoS routing in multihop mobile networks, in *Twentieth Annual Joint Conference of the IEEE Computer and Communications Societies. Proceedings. IEEE*, Vol. 3, 2001, pp. 1735–1744,.3 INFOCOM 2001.

45. Aggelou, G. N. and Tafazolli, R., QoS support in 4th generation mobile multimedia ad hoc networks, in *3G Mobile Communication Technologies, 2001. Second International Conference* on (Conf. Publ. No. 477), 2001, pp. 412–416.

46. Aggelou, G. N. and Tafazolli, R., A simulation analysis on reactive route repair techniques for QoS sensitive applications in mobile ad hoc networks, in *Proceedings of the 1st ACM International Symposium on Mobile Ad hoc Networking and Computing (Q2SWinet)*, 2000, pp. 139–140.

47. Li, Y., Chen, X., and Yu, D., Disjoint multi-path QoS routing in ad hoc networks, in *International Conference on Wireless Communications and Networking*, Vol. 2, 2005, pp. 739–742.

48. Zhang, B. and Mouftah, H. T., *QoS Routing for Wireless Ad Hoc Networks: Problems, Algorithms, and Protocols*, IEEE Communications Magazine, Vol. 43 (10), pp. 110–117, 2005.

49. Ohba, S., Barolli, L., Ikeda, M., De Marco, G., Durresi, A., and Iwashige, J., An effective topology extraction algorithm for search reduction space of a GA-based QoS routing method in ad-hoc networks, in *Proceedings of the 8th International Symposium on Parallel Architectures, Algorithms and Networks (ISPAN'05)*, 2005, pp. 6–12.

50. Islam, M., Pose, R., and Kopp, C., A hybrid QoS routing strategy for suburban ad-hoc networks, in *The 11th IEEE International Conference on Networks (ICON)*, 2003, pp. 225–230.

51. Zhang, Y. and Gulliver, T. A., Quality of service for ad hoc on-demand distance vector routing, in *IEEE International Conference on Wireless and Mobile Computing (WiMob'2005)*, Vol. 3, 2005, pp. 192–196.

52. Labiod, H. and Quidelleur, A., QoS-ASR: an adaptive source routing protocol with QoS support in multihop mobile wireless networks, in *IEEE 56th Vehicular Technology Conference (VTC02)*, Vol. 4, 2002, pp. 1978-1982.

53. Aggelou, G., On the performance analysis of the minimum-blocking and bandwidth-reallocation channel-assignment (MBCA/BRCA) methods for quality-of-service routing support in mobile multimedia ad hoc networks. *IEEE Trans Vehicular Technol* 2004; 53(3): 770–782.

54. Punde, J., Pissinou, N., and Makki, K., On quality of service routing in ad-hoc networks, in *Proceedings of the 28th Annual IEEE International Conference on Local Computer Networks (LCN)*, 2003, pp. 276–278.

55. Munaretto, A., Badis, H., Al Agha, K., and Pujolle, G., A link-state QoS routing protocol for ad hoc networks, in *4th International Workshop on Mobile and Wireless Communications Network*, 2002, September 2002, pp. 222–226.

56. Badis, H., Gawedzki, I., and Al Agha, K., QoS routing in ad hoc networks using QOLSR with no need of explicit reservation, in *IEEE 60th Vehicular Technology Conference*, Vol. 4, 2004, pp. 2654–2658.

57. Ge, Y., Kunz, T., and Lamont, L., Quality of service routing in ad-hoc networks using OLSR, in *36th Annual Hawaii International Conference on System Sciences (HICSS'03)*, Big Island, Hawaii.

58. Ziane, S. and Mellouk, A., A swarm intelligent multi-path routing for multimedia traffic over mobile ad hoc networks, in *1st ACM International Workshop on Quality of Service and Security in Wireless and Mobile Networks (Q2SWinet)*, 2005, pp. 55–62.

59. Zhou, Z. and Jiang, Z., A novel QOS routing scheme in mobile ad hoc network-LAQR, in *10th Asia-Pacific Conference on Communications and 5th International Symposium on Multi-Dimensional Mobile Communications*, Vol. 2, 2004, pp. 863–867.

60. Brayer, K., Achieving timely message delivery quality-of-service in fixed and variable connectivity distributed routing networks. *IEEE J Select Areas Commun* 2004; 22(7): 1183–1196.

61. Kunavut, K. and Sanguankotchakorn, T., QoS-aware routing for mobile ad hoc networks based on multiple metrics: Connectivity Index (CI) and delay, in Electrical Engineering/Electronics Computer Telecommunications and Information Technology (ECTI-CON), 2010 International Conference on, 19–21 May 2010, pp. 46–50.

62. Hartenstein, H., Mauve, H., and Widmer, A., A survey on position-based routing in mobile ad hoc networks. *IEEE Netw* 2001; 15(6): 30–39.

63. Basagni, S., Chlamtac, I., and Syrotiuk, V., Dynamic source routing for ad hoc networks using the global positioning system, in *Wireless Communications and Networking Conference, 1999. WCNC. 1999 IEEE,* 1999, Vol. 1, pp. 301–305.

64. Li, J. and Mohapatra, P., PANDA: a positional attribute-based next-hop determination approach for mobile ad hoc networks, Technical Report, Department of Computer Science, University of California, Davis, 2002.

65. Park, T. and Shin, K. G., Optimal tradeoffs for location-based routing in large-scale ad hoc networks, in *IEEE/ACM Transactions on Networking,* Vol. 13, no. 2, April 2005.

66. Huang, C., Dai, F., and Wu, J., On-demand location-aided QoS routing in ad hoc networks, in *Proceedings of the International Conference on Parallel Processing (ICPP'04),* IEEE Computer Society, Washington, DC, USA, 2004, pp. 502–509.

67. Abdrabou, A. and Zhuang, W., Statistical QoS routing for IEEE 802.11 multihop ad hoc networks, in *IEEE Transactions on Wireless Communications,* Vol. 8, no. 3, March 2009, pp. 1542–1552.

68. Calafate, C. T., Malumbres, M. P., Oliver, J., Cano, J. C., and Manzoni, P., QoS support in MANETs: a modular architecture based on the IEEE 802.11e technology, in Circuits and Systems for Video Technology, IEEE Transactions on, vol. 19, no. 5, May 2009, pp. 678–692.

69. Papadimitratos, P., Haas, Z.J., and Sirer, E., Path set selection in mobile ad hoc networks, in 3rd ACM MOBIHOC'02, 2002, pp. 1–11.

70. Novatnack, J., Greenwald, L., and Arora, H., Evaluating ad hoc routing protocols with respect to quality of service, in *IEEE International Conference on Wireless and Mobile Computing (WiMob'2005),* Vol. 3, 2005, pp. 205–212.

71. Barolli, L., Koyama, A., and Norio, S. QoS routing method for ad-hoc networks based on genetic algorithm, in *Proceedings of the 14th International Workshop on Database and Expert Systems Applications (DEXA'03),* IEEE Computer Society, Washington, DC, USA, 2003, p. 175.

72. Maleki, M., Dantu, K., and Pedram, M., Power-aware source routing protocol for mobile ad hoc networks, in ACM ISLPED'02, 12–14 August, 2002.

73. Singh, S., Woo, M., and Raghavendra, C., Power-aware routing in mobile adhoc networks, in *Proceedings of Mobicom 98 Conference,* Dallas, October 1998.

74. Lindsey, S., Sivalingam, K., and Raghavendra, C. S., Power aware routing and MAC protocols for wireless and mobile networks. *Wiley Handbook on Wireless Networks and Mobile Computing*, Edited by Stojmenivic, I., Monterey, CA: John Wiley & Sons, 2001.

75. Stojmenovic, J. and Lin, X., Power-aware localized routing in wireless networks, in Proceedings IEEE IPDPS, Cancun, Mexico, May 2000.

76. Toh, C. K., *Maximum Battery Life Routing to Support Ubiquitous Mobile Computing in Wireless Ad hoc Networks*, IEEE Communication Magazine, June 2001.

77. Akyildiz, I. F., Su, W., Sankarasubramaniam, Y., and Cayirci, E. Wireless sensor networks: a survey. *Computer Networks* 38(4): 393–422, 2002.

5

Power-Aware Video Compression for Mobile Environments

Ishfaq Ahmad
University of Texas

Victor Yongfang
Liang

Zhihai (Henry) He
University of Missouri

5.1 Introduction

A raw video stream tends to be quite demanding when it comes to storage requirements and the need for high network capacity when being transferred. For example, an uncompressed HDTV picture with 2.2 million pixels and raw coding with 24 bits per pixel (8 per color component) would require 1.5–3 Gbits/s depending on the picture frequency. Therefore, before being stored or transferred, the raw stream is usually transformed to a representation using compression.

To ensure interoperability between different terminal devices when sharing video content and compliance when creating (or personalizing) video bitstreams, different video coding standards/recommendations have been proposed targeting various application areas, including H.261/H.263/H.264 by ITU-T [1–3] and MPEG-1/MPEG-2/MPEG-4 by ISO/IEC [4–6]. The H.26x recommendations have been designed mostly for real-time video communication applications, such as video conferencing and video telephony. On the other hand, the MPEGx standards are designed to address the needs of video storage (DVD), broadcast video (broadcast TV), and video streaming. These standards/recommendations have been the engines behind the commercial success of various video services and applications summarized in Table 5.1; each of those was designed to fit a specific application and best suited to particular requirements.

MPEG-1 [4] is a standard for the compression of moving pictures and audio up to 1.5 Mbits/s. It is the standard of compression for VideoCD, the most popular video distribution format throughout much of Asia. MPEG-2 [5] can be used for application between 1.5 and 15 Mbits/s such as Digital Television set-top boxes and DVD movies. MPEG-2 scales well to HDTV resolution and bit rates, obviating the need for an MPEG-3. The focus and scope of MPEG-4 [6] was defined as the intersection of the traditional separate industries of telecommunications, computer, and file where audio-visual applications exist. It aims at application such as Internet and intranet video, video e-mail, home movies, virtual reality games, simulation, and training. H.261 [7] is an ITU standard designed for two-way communication over ISDN lines (video conferencing) and supports data

TABLE 5.1   Video Codecs in Various Video Services and Applications

Services and Applications		Video Codec
3GPP (third-generation partnership project)	MMS	H.263 profile 0 level 10
	PSS Release 5	H.263 profile 0 level 10
	PSS Release 6	N/A
3GPP2	Circuit switched video conferencing services	MPEG-4 simple profile level 0 and H.263 baseline
	Packet switched video conferencing services	MPEG-4 Visual or ITU-T H.263 (or both) shall be supported
3G-324M mobile videoconferencing		H.263 baseline level 10
H.320 videoconferencing		H.261 QCIF
H.323 videoconferencing		H.261 QCIF (if have video service)
HD-DVD, Blue-ray DVD		Microsoft VC-9 (VC-1), MPEG-2, MPEG-4 AVC (H.264)
DVD		MPEG-2 Main profile at main level
SVCD (Super Video CD)		MPEG-2 MP at low level MPEG1
VCD (Video CD)		MPEG1

rates that are multiples of 64 kbit/s. H.263 [1] was developed for low-bit rate video coding between 20 and 64 kbps. H.263 supports CIF, QCIF, SQCIF, 4CIF, and 16CIF resolutions. It has widely been used in videoconferencing and video-telephony applications. H.264 (also known as MPEG-4 AVC) [2] was jointly developed by the video coding experts group (VCEG) of the ITU-T and the moving picture experts group (MPEG) of ISO/IEC with an objective to create a single video-coding standard, which simultaneously resulted in advanced video coding (AVC) of MPEG-4 Part 10 and ITU-T H.264 recommendations. H.264/AVC achieves a significant improvement in compression efficiency and can be used in a wide range of applications, including video telephony, video conferencing, TV, and storage (DVD and/or hard disk based, especially high-definition DVD).

Mobile video applications arise in a wide range of environments, such as battlefield intelligence, surveillance, reconnaissance, security monitoring, emergency response, disaster rescue, environmental tracking, tele-medicine, and multimedia systems in consumer electronics. One example is the wireless camera flashlight used by police officers as part of a pervasive video surveillance system. The officers can use the flashlight to capture the crime scene, process it, and communicate with the department for the purpose of criminal identification or archiving.

Video compression in wireless mobile environments has to meet the following challenges:

- *Power Awareness:* The nickel–cadmium and lithium–ion batteries generally used in mobile devices have increased their energy capacity roughly by 10–15% per year. However, it is conjectured that only another 15–25% increase is possible [8]. The implication of limited power on video encoding is two-fold. First, efficient video compression significantly reduces the amount of the video data to be transmitted, which, in turn, saves a significant amount of energy in data transmission. Second, more efficient video compression often requires higher computational complexity and larger power consumption in computing. To prolong the lifetime of the battery, a video encoding system capable of adjusting its energy consumption as demanded by the situation and its environment is highly desirable.
- *Ensuring Quality of Service:* The quality of service (QoS) in the context of video encoding is measured from two perspectives: First, the picture quality of motion picture must be optimized. One can always aim for better than HDTV quality on a handheld device, and the optimization pursuit may be without limit. Second, the available coding bits are best used when and where needed the most at a limited transmission bandwidth.
- *Content Adaptive:* Pervasive environments are far from ideal. They contain highly obscure and unintelligible data, which exert tremendous stress on an encoder, in terms of the computational requirements as well as the bit budget to encode the scene complexity. Therefore, we also need to develop effective adjustment techniques to consume minimal power while preserving desired video quality, according to the changes in the environment as well as in the power supply level.
- *Fast Encoding Speed:* Real-time video compression remains a computationally demanding problem. With limited CPU power available in pervasive computing devices, such as miniature cameras and sensors, another research challenge is to

design fast algorithms without compromising on the solution and at the same time invent techniques to speed up the execution of the algorithms.

In this chapter, we explain a methodology to meet the PQRS goals in ubiquitous video environment under power constraints, and techniques for efficiently utilizing the energy supply while preserving desirable video quality.

5.2 Related Research Works

The source coding schemes recommended by the international standardization groups are very detailed for achieving high-bit rate reduction under the constraint of the highest possible picture quality. Therefore, these coding schemes will result in high complexity, which usually lead to high power consumption.

5.2.1 Design of Complexity Scalable Video Coding

To allow a flexible control on the video encoder, the complexity scalability must be introduced. In complexity scalable video coding, the computational complexity of the video encoder can be adjusted by changing its complexity parameters. In [9], a hierarchical block motion estimation (ME) algorithm based on the partial distortion measure is proposed. It uses a coarse to fine approach to refine the search for each higher level, by dividing the motion vector (MV) search into several levels in a way that lower levels use partial distortions with higher decimation ratios. This algorithm can provide different computational complexities, that is, different levels of power consumption, with different ME precisions. In [10], a flexible framework is presented for DCT-based video encoding. In this framework, each of the encoding components (DCT and ME) features a set of parameters that can be used to control the computational complexity and performance and allow the encoder to run in real-time on machines with different computing power levels. A partially predefined configuration architecture for DCT and ME is proposed in [11]. A low power video encoder with power, memory, and bandwidth scalability for use in portable video applications is presented in [12]. Scalable compression is achieved by using block transform-based vector quantizer encoders implemented with table lookups. It can change its power consumption depending on the available channel bandwidth and can also trade-off bandwidth for power. Recently, in [13], depending on the battery power level, a number of B, P, I frames are discarded to reduce the transmission bits. A multi-stage coded modulation to accommodate rates in different modes is utilized in [14]. However, the complexity parameters in the reviewed methods are obtained empirically and are lack of adaptability to the coding environment. Moreover, the techniques are designed to meet specific requirements for specific applications and video-coding standards.

5.2.2 Modeling Power Consumption and Rate-Distortion

To better understand the power consumption behavior of a video encoder, a power-consumption model is desirable. The model can provide an estimation of the actual consumed power, which helps system analysis and system design. In [15], a set of experiments is performed to understand the energy usage pattern of handheld devices while decoding and displaying MPEG compressed video in software. Experiments are designed so as to bring forth parameters that can be used to predict the energy requirement for MPEG playback. In [16], a linear model is used to model the power consumption required for computing ME, DCT, and quantization, of a H.263 video encoder, running with different ME algorithms.

For video coding under bit rate constraints, the goal is to maximize the picture quality, that is, minimize the coding distortion, for a given bit rate budget. Thus, it is necessary to model the rate–distortion (R–D) relationship. R–D modeling based on statistical properties of the source data [17], empirical analysis on the observed data [18] resulted in several other techniques proposed in [19–21].

5.2.3 Distortion Optimized Video Coding

The problem in this category is to minimize the distortion or maximize the quality while the power supply is limited. The goal in [22] is to minimize the amount of distortion in the reconstructed video sequence under certain channel bandwidth and transmission power constraints, with transmission power allocated across packets. In [23], a joint source coding and power control approach to simultaneously maximize the per-cell capacity while maximizing the quality of the delivered video to individual users subject to a constraint on the total available bandwidth is proposed. An interesting work is reported in [24], where a power–distortion optimized coding mode selection scheme is proposed for variable bit rate videos in wireless code-division multiple-access (CDMA) systems.

5.2.4 Power Optimized Video Coding

In [25], a joint source coding and transmission power allocation scheme is proposed. In this scheme, no channel coding is used, and the transmission power is allocated adaptively to different video segments based on their relative importance. Alternatively, [26] formulates an optimization problem that corresponds to minimizing the energy required to transmit a video frame with an acceptable level of distortion. In [27], the transmitter wants to obtain a joint power and coding-rate selection in order to maximize a video quality metrics within an analytical model. In [28,29], the authors consider the processing power for source coding and channel coding as well as transmission power for a given video quality and propose a power-minimized bit-allocation scheme. This work is extended in [30], considering the interference to other users when performing the optimization of power consumption. In [31], the authors introduce an approach for minimizing the total power consumption of a mobile transmitter due to source compression, channel coding, and transmission subject to a fixed end-to-end source distortion.

5.3 Complexity Scalable Video Coding Design

As the first step of power and distortion-optimized video coding, the encoder needs to be parameterized to enable flexible control on power consumption and video distortion.

In other words, we need to introduce some complexity parameters into the encoder to control the coding behavior of the major encoding modules. Typical video encoders, such as MPEG-2, H.263, and MPEG-4, employ a hybrid motion-compensated DCT-encoding scheme, which is summarized as follows. A video frame is divided into macroblocks (MBs). Each MB consists of four blocks (8×8 pixels) of luminance and one block for each of the two-chroma components. To exploit the temporal dependencies of MBs between successive frames, ME/MC is done through interpicture prediction. Transform coding of the residual prediction error signal, such as the DCT, is used to exploit the spatial redundancy. After DCT is performed, the coefficients are numbered in a zig-zag order from the top left to bottom right. Scalar quantization is then applied to the resultant DCT coefficient matrix. Quantization is the lossy component of video compression. It simply reduces the number of bits needed to store the transformed coefficients by reducing the precision of those values. The resulting data are entropy encoded using a Huffman variable word length scheme in a lossless manner to give better overall compression gain. To decompress the image, the process is carried out in reverse. Among all the coding operations of video coding, the ME and DCT have been shown to have high complexity.

DCT/IDCT is typically done on each 8×8 block. (Note that in the new H.264/AVC, DCT is done on 4×4 blocks.) One-dimensional DCT requires 64 multiplications and for an 8×8 block eight 1-D DCTs are needed. Two-dimensional DCT requires 54 multiplications and 468 additions and shifts. IDCT has similar complexity as DCT.

ME aims to find the best match of a reference block in the reference frame that yields the minimum block distortion measure (BDM) within a given search window. The sum of absolute difference (SAD) between the reference block and current block is popularly used as the BDM. For the SAD-optimal ME problem, the search for the optimal MV can be expressed as:

$$(u_0, v_0) = \text{argmin SAD}(u, v), \tag{5.1}$$

where (u_0, v_0) is the optimal MV, representing the horizontal and vertical displacement, respectively, and $\text{SAD}(u, v)$ is the SAD value of the candidate MV (u, v) within the search window, given by

$$\text{SAD}(u,v) = \sum_{i=0}^{M-1}\sum_{j=0}^{N-1}\left|f_{t-r}(i+u,j+v) - f_t(i,j)\right| \tag{5.2}$$

where $f_{t-r}(\cdot,\cdot)$ and $f_t(\cdot,\cdot)$ refer to the blocks with size of $M \times N$ in the reference frame and current frame, respectively. One SAD computation requires $M \times N$ subtractions, $M \times N$ absolute value operations, and $(M \times N - 1)$ additions. The MV has the minimum distortion, chosen from a certain number of MV candidates. In full search ME, all the candidates within the search window need to be evaluated. Remember that evaluating each MV candidate in the search window requires one SAD computation, which makes the MV search quite computationally complex.

Figure 5.1 illustrates the structure of a typical hybrid block-based video encoder. There are several major encoding modules: ME and compensation (COMP), DCT,

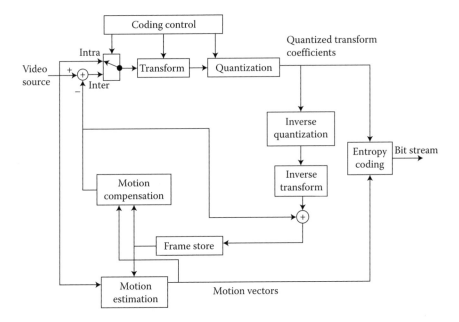

FIGURE 5.1 Video-coding structure.

quantization (QUANT), entropy encoding (ENC) of the quantized DCT coefficients, inverse quantization (DQUANT), inverse DCT (IDCT), picture reconstruction (RECON), and interpolation (INTERP) [32]. For the ease of exposition, the DCT, IDCT, QUANT, DQUANT, and RECON modules are collectively referred to as PRECODING, which can be considered as the data representation module. In this way, the video encoder has only three major modules: ME, PRECODING, and ENC.

It is widely known that the ME process, which includes the SAD computation, is the most computation-intensive module, consuming most of the processor cycles. Following ME, the PRECODING modules collectively take the second place of the total processor cycles consumption. The ENC module, which is basically a bit splicing engine, uses a relatively small amount of the total CPU time. In addition, its computational complexity mainly depends on the coding bit rate. Moreover, since the output bitstream must be conformed to bitstream syntax defined by the specific video standard, parameterized coding is not performed in the component of entropy coding.

5.3.1 Complexity Scalable ME Design

Since the computational complexity of each SAD computation is a constant, the overall computational complexity of the ME module is linearly proportional to the number of SAD computations, denoted by λ_{ME}. In the proposed energy scalable framework, λ_{ME} is determined by system-level power management and quality optimization. Since ME is performed on the MB level, therefore at the frame-level, we need to allocate the current available λ_{ME} SAD computations among the MBs in the video frame.

Initially, Nsad is set to be λ_{ME}. Suppose the motion search range is SR. If $nsad_{ij} \geq (2 \times SR + 1)^2$; it means the computational power is enough to perform a full search for this block. Otherwise, the complexity controllable ME scheme described in the following is used to find the MV, whose complexity is controlled by $nsad_{ij}$.

5.3.2 Complexity Controllable ME Scheme

The proposed adaptive search scheme is based on the popular media-bias diamond search wherein four points around the current minimum are searched to find the next minimum in every iteration (Figure 5.3a). For a given number of pre-allocated SAD calculations $nsad_j$, we denote SI_j the number of search iterations during the MV search, that is, the number of diamond search patterns we need to perform. It is approximated by a linear relationship, given by

$$SI_j = \psi \times nsad_j, \tag{5.3}$$

The parameter ψ is adaptively updated for each block, since the motion characteristics and the SAD error surface are correlated to those of the adjacent blocks. Now, the proposed search strategy works as follows. After the allocation of SAD computation, if $nsad_j$ is 0, which means we have no available computation power at all, the MV of the collocated MB in the previous coded frame is chosen as the MV for current block because of the high temporal correlation between the current frame and the previous encoded frame. If $nsad_j$ equals to 1, the MV is chosen from the median MV predictor and the temporal previous MV that yields the smaller SAD value. In other cases, the number of search iteration is determined by Equation 5.3 and a parameterized search is used to find the MV, summarized as following:

- *Step* 1: We start with the diamond search pattern with size 1, showed at Figure 5.2a, and continue the search until the search center is found with the minimum SAD. If the number of search iteration exceeds SI_j, we stop the search.

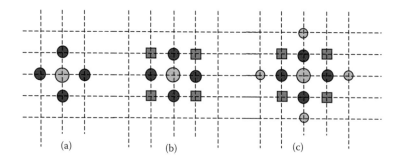

(a)   (b)   (c)

FIGURE 5.2 The biggest round point has the minimum distortion. (a) The diamond search with size equal to 1. (b) The diagonal search with size 1, the square points are checked. (c) The diamond size is changed to 2. The outermost round points are checked.

FIGURE 5.3 Coded video quality comparison for Frame 100 of "Foreman" and Frame 80 of "Carphone" when (a) 100% blocks are encoded; (b) 20% blocks are encoded.

- *Step* 2: We switch to the diagonal search with the size of the current diamond size, showed at Figure 5.2b, and continue the search. If the search center is found with minimum SAD, we increase the diamond size by 1. Otherwise, the diamond size is reset to 1. Go back to Step 1. During the search, whenever the number of search iteration exceeds SI_j, we stop the search.
- *Step* 3: The MV is returned and the parameter ψ is updated.

Note that like other conventional fast ME algorithms, the partial distortion computation technique is also applied. In addition, the checked points during the search procedure are tracked to avoid the unnecessary checking when the search pattern moves.

5.3.3 Complexity Scalable PRECODING Design

In this section, we present a parametric complexity scalability scheme to collectively control the computational complexity of the PRECODING modules, namely the DCT, QUANT, DQUANT, IDCT, and RECON modules.

In typical video encoding as illustrated in Figure 5.1, DCT is applied to the difference MB after ME and compensation, or the original MB if its coding mode is INTRA. After the DCT coefficients are quantized, DQUANT, IDCT, and RECON are performed to reconstruct the MB for motion prediction of the next frame. In transform coding of videos, especially at low coding bit rates, the DCT coefficients in the MB might become all zeros after quantization. We refer to this MB as an all-zero MB (AZMB). Otherwise, it is called a non-zero MB (NZMB). In international standards for video encoding, such as MPEG-2, H.263, and MPEG-4, "non-zeros" also means the CBP (coded block pattern) value of the MB is non-zero. If we can predict an MB to be AZMB, all the above PRECODING operations can be skipped, because the output of DQUANT and IDCT of an AZMB is still an AZMB, and the reconstructed MB is exactly the reference MB used in ME and compensation. Therefore, the encoder can simply copy over the reference MB to reconstruct the current MB.

In this work, the unique property of the AZMB is used to design a complexity scalability scheme for the PRECODING modules. Let $\{x_{nk} \mid 0 \leq n, k \leq 7\}$ be the coefficients in the different MB after ME. For INTRA MBs, $\{x_{nk}\}$ are the original pixels in the video frame. Let $\{y_{ij} \mid 0 \leq n, k \leq 7\}$ be the DCT coefficients. According to the definition of DCT, we have

$$y_{ij} = \frac{1}{4} C_i C_j \sum_{n=0}^{7} \sum_{k=0}^{7} x_{nk} \cos\left(i\pi \frac{2n+1}{16}\right) \cos\left(j\pi \frac{2k+1}{16}\right), \tag{5.4}$$

where

$$C_i = \begin{cases} \dfrac{1}{\sqrt{2}} & \text{if } i = 0 \\ 1 & \text{else} \end{cases}, \quad C_j = \begin{cases} \dfrac{1}{\sqrt{2}} & \text{if } j = 0 \\ 1 & \text{else} \end{cases}. \tag{5.5}$$

We can see that

$$\left| y_{ij} \right|^2 \leq \sum_{n=0}^{7} \sum_{k=0}^{7} \left| x_{nk} \right|^2, \tag{5.6}$$

Note that the right-hand side is the SAD of the difference MB, which is already computed during the ME. This suggests us that the SSD could be an efficient and low-cost measure to predict the AZMB. After ME and compensation, let SSD$_i$| $1 \leq i \leq M$} be the SSD values of the M MBs in the video frame sorted in an ascending order. In the proposed complexity scalability scheme for PRECODING, we force the first M-λ_{PRE} MBs to be AZMBs, and treat the remaining λ_{PRE} MBs as NZMBs to which the PRECODING operations are applied. Let C_{NZMB} be the number of processor cycles needed by the PRECODING operations to finish one NZMB. The value of C_{NZMB} can be obtained either by theoretical cycle estimation of the PRECODING modules, or from simulation statistics.

In practice, the value of C_{NZMB} may vary slightly from MB to MB. Note that the power management and energy consumption control operate on a level much higher than the MB. The overall complexity of the PRECODING modules, denoted by C_{PRE} is then given by:

$$C_{\text{PRE}} = \lambda_{\text{PRE}} \times C_{\text{NZMB}}, \tag{5.7}$$

We refer to this type of complexity scalability scheme as λ_{PRE} scalability. Figure 5.3 shows the 100th frame of "Foreman" encoded at 192 kbps, and the 80th frame of "Carphone" encoded at 64 kbps with 100 and 20% PRECODING complexity. Perceptually, we can hardly see much difference between them.

5.4 Power Rate Distortion Analysis

Rate–distortion (R–D) analysis has been one of the major research focus in information theory and communication for the past few decades, from the early Shannon's source coding theorem for asymptotic R–D analysis of generic information data [33], to recent R–D modeling of modern video encoding systems [19,34]. However, very little work is reported on P–R–D analysis. In energy-aware video encoding, the coding distortion is not only a function of the encoding bit rate as in the traditional R–D analysis, but also a function of the power consumption P. In other words,

$$D = D(R; P), \tag{5.8}$$

which describes the P–R–D behavior of the video encoding system. The P–R–D model provides a theoretical basis, as well as a practical guideline, for a system design and performance optimization. Using the P–R–D model, we can perform energy consumption control on the underlying device at the system level. For example, in a wireless sensor network, we can perform across-node energy optimization and network lifetime maximization.

5.4.1 Power Consumption Analysis

We consider the complexity control parameter for the ME module to be the number of SAD computations per frame, denoted by λ_{ME}. The computational complexity of ME, denoted by C_{ME}, is simply given by:

$$C_{ME} = \lambda_{ME} \times C_{SAD,} \tag{5.9}$$

where C_{SAD} represents the complexity of one SAD computation between the current MB and its reference MB. The computational complexity is measured by the number of processor cycles used by the operation.

The computational complexity of all the PRECODING modules is controlled using one single parameter λ_{PRE}, which is the number of non-zero MBs in the video frame. Let C_{NZMB} and C_{PRE} be the PRECODING computational complexity of one NZMB and the whole video frame, respectively. As indicated above,

$$C_{PRE} = \lambda_{PRE} \times C_{NZMB} \tag{5.10}$$

The ENC module, as a variable length-coding (VLC) engine, mainly consists of VLC table look-up and bit splicing of the code words. The computational complexity of the ENC module, denoted by CENC, is approximately proportional to R. Therefore, we have

$$C_{\text{ENC}} = S \times R \times C_{\text{BIT}}, \tag{5.11}$$

where C_{BIT} is the per bit ENC complexity and S is the size of the picture. Here, S is needed because R represents the coding bit rate in the unit of bits per pixel. The computational complexity of the video encoder, denoted by C and measured by the number of processor cycles per second, is given by:

$$C(R; \lambda_{\text{ME}}, \lambda_{\text{PRE}}, \lambda_{\text{F}}) = \lambda_{\text{F}} \times (C_{\text{ME}} + C_{\text{PRE}} + S \times R \times C_{\text{BIT}}) \tag{5.12}$$

where λ_{F} is the encoding frame rate. This model presents a complexity-scalable architecture for video encoding, whose computational complexity is mainly controlled by the parameter set $\{\lambda_{\text{ME}}, \lambda_{\text{PRE}}, \lambda_{\text{F}}\}$.

Let Φ be a mapping function to translate the computational complexity of the video coding system into corresponding power consumption. We can derive the relationship between the power consumption and the complexity control parameters,

$$P = \Phi(\lambda_{\text{F}} \times (C_{\text{ME}} + C_{\text{PRE}} + S \times R \times C_{\text{BIT}})). \tag{5.13}$$

5.4.2 R–D Analysis

5.4.2.1 ME Module R–D Analysis

To analyze the R–D behavior of the complexity control parameter λ_{ME}, we need to investigate the relation between λ_{ME} and the frame SAD S_f, which is the average SAD per pixel in the motion compensated difference frame. To this end, we collect the frame SAD statistics for different λ_{ME} from several test video sequences. Figure 5.4 plots the frame SAD S_f as a function of λ_{ME} for two QCIF video sequences: "Akiyo" and "Foreman." The simulation results suggest the following relation between λ_{ME} and S_f:

$$S_f\left(\lambda_{\text{ME}}\right) = \beta_0 + \beta_1 \times e^{-\beta_2 x}, \quad x = \lambda_{\text{ME}}/\lambda_{\text{ME}}^{\max}, \tag{5.14}$$

where the model parameters β_0, β_1, and β_2 are estimated by the statistics of previous frames; and $\lambda_{\text{ME}}^{\max}$ is the maximum value of λ_{ME}. Besides the SAD, another operation called SSD (sum of square difference), which is the square difference between the current MB and its reference, is often used in ME. In hardware design, the SSD is more advantageous than the SAD because the subtraction and multiplication operations can be completed by a single instruction.

5.4.2.2 Precoding R–D Analysis

Using the mathematical framework for optimal bit allocation, we analyze the R–D behavior of the complexity control parameter λ_{PRE}, as discussed above. The dynamic rate

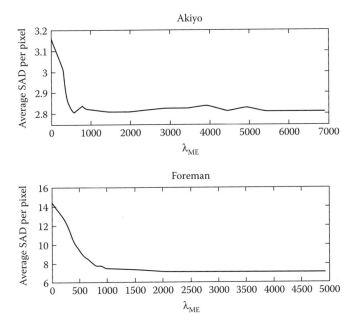

FIGURE 5.4 Frame SAD as a function of λ_{ME}.

control is a near-optimal bit allocation process. Let $\{\sigma_i^2 \mid 1 \leq i \leq M\}$ be the variance in the MBs in the video frame sorted in an ascending order. Let R be the target coding bit rate in bits per pixel (bpp). According to the classic R–D distortion formula, the distortion of the *i*th MB is given by

$$D_i(R_i) = \sigma_i^2 \cdot 2^{-2\gamma R_i}, \tag{5.15}$$

where R_i is the bit rate of the *i*th MB, and is a model constant. The optimal bit allocation can be then formulated as

$$D = \min_{\{R_i\}} \frac{1}{M} \sum_{i=1}^{M} \sigma_i^2 \cdot 2^{-2\gamma R_i} \tag{5.16}$$

$$\text{s.t. } R = \frac{1}{M} \sum_{i=1}^{M} R_i. \tag{5.17}$$

The minimum distortion obtained by the optimal bit allocation is

$$D = \left(\prod_{i=1}^{M} \sigma_i^2 \right)^{\frac{1}{M}} \cdot 2^{-2\gamma R}, \tag{5.18}$$

In our complexity scalability scheme, the first M-λ_{PRE} MBs are encoded as AZMBs, while the remaining λ_{PRE} MBs are encoded as NZMBs. In this case, the bit rate of each AZMB is zero, and its coding distortion, denoted by D_i^z, is exactly the variance of the difference MB, that is,

$$D_i^z = \sigma_i^2 * 2^{-2\gamma*0} = \sigma_i^2, \quad 1 \le i \le M - L, \tag{5.19}$$

where $L = \lambda_{PRE}$ is introduced to simplify the notation. Since all the coding bits are allocated among the NZMBs, according to Equation 5.19, the coding distorting of each NZMB, denoted by D_i^{nz}, is given by

$$D_i^Z = \left(\prod_{i=M-L+1}^{M} \sigma_i^2 \right)^{\frac{1}{L}} 2^{-2\gamma \frac{MR}{L}}, \quad M - L + 1 \le i \le M \tag{5.20}$$

The overall distortion D of the video frame, which is an average distortion of the AZMBs and NZMBs, is given by

$$
\begin{aligned}
D = D(L) &= \frac{1}{M} \left[\sum_{i=1}^{M-L} D_i^2 + \sum_{i=M-L+1}^{M} D_i^{nz} \right] \\
&= \frac{1}{M} \left[\sum_{i=1}^{M-L} \sigma_i^2 + L \left(\prod_{i=M-L+1}^{M} \sigma_i^2 \right)^{1/L} 2^{-2\gamma MR/L} \right].
\end{aligned}
\tag{5.21}
$$

To derive the expression for $D(L)$, we consider the continuous-time version of Equation 5.21. Note that $\{\sigma_i^2 \mid 1 \le i \le M\}$ is an increasing series. Figure 5.5 shows $\{\sigma_i^2\}$ for the 100th frame of the "Foreman." Experiments on other video frames and other video sequences yield similar results.

This suggests that it is reasonable to model $\{\sigma_i^2\}$ with the following linear function

$$\xi(t) = A \cdot t, \quad t \in [0,1], \tag{5.22}$$

such that

$$\sigma_i^2 = \xi\left(\frac{i}{M}\right), \quad 1 \le i \le M \tag{5.23}$$

here A is a positive constant. It should be noted that at the right end of the curve, the linear approximation is not accurate. However, since the R–D modeling is a statistical procedure to model the behavior of the whole frame, which has a large number of MBs, the approximation error within this small region will not affect much the performance of the whole model. Our simulation results that will be presented later confirm this

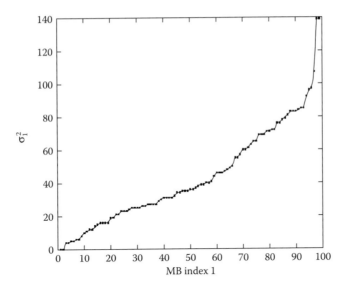

FIGURE 5.5 The MB variances sorted in an ascending order for the 100th frame of "Foreman."

observation. Similarly, we define $y = L/M$, and consider $D(y)$ as the continuous-time version of $\{D(L)\}$, that is,

$$D(y) = D(L/M) \tag{5.24}$$

Note that the first term on the right-hand side of Equation 5.28 can be written as

$$\frac{1}{M} \sum_{i=1}^{M-L} \sigma_i^2 = \int_0^{1-y} \varsigma(t)\, dt = \int_0^{1-y} A \cdot t\, dt = \frac{A}{2}(1-y)^2, \tag{5.25}$$

where $y = L/M$ represents the fraction of NZMBs in the video frame. Let

$$z = \left(\prod_{i=1}^{M} \sigma_i^2 \right)^{\frac{1}{L}}$$

,

we have

$$\ln(Z) = \frac{M}{L} \cdot \frac{1}{M} \sum_{i=M-L+1}^{M} \sigma_i^2 = \frac{1}{y} \int_{1-y}^{1} \ln(At)\, dt = \ln A - \frac{1}{y}[y + (1-y)\ln(1-y)]. \tag{5.26}$$

Therefore,

$$D(y) = A(\frac{1}{2}(1-y)^2 + ye^{-(1/y)[y+(1-y)\ln(1-y)]} \cdot 2^{-2\gamma(R/y)}$$ (5.27)

5.4.3 The P–R–D Model

5.4.3.1 Parameters Estimation and Model Simplification

The R–D model for the PRECODING modules given by Equation 5.27 has one parameter A. Note that

$$\frac{1}{M}\sum_{i=1}^{M}\sigma_i^2 = \int_0^1 \varsigma(t)\,dt = \frac{A}{2}.$$ (5.28)

Therefore, A can be estimated by

$$A = \frac{2}{M}\sum_{i=1}^{M}\sigma_i^2 = \frac{2}{M}\sum_{i=1}^{M}SSD_i$$ (5.29)

The R–D model in Equation 5.27 will be used for energy consumption control and picture quality optimization. Since the model is highly nonlinear, it is not suitable for mathematical optimization. Therefore, we need to simplify the formulation, specifically the exponential term. Taylor expansion yields the following linear approximation:

$$e^{-\frac{1}{y}[y+(1-y)\ln(1-y)]} \approx (e^{-1} + e^{-3}) + (1 - e^{-1} - e^{-3})(1 - y).$$ (5.30)

Figure 5.6 shows the nonlinear exponential function (solid line) and its linear approximation (dashed line). It can be seen that an approximation error is relatively small. With the linear approximation, the PRECODING C–R–D model becomes

$$D(y) = A \cdot \left[\frac{1}{2}(1-y)^2 + y(1 + a_0 y) \cdot 2^{-2\gamma\frac{R}{y}}\right],$$ (5.31)

where $a_0 = e^{-1} + e^{-3} - 1$.

5.4.3.2 Integrated P–R–D Model

For a complexity target of λ_{ME} SSD computations, the average MB variance is given by

$$\frac{1}{M}\sum_{i=1}^{M}\sigma_i^2 = \beta_0 + \beta_1^{-\beta_2 x}, \quad x = \frac{\lambda_{ME}}{\lambda_{ME}^{\max}}.$$ (5.32)

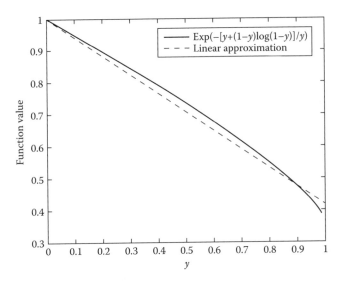

FIGURE 5.6 Linear approximation of $\exp(-(1/y)[y + (1 - y)\ln(1 - y)])$.

According to Equation 5.29 and the PRECODING C–R–D model in Equation 5.38, we have

$$D = D(R; x, y) = 2\left(\beta_0 + \beta_1^{-\beta_2 x}\right)\left[\frac{1}{2}(1-y)^2 + y(1+a_0 y) \cdot 2^{-2\gamma(R/y)}\right], \tag{5.33}$$

where x and $y = \lambda_{\text{PRE}}/M$ are the normalized complexity control parameters. Both x and y range from 0 to 1, with 0 and 1 representing the lowest and highest computational complexity, respectively. It should be noted that the distortion in Equation 5.33 measures only the quality for a single frame. The research in video quality evaluation suggests that the video presentation quality should be measured not only by the spatial quality of a single frame, but also by the temporal quality in motion smoothness [35]. Therefore, the encoding frame rate λ_F plays a very important role in quality evaluation. It is also a key parameter in energy consumption control. For example, at lower frame rates, more energy can be allocated to each frame to improve the spatial quality. However, in this case, the temporal video quality degrades. Although many results have been published in subjective video quality evaluation, most of them focus on experimental studies. For quality optimization of video coding, we need an analytic, mathematically tractable model to describe the video presentation quality. The experimental results in [35] suggest that the video presentation quality D_v should consist of two parts: the spatial quality of a single picture D_{spatil} and the temporal motion quality D_{temporal}. D_{spatil} given by Equation 5.33. D_{temporal} depends on the encoding frame rate. In typical video decoding and display, if a video frame is skipped, the previous decoded picture stays on the screen until the next frame is decoded. In other words, the decoder reconstruction of the skipped frame is the copy of its previous decoded frame. From the video encoder point of view, the ME complexity x, the PRECODING complexity y, and the bit rate R of the skipped video frame are all zeros. Therefore, from Equation 5.33, we can see that its coding distortion is given by

$$D_{\text{temporal}} = D(R;x,y)|_{R=0,x=0,y=0} = \beta_0 + \beta_1, \qquad (5.34)$$

which is the MSE between the skipped frame and its previous reconstruction. Let ω_s and ω_t be the perceptual weight on the spatial quality and temporal quality, respectively. The experimental results in [35] suggest that ω_s and ω_t should be a function of the frame rate. For example, if the video encoder encodes only one frame per minute, although each picture has very high quality, the viewer will complain about the bad video streaming service because he has missed a lot of important motion information and the spatial information in between. In this work, we choose the perceptual weight as follows:

$$\omega_t = (1-z)^2; \quad \omega_s = 1 - \omega_t; \qquad (5.35)$$

where $z = \lambda_F/f_{\text{max}}$, and f_{max} is the maximum frame rate with a default value of 30 fps. Therefore, the video presentation quality is defined as

$$D_v = \omega_s D_{\text{spatial}} + \omega_t D_{\text{temporal}}$$

$$= (1-z)^2(\beta_0 + \beta_1) + 2(2z - z^2)(\beta_0 + \beta_1^{-\beta_2 x}) \left[\frac{1}{2}(1-y)^2 + y(1 + a_0 y) \cdot 2^{-2\gamma(R/y)} \right]. \qquad (5.36)$$

Let C_1, C_2, and C_3 be the constants in (5.16), we have the power consumption computed as:

$$P = \Phi \Big(z \big(C_1 x + C_2 y + C_3 R \big) \Big), \qquad (5.37)$$

For a given power supply level P and a given rate R, we need to find the best configuration of the complexity parameters for the ME and PRECODING modules to maximize the picture quality. Mathematically, this can be formulated as:

$$\begin{aligned} &\min_{\{x,y,z\}} D_v(R;x,y,z) \\ &\text{s.t. } P = \varphi(z(c_1 x + c_2 y + c_3 R)) \end{aligned} \qquad (5.38)$$

The minimization parameters (x, y, z) can be obtained using binary search of the minimum point.

5.4.3.3 R–D Optimized Power-Scalable Video Encoding

The R–D optimized power-scalable video encoder system operates as follows:

Step 1: Determining the model parameters: In Equation 5.38, the ME model parameters β_0, β_1, β_2 are estimated from the statistics of previous frames using linear regression. a_0 is a constant determined by Equation 5.33. The model parameter is also determined from the R–D statistics of the previous frames. At the beginning stage, no power control is applied, because the system has sufficient power supply.

Step 2: Optimization: Find the optimal complexity control parameters $\{x,y,z\}$ using Equation (5.38). This step is executed only if the power control is triggered according to the adjustment frequency, for example, once per 5 s.

Step 3: Frame rate and ME complexity control: Set the encoding frame rate to be $\lambda_F = z \cdot f_{max}$. The available SSD computations for ME is given by $\lambda_{ME} = x \cdot \lambda_{ME}^{max}$. Using the MHI-based allocation scheme presented above to allocate the SSD computation among the MBs. Using the fast and efficient diamond ME scheme to find the MV and the minimum SSD for each MB. The number of diamond search layers is controlled by the allocated SSD computations.

Step 4: PRECODING complexity control: Find the $(1 − y) \cdot M$ MBs with the smallest SSD values and forces them to be AZMBs. The PRECODING operation is applied to the remaining NZMBs. Dynamic rate control is used to reallocate the bits from the AZMBs to the NZMBs.

5.4.4 Empirical Data

To evaluate the performance of the P–R–D model and the power-scaling video encoding system, we implemented the proposed P–R–D model and power scalability scheme in the public domain H.263+ encoder. Similar performance is expected for other coding systems, such as MPEG-2 and MPEG-4. In our simulations, the maximum search points for each MB λ_{ME}^{max} is 50, and the maximum frame rate $f_{max} = 30$ fps. To test the accuracy of the P–R–D model, we run the video encoder over the "Foreman" QCIF sequence at 128 kbps and 15 fps for different complexity control parameters (x, y) and measure the corresponding distortion. Figure 5.7 shows the actual distortion function $D(x, y)$.

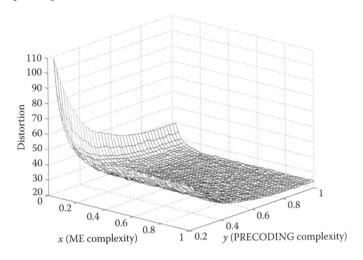

FIGURE 5.7 Actual complexity-distortion surface $D(x, y)$.

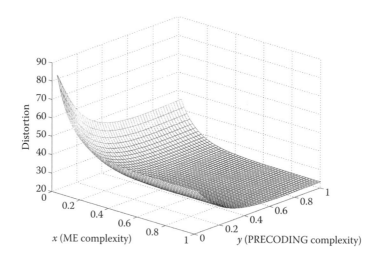

FIGURE 5.8   The complexity–distortion surface estimated by the P–R–D model.

The estimation given by the P–R–D model is shown in Figure 5.8. We can see that model estimation is quite accurate. Simulations over other test videos yield similar results. For a given bit rate R and the device power supply level, using Equation (5.38) the encoder can find the best configuration of parameters to maximize the video quality.

Figures 5.9 through 5.11 show the picture distortion, and the optimal control parameters $\{x, y, z\}$ as functions of the percentage of power consumption for different coding bit rates R. Some interesting observations can be made: (i) As the encoder scales down its power consumption, as a percentage of its maximum power consumption level, the video quality degrades. The video encoding automatically changes from high-quality motion video coding (when the energy supply is plenty) to still image coding (when the device is running out of energy). (ii) At lower bit rates, the ME wins over the PRECODING in power allocation, because the ME is computation-hungry but the PRECODING is bit-rate-hungry; hence, as shown in Figure 5.9, the complexity for the ME is high but the complexity for the PRECODING is low. Figure 5.12 shows the achievable minimum distortion D as a function of R and the power P. Figure 5.13 shows the "Carphone" QCIF video coded at 64 kbps and 15 fps for different power consumption levels. We can see the picture quality degradation is very graceful. We can see that the P–R–D model has direct applications in energy management, resource allocation, and QoS provisioning in wireless video communication, especially over wireless video sensor networks.

5.5 Power and Distortion Optimized Video Coding

Traditional power-conserving video processing techniques usually focus on low-power device and circuit design, as discussed in [36]. Recently, some research works on high-level management of algorithms and architectures have been reported, including power-aware design techniques that attempt to adjust the source-coding parameters to maximize the performance under power dissipation constraints [14] and low-power

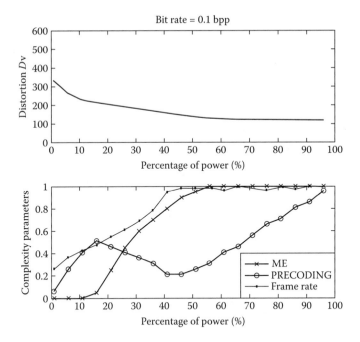

FIGURE 5.9 R–D optimized power control for the "Football" CIF video at $R = 0.1$ bpp.

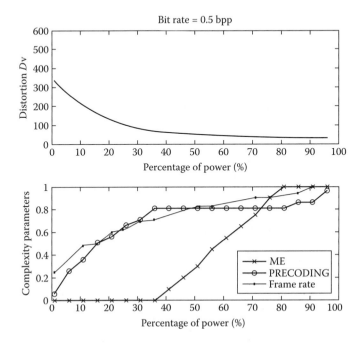

FIGURE 5.10 R–D optimized power control for the "Football" CIF video at $R = 0.5$ bpp.

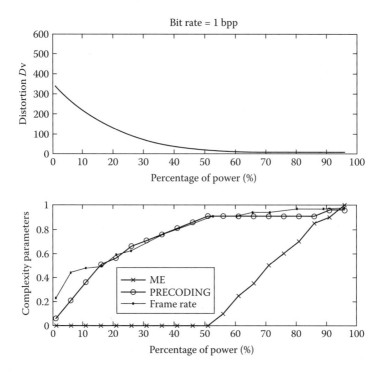

FIGURE 5.11   R–D optimized power control for the "Football" CIF video at $R = 1$ bpp.

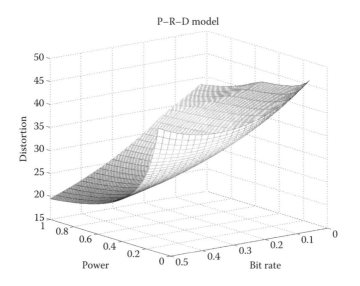

FIGURE 5.12   The P–R–D model.

| 100% power, PSNR = 33.76 | 75% power, PSNR = 33.72 |
| 25% power, PSNR = 32.05 | 5% power, PSNR = 29.41 |

FIGURE 5.13 The encoded "Carphone" QCIF sequence at 64 kbps and 15 fps for different power supply level.

design techniques that try to lower the intensive complexity of video encoding with or without a desired performance target [30,37].

In power-distortion optimized video coding, the coding complexity needs to be taken into account since it affects both the power consumption and the video distortion. This in turn requires developing an appropriate R–C–D model to study and understand the interaction and tradeoffs between rate, complexity, and distortion. An appropriate model helps to provide valuable insights into the power–distortion optimization problem. Some tradeoffs of power and complexity have been addressed. Research works on R–C–D modeling are also reported in [14]. In our previous work in [17], based on the analysis of power consumption, bitrate, and distortion behavior, a video system that is able to minimize the distortion under a given power consumption constraint is developed. However, the little reported works are either ad hoc, in that they depends on the coding behavior of a specific coding algorithm, or they propose complex models so that estimation of model parameters might become a formidable task. Up until now, no general R–C–D model has been proposed.

5.5.1 The Power–Distortion Optimization Problem

The coding procedure can simply be decomposed into three consecutive parts: (i) ME/MC; (ii) mode selection; (iii) entropy coding. The first two parts produce the quantized transform coefficients. The third part is responsible for converting the symbols of quantized transform coefficients into a standard compliant bitstream. Parameterization can be introduced into these parts, since video coding standards only define the bitstream syntax, or decoder operation, parametric video coding causes no conflict. Parametric ME can be achieved by controlling the ME precision [9], the size of the MV search window or the number of search points during the searching for the MVs [17]. Parametric mode selection can be achieved by controlling the λ_{PRE} parameter, which is the fraction of non-SKIP MBs in the video frame [17], or limiting the number of available coding modes, or the INTRA ratio parameter, which is the fraction of INTRA MBs in the video frame. Since the output bitstream must be conformed to bitstream syntax defined by the specific video standard, parameterized coding is not performed in the component of entropy coding.

Let $C = \{c_1, c_2, \ldots, c_N\}$ be the parameter set of the parametric encoder where N is the total number of parameters. For example, we may have parameter c_1 for ME, which is the size of the search window and c_2 for mode selection, which is the INTRA ratio parameter. Without loss of generality, each element c_i is normalized to lie between 0 and 1, measuring the level of coding complexity, wherein 0 and 1 corresponds to no and full coding complexity, respectively. The larger the value, the higher the coding complexity is. In the following, c_i is referred as the complexity parameter.

5.5.2 The Problem Formulation

The complexity parameters affect the compression performance in terms of power consumption and video distortion. Let $P(R,C)$ and $D(R,C)$ denote the power consumption and video distortion with parameter set C at coding bit rate R; detailed analysis will be given in the next section. The objective is to find a parameter set that minimizes both the power consumption and the video distortion, which can be formulated as a multiple objective optimization (MOO) problem, given by

$$\min \begin{bmatrix} P(R,C) \\ D(R,C) \end{bmatrix},$$

(5.39)

$$\text{s.t.} \quad P(R,C) \leq P_c$$

where P_c is the power supply constraint for video coding. We assume P_c is attainable either through low-level circuits design or pre-determined by system-level power allocation. For simplicity, we only consider the case that the bit rate is given as a constant and controlled by a rate control scheme.

In general, higher coding complexity results in smaller distortion, but consumes more energy. In contrast, lower coding complexity has lower power consumption, but at the expense of larger distortion. Thus, the objective functions in Equation 5.39 are incommensurate and in conflict with one another with respect to their minimum goals. For such an MOO problem, there is no unique solution [38]. Figure 5.14 illustrates the concept of the MOO problem with conflict objective functions. The curve AB is the pareto-optimal frontier, along which no further improvement can be done on power consumption P or video distortion D without sacrificing the other one.

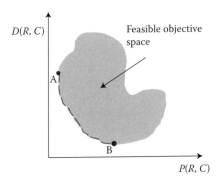

FIGURE 5.14 Illustration of the MOO problem.

5.5.3 Problem Analysis and Optimization Strategy

5.5.3.1 Power Consumption Analysis

The video coding operations are decomposed into two parts: (1) Residual error prediction and transform; (2) entropy coding. The first part includes coding operations such as ME, DCT, IDCT, MC, and so on, wherein the complexity parameter is usually introduced, as discussed in the previous section. Obviously, the energy consumed by this part depends on the embedded complexity parameters. The energy consumption of entropy coding is approximately proportional to the coding bit rate R [46]. We model the energy consumption (with units of Joules) of encoding one frame by

$$E_f = E_0 + E_r \cdot R + E_c, \tag{5.40}$$

where E_0 accounts for miscellaneous energy overheads, such as energy cost for I/O and E_r is a linear factor constant. E_c is the energy consumption of the parametric coding operations. In this work, we consider E_c as the summation of the energy consumption of individual modules:

$$E_c(C) = \sum_{i=1}^{N} E_i c_i, \tag{5.41}$$

where E_i is the energy consumption for module i. Let f_s denote the coding frame rate (frames/second), the power consumption (with units of Watts) is given by

$$P(R,C) = f_s \cdot E_f = f_s \cdot (E_0 + E_c(C) + E_r \cdot R) = P_0 + P_c(C) + P_r R. \tag{5.42}$$

Note that changing the complexity parameters and bit rate affects the miscellaneous energy overheads. We observe, however, even if efficient models of energy overheads are available, the dependence is not obviously making a comprehensive compositional model potentially complex. Thus, in this work, we ignore this effect and E_0 is treated as a constant. Because the complexity parameters are normalized values and bounded by 1, P is also bounded. When the video encoder runs at full complexity ($c_i = 1$), P is the maximum, denoted by P_{max}.

5.5.3.2 R–C–D Modeling

For a parametric video encoder, D is not only a function of R, but also the complexity parameters. To model the R–C–D behavior, we begin the analysis by considering only one complexity parameter c in the video encoder. In this case, $D(R, C) = D(R, c)$. Two implementations of parametric video coding are investigated. In the first implementation, we use the search range of full search ME as the complexity parameter with maximum value 15. When the search range decreases, the complexity decreases since we have less search points for MV search. In the second implementation, we use the INTRA refresh rate as the complexity parameter with maximum value 10. In this approach, if the consecutive times of one MB being coded as an INTER MB are bigger than the INTRA refresh rate, this MB will be forced as INTRA coded. As the refresh rate decreases, the complexity of the encoder decreases since the MBs are more frequently coded as INTRA MBs for which ME is not required. In Figures 5.15 and 5.16, the R–D curves of "Stefan" sequence at different constant values of c ($c_1 < c_2 < c_3 < c_4$) are plotted. Note that c is a normalized value.

As can be seen from Figures 5.15 and 5.16, the R–D performance deteriorates as c decreases. Based on the assumption of independent identically distributed (*i.i.d.*) memoryless source for the quantized transform coefficients, typical R–D model under the mean square distortion criterion is

$$D(R) = \varepsilon^2 \cdot \sigma_x^2 \cdot e^{-\alpha R}, \tag{5.43}$$

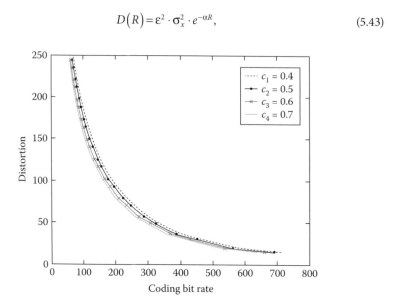

FIGURE 5.15   R–D curves of frame 0 ~ 99 of "Stefan" sequence with ME search range control.

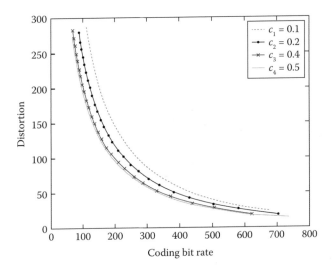

FIGURE 5.16   R–D curves of frame 0 ~ 99 of "Stefan" sequence with INTRA refresh rate control.

where ε^2 is a source-dependent parameter equal to 1 for uniform distribution, 1.4 for Gaussian distribution, and 1.2 for Laplacian distribution. σ_x^2 denotes signal variance. The parameter α equals to 1.386 for uniform, Gaussian and Laplacian distribution. Similar to the R–D model, but taking into account of the impact of c on the R–D performance, we model the R–C–D behavior with

$$D(R,c) = \varepsilon^2 \cdot \sigma_x^2 \cdot c^{-\beta} \cdot e^{-\alpha R} = \gamma \cdot c^{-\beta} \cdot e^{-\alpha R}, \quad 0 < c \leq 1, \ \alpha > 0 \ \text{and} \ \beta > 0, \quad (5.44)$$

where α, β, and γ are model parameters and can be obtained by using linear regression techniques. When we have constant complexity, the R–C–D model becomes the classical R–D model. It is worth to mention that different implementation of parametric video coding has different impact on the video distortion. For example, the video distortion is less sensitive to parametric ME control than to parametric DCT/quantization control where the distortion mainly comes from. Parameter β accounts for this difference. For a specific video encoder, it is possible to have a more accurate (but potentially more complex) R–C–D model through theoretic R–D analysis. However, this theoretic R–C–D model is only valid for the specific video encoder, which limits its application. In contrast, the proposed R–C–D is a general model. As we show in the following, it remains valid for different video encoders. The simplicity of the model significantly increases its usability and provides valuable insights into the R–C–D tradeoffs in power–distortion optimization.

We empirically evaluated the accuracy of the proposed R–C–D model, using the ME search range control and INTRA refresh rate control. For each implementation, we varied the complexity parameter and the quantization factor to obtain the output bit rate and distortion. We performed data fitting by minimizing the sum of squared MSE differences between the model and the measured data. Figures 5.17 through 5.20 present

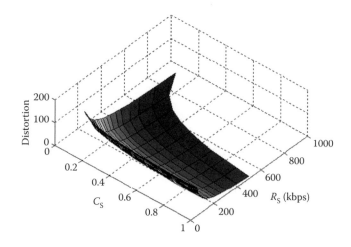

FIGURE 5.17 R–C–D surface using INTRA refresh rate control for "Foreman" QCIF sequence.

the results of the actual R–C–D surfaces and the results estimated from the model. We note that different implementations of parametric video coding have significantly different forms of R–C–D surfaces. For example, the surface of complexity control using ME search range is more flat than the one using the INTRA refresh rate. Nevertheless, the proposed R–C–D model still fits the actual data well. Simulations over other test video sequences yield similar results. The average prediction accuracy is given in Table 5.2. The proposed model is accurate enough to appropriately describe the R–C–D behavior of video coding.

Now we consider the case when the parametric encoder has multiple complexity parameters: $C = \{c_1, c_2, \ldots c_N\}$. Since the encoder can be regarded as a nonlinear system,

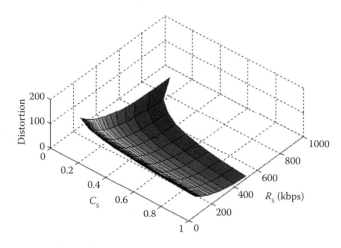

FIGURE 5.18 Data generated by the proposed R–C–D model using INTRA refresh rate control for "Foreman" QCIF sequence.

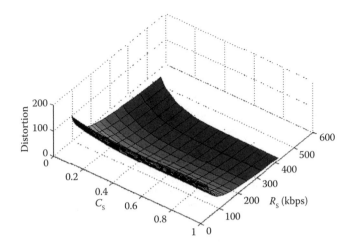

FIGURE 5.19 R–C–D surface using ME search range control for "Foreman" QCIF sequence.

the overall distortion is not simply a linear addition of the individual effects of c_i. As discussed above, the complexity parameters are usually separately applied to different coding modules. We observe that the effects of various complexity parameters can be described in a separable manner. Based on the observation, we model the R–C–D using

$$D(R,C) = \gamma \cdot e^{-\alpha R} \prod_{i=1}^{N} c_i^{-\beta_i}, \qquad 0 < c_i \leq 1, \quad \alpha > 0, \quad \beta_i > 0, \qquad (5.45)$$

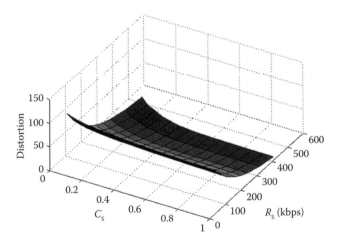

FIGURE 5.20 Data generated by with the proposed R–C–D model using ME search range control for "Foreman" QCIF sequence.

where N is the number of complexity parameters. A larger model parameter β_i indicates c_i has more contribution/effect on the overall distortion.

5.5.4 Optimization Strategy

Because the energy drawn from a battery is not always equivalent to the energy consumed in device circuits, understanding the battery discharge behavior is essential for optimal system design. Figure 5.21 gives an example of the discharge characteristic of the lithium–ion battery [39], which is used widely in today's mobile devices because of its high energy density and capacity.

> *Observation* 1: As the battery discharges, its voltage drops. There is an inflexion point on the discharge curve, after this point the power will run out quickly. In this case, the video quality will degrade quickly due to insufficient power supply.
>
> *Observation* 2: The effective battery capacity is increased if the average discharge current from the battery decreases, which suggests that reducing the discharge current, that is, lowering the complexity is essential for battery lifetime extension.

According to these observations, we have different preferences for the distortion versus power consumption during the whole battery lifetime. Thus, we apply the constraint-oriented strategy to solve the above MOO problem. In this strategy, one objective function is used as the main objective and the other is treated as the secondary objective.

5.5.4.1 Distortion Preference

When we do not have enough power supply to perform full complexity encoding, that is, $P_c < P_{max}$, we substitute problem (5.39) by

$$\min D(R, C), \quad \text{s.t. } P(R, C) < P_c, \tag{5.46}$$

which is a distortion optimization problem with a power consumption constraint. Using the Lagrange multiplier method to solve the problem, we have the Kuhn–Tucker conditions as

$$\frac{\partial D}{\partial c_i} + \lambda \frac{\partial P}{\partial c_i} = 0, \tag{5.47a}$$

$$\lambda \geq 0, \quad P(R,C) \leq P_c, \quad \text{and} \quad \lambda(P(R,C) - P_c) = 0 \tag{5.47b}$$

TABLE 5.2 Average Prediction Accuracy of the R–C–D Model

Parametric Video Coding	Foreman (%)	Carphone (%)	Akiyo (%)
ME search range control	83	82	89
INTRA refresh rate control	82	81	80

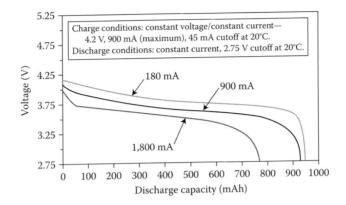

FIGURE 5.21 Lithium-ion battery discharge characteristic.

where λ is the Lagrange multiplier controlling the P–D tradeoffs. From condition (5.47b), either $\lambda = 0$ or $P(R, C) = P_c$. If $\lambda = 0$, from condition (5.47a), we should have

$$\frac{\partial D}{\partial c_i} = 0, \tag{5.48}$$

which is conflict with the following equation derived from the R–C–D model:

$$\frac{\partial D}{\partial c_i} = -\gamma\beta_i c_i^{-\beta_i - 1} \prod_{j=1, j\neq i}^{N} c_j^{-\beta_j} e^{-\alpha R} < 0, \quad i \neq j. \tag{5.49}$$

As a result, we have

$$P(R, C) = P_c. \tag{5.50}$$

Remark 1: In the distortion-preferred optimization problem, the optimal solution is at the boundary of the constraint where the power consumption is maximized to the given upper bound. That is, the video encoder has to consume all the available energy to get the minimal distortion.

The solution can be obtained by solving Equations 5.47a and 5.50. As an example, suppose we have two complexity parameters, c_1 and c_2, we have

$$D = \gamma c_1^{-\beta_1} c_2^{-\beta_2} e^{-\alpha R}, \quad P = P_0 + P_1 c_1 + P_2 c_2 + P_r R. \tag{5.51}$$

The solution is

$$c_1 = \frac{P_c - P_0 - P_r R}{P_1(1 + (\beta_2/\beta_1))}, \quad c_2 = \frac{P_1 \beta_2 c_1}{P_2 \beta_1} \tag{5.52}$$

When only one complexity is used, the solution can be easily obtained without solving the unconstrained problem. The optimal complexity parameter is simply given by

$$c_{opt} = \frac{(P_c - P_0 - P_r R)}{P_1}.$$

$$(5.53)$$

5.5.4.2 Power Consumption Preference

When $P_c \geq P_{max}$, to preserve power, we substitute problem (5.39) by

$$\min P(R, C), \quad \text{s.t. } D(R, C) \leq D^*,$$

$$(5.54)$$

where D^* is a given upper bound distortion. Notice that different from Equation 5.1, here the power constraint is released. Analysis on the Kuhn–Tucker conditions results in a similar conclusion.

Remark 2: Minimal power consumption is achieved when the video distortion is maximized to the upper bound value, in other words, when

$$D(R, C) = D^*.$$

$$(5.55)$$

Figure 5.22 gives one example of how the upper bound distortion affects the solution. The power consumption is measured by the percentage of power compared with that is consumed by running at full coding complexity.

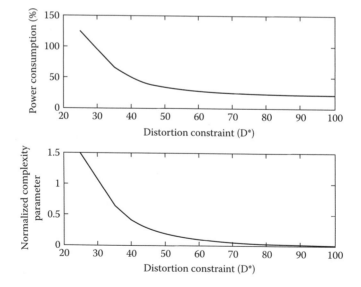

FIGURE 5.22 P–D curves for "Foreman" QCIF sequence using INTRA refresh control under different distortion constraints.

```
if(SSAD_i - SSAD_i' < Thresold)
/* We may get improvement by increasing the complexity */
       N++ ;
       if (C_a^i > C_e^i)
            C_e^{i+1} = C_e^i + 2^N · (C_a^i - C_e^i)
       else
            C_e^{i+1} = C_e^i
else
       N = 0;
       C_e^{i+1} = min(C_a^i, C_e^i)
```

FIGURE 5.23   Greedy ME control.

Figure 5.23 indicates that as the distortion constraint increases, the video encoder can run at lower coding complexity for smaller power consumption. Also notice that as the expected distortion decreases constraint more, bigger complexity parameters, and thus higher power consumption is required. In some cases, it is even not feasible, as shown in the figures, the complexity parameter bigger than 1 and Power Consumption higher than 100%.

Considering the human being can tolerate some video distortion, this degradation in perceived video quality may be negligible with respect to the improvement in power saving. As the curve shows, this improvement in power consumption by increasing the distortion constraint will get saturate. Therefore, the selection of an appropriate distortion constraint is crucial for power consumption saving. However, the constraint of distortion is very subjective and application dependent, which makes it quite difficult to pre-determine the distortion constraint before encoding the sequence. It is desirable to have a dynamic control to attack this challenge.

Remark 3: From Equation 5.49, each complexity parameter has different influence on the overall distortion according to their importance, reflected by their model parameter β_i. For a same amount of complexity change, the larger the value of β_i, the more changes in the distortion.

Remark 4: For a given amount of complexity change, the smaller the value of c_i, the larger the decrement of distortion is. Thus, as the encoder slows down (c_i becomes smaller), the distortion becomes more sensitive to the change of complexity.

Based on Remarks 3 and 4, instead of pre-determining the distortion constraint, we apply a progressive control approach. Each parameter is adjusted individually according to their importance. Note that for a given bit rate, the minimum distortion can always be achieved by running at full complexity from Remark 1. We start running the video encoder at full coding complexity, and progressively lower the complexity to preserve power. According to Remark 4, it is desirable to promptly increase the complexity when the video quality begins to deteriorate. The advantage of the progressive control is that we do not need accurate model parameters to solve the minimization problem. Also, since a predefined upper bound distortion is not required, this approach is adaptive to various coding content.

5.5.5 Power and Distortion Optimized Video Coding

In this section, based on the optimization strategies presented in the previous section, we demonstrate how to achieve power–distortion optimized video coding in a practical parametric video encoder. We use ME and model selection as the candidates for complexity control, since they significantly affect both the video distortion and power consumption.

5.5.5.1 Progressive Complexity Adjustment

After applying the ME and mode selection control, which will be described below, the encoder might be in "sleep" (if the scene has been idle for a while), therefore it is necessary to "wake up" the encoder when with rapid motion or sudden scene change. A variety of techniques have been proposed to perform the detection of scene changes. Considering the requirement of low computational complexity in the research, we use the percentage of INTRA coding MBs to detect the scene change. This approach has low computational overhead, but still has pretty good performance. When the scene change occurs, the video coding will restart at the highest complexity level in order to re-understand the content and choose another optimal parameter set.

5.5.5.2 Greedy ME

We propose a greedy adjustment on the complexity scalable ME. We start with the highest complexity parameter value and keep reducing the value until we obtain a performance level that is no longer acceptable. Because of its simplicity, the sum of SADs of all the MBs is used as the distortion measure.

Note that the process of ME can be simply considered as a sequence of SAD computations. For each MV candidate, one SAD value is calculated and compared with the previous minimal SAD. A smaller value of SAD indicates a better MV. As the search for the optimal MV continues, more SAD computations are required and smaller SAD may be obtained. Denote SAD_m and SAD'_m the last SAD value and the second last SAD during the MV search for the mth MB. At frame i, let $SSAD_i = \sum_m SAD_m$ and $SSAD'_i = \sum_m SAD'_m$ be the sum of SAD and the sum of second last SAD, respectively. Clearly, $SSAD_i$ needs more computation than $SSAD'_i$. If $SSAD_i$ is smaller than $SSAD'_i$, it implies that we may get smaller distortion if we increase c_1 at frame $i + 1$. Let c_e^i and c_a^i denote the estimated and actual complexity parameter for frame i, respectively. The greedy ME control is given in Figure 5.23. We exponentially increase the complexity when we can get a performance improvement.

The threshold in Figure 5.23 controls the C–D tradeoff. In the experiments, we use the extreme value threshold = 0. Figure 5.24 shows the results of greedy ME complexity control for three different sequences, representing motion characteristic from low motion to high motion, respectively. The "Akiyo" sequence is a typical head-and-shoulder sequence and has very little motion. The camera is assumed to be stationary. From Figure 5.24a we can observe that the ME complexity parameter c_1 is gradually lowered to preserve power and finally stays in constant when the encoder only consumes computation on the active region (the news reporter). The "Foreman" sequence has a medium motion with the

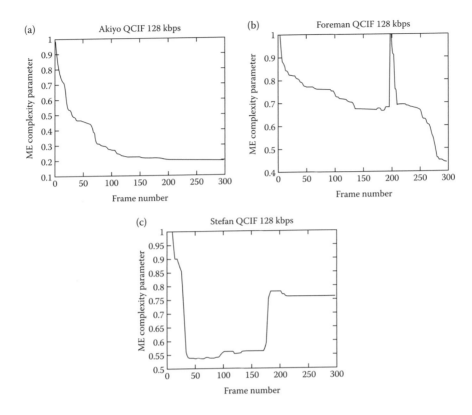

FIGURE 5.24   Results of greedy ME complexity control.

camera motion. In Figure 5.24b, the encoder resets the complexity parameter to the highest value when a scene change is detected and starts another greedy search. "Stefan" represents a typical sports sequence with fast motion. In "Stefan," from the 180th frame, the camera moves very quickly to focus on the sportsman. The proposed control is able to increase the complexity parameter promptly when the scene switches to such a fast motion, as shown in Figure 5.24c.

5.5.6 Optimal Mode Selection

Offline simulation shows that the model parameter of model selection is much bigger than that of ME, that is, $\beta_2 > \beta_1$. This is valid because in video coding, the quantization process, affected by the procedure of mode selection, mainly causes the distortion. Instead of using progressive control as ME, we try to find the optimal value for c_2.

In a block based video encoder, the residual error after motion compensation is transformed into the frequency domain using certain type of transform, such as DCT. After that, the transform coefficients are quantized by a predetermined quantization parameter for further entropy compression. We denote r, R as the residual error and the transform coefficients respectively. r can be the input MB for INTRA coding. The transform can be described by

$$R = ArA^{\mathrm{T}}, \tag{5.56}$$

where A is the transform matrix. The distribution of the residual error can be approximated by a Laplacian distribution with zero mean and a separable covariance:

$$r(m,n) = \sigma_r^2 \, \rho^{|m|} \, \rho^{|n|}, \tag{5.57}$$

where m and n are the horizontal and vertical distances between two pixels, respectively, σ_r is the variance of the residual errors. ρ ($|\rho| < 1$) is the correlation coefficient. Typical value of ρ ranges from 0.6 to 0.75. Let L denote the matrix of the correlation coefficients, given by

$$L = \begin{bmatrix} 1 & \rho & \rho^2 & & \rho^M \\ \rho & 1 & \rho & \cdots & \\ \rho^2 & \rho & 1 & & \\ \vdots & & & \ddots & \\ \rho^N & & & & 1 \end{bmatrix}. \tag{5.58}$$

Let Q be the quantization parameter and EM be an estimation matrix, of which the element $EM(u,v)$ is given by

$$EM(u,v) = \frac{K \times Q}{n \times \sqrt{[ALA^{\mathrm{T}}]_{u,u}[ALA^{\mathrm{T}}]_{v,v}}}, \tag{5.59}$$

where K is a constant and n is a confidence parameter. We can estimate the probability of $R(u, v)$ being zero by:

$$SAD < EM(u, v), \tag{5.60}$$

where SAD is the SAD value after ME. When Equation 5.60 is satisfied, $R(u,v)$ will be quantized to zero with probability 68%, 94%, and 99% with $n = 1$, $n = 2$, and $n = 3$, respectively [??].

Since the DC coefficient dominates the transform coefficients, when the quantized DC coefficient becomes zero, we can assume that all the quantized transform coefficients are zero. This assumption will not influence the reconstructed video quality apparently, because the human eyes are more sensitive to low frequency coefficients. Therefore, if $SAD < EM(0,0)$, then $R(0,0)$, which is the DC coefficient, all the quantized transform coefficients are zero, which implies the MB will be a zero MB. After doing this for each MB, we can calculate the percentage of zero MBs and determine c_2. Being able

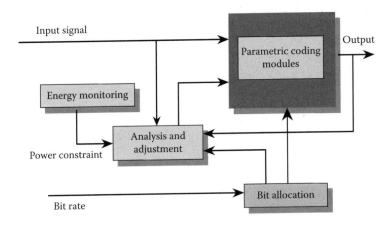

FIGURE 5.25 Power-distortion optimized video encoder.

to determine c_2, the number of complexity parameter is reduced to one, which can be calculated by Equation 5.53. This significantly reduces the complexity of solving the distortion-preferred optimization problem, as discussed in the previous section.

5.5.7 System Architecture

Figure 5.25 shows the architecture of the proposed video coding system, which includes four major modules: the "parametric coding modules," the "analysis and adjustment" module, the "energy monitoring" module, and the "bit allocation" module. In the system, the coding modules demanding high computational power are complexity scalable and can be adjusted by the embedded parameter c_1 and c_2. The "analysis and adjustment" modules model the R–C–D behavior of the system and apply the progressive control, aiming to determine a control parameter set for the coding modules. The "energy-monitoring" module provides the current power supply of the system, that is, the power constraint of this power-sensitive platform. The "bit allocation" module determines the number of bits to encode the next frame or group of frames according to the available bit rate determined by the transmission bandwidth.

5.6 Experimental Results

In this section, we evaluate the performance of the power–distortion optimized vide coding. The public domain H.263+ encoder is tested in the experiments. The approach is generally applicable to other standards and similar performance is expected. The video distortion is measured by peak signal-to-noise ratio (PSNR). To compare with the traditional fast video coding approach, the fast ME method in the reference software is tested in the experiments, referred as the FastME approach in the following. The TMN8 rate control algorithm is used. In the simulations, the power consumption is measured using the linear model, given by Equation 5.40. The complexity ratios are obtained through run-time complexity profile analysis. The maximum ME complexity corresponds

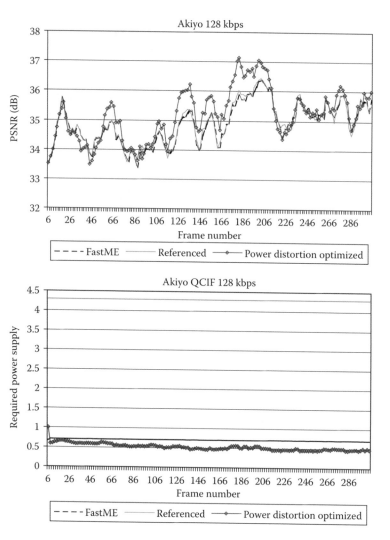

FIGURE 5.26 PSNR and power consumption comparison for the "Akiyo" sequence.

to 50 search points for each MB and therefore the full search ME will use an ME complexity parameter much higher than 1.

Figures 5.26 through 5.28 illustrate the comparisons of reconstructed video quality and power consumption for the tested sequences. We can observe that the proposed PDO (power–distortion optimization) video coding achieves significant power saving while maintaining similar video quality compared with the referenced video encoder. The proposed method has much better performance than the referenced encoder in terms of power saving. For the "Akiyo" sequence, these methods achieve similar video quality performance. From the power saving perspective, the PDO has the best perfor-

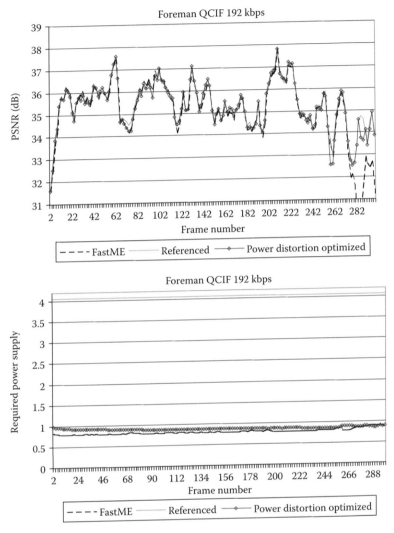

FIGURE 5.27 PSNR and power consumption comparison for the "Foreman" sequence.

mance, followed by the FastME method. This is because in the PDO control method, the video encoder eventually only spends some computation on the talking women, saving a great amount of energy. In Figures 5.27 and 5.28, the FastME and the PDO approaches have similar performance on power saving. However, the PDO approach achieves much better video quality, close to that of the ref video encoder, especially for fast motion sequence "Stefan." The FastME method reduces the power consumption at the expense of degraded video quality. From the results, we can see that the PDO video coding not only significantly reduces the power consumption, but also maintains the video quality for different motion sequences.

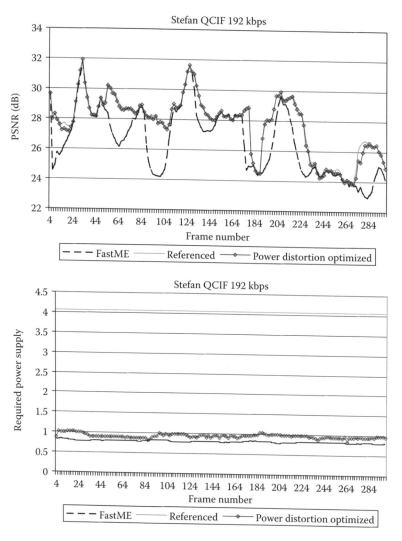

FIGURE 5.28   PSNR and power consumption comparison for the "Stefan" sequence.

Figure 5.29 shows the results of the PDO approach when the power supply changes. In Figure 5.29a, the *x*-axis is the time and the *y*-axis is the normalized power supply. One can observe that the proposed PDO approach is able to automatically adjust the system complexity according to the power supply level, which helps to extend the battery lifetime. As the power supply decreases, the video quality decreases gracefully. Note that for the referenced encoder and the FastME approach, they cannot adapt to the power supply level. Frames have to be dropped when the power supply is insufficient.

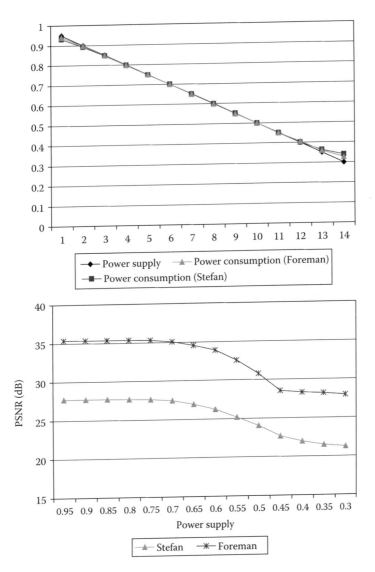

FIGURE 5.29 PSNR and power consumption with insufficient power supply.

References

1. ITU-T, Video Codec for Audiovisual Services at p × 64 kbits/s, Recommendation H.261, Geneva, 1990.
2. ITU-T, Video Codec for Low Bit-Rate Communication, Recommendation H.263, December 1995.

3. ITU-T Rec. H.264 | ISO/IEC 14496-10 AVC, Advance Video Coding, Draft, October 2002.

4. ISO/IEC CD 11172/2 (MPEG-1 Video), Information technology—Coding of Moving Pictures and Associated Audio for Digital Storage Media at Up to About 1.5 Mbits/s, Video, 1993.

5. ISO/IEC CD 13818/2 (MPEG-2 Video), Information technology—Generic Coding of Moving Pictures and Associated Audio Information, Video, 1995.

6. ISO/IEC CD 14496/2 (MPEG-4 Video), Information technology—Coding of Natural/Visual Objects, Part 2: Visual, 1999.

7. ISO/IEC, Information Technology—Generic Coding of Audio Visual Object, ISO/IEC JTC 1/SC 29/WG 11 N 2502, Atlantic City, 1998.

8. Paulson, L. D., Will fuel cells replace batteries in mobile devices. *IEEE Comput*, 2003; 36(11): 10–12.

9. Cheung, C. K. and Po, L. M., A hierarchical block motion estimation algorithm using partial distortion measure, in *Proceeding of ICIP'97*, vol. 3, pp. 606–609, 1997.

10. Ismaeil, I., Docef, A., Kossentini, F., and Ward, R., Computation-performance control for DCT-based video coding, in *IEEE International Conference on Acoustics, Speech, and Signal Processing, 2000. ICASSP '00. Proceedings. 2000*, vol. 6, 5–9 June 2000.

11. Burleson, W., Jain, P., and Venkatraman, S., Dynamically parameterized architectures for power-aware video coding: motion estimation and DCT, in *Second International Workshop on Digital and Computational Video*, Tampa, FL, USA, 8–9 February 2001.

12. Chaddha, N. and Vishwanath, M., A low power video encoder with power, memory and bandwidth scalability, in *Ninth International Conference on VLSI Design*, pp. 358–363, Bangalore, India, 3–6 January 1996.

13. Agrawal, P., Chen, J., Kishore, S., Ramanathan, P., and Sivalingam, K., Battery power sensitive video processing in wireless networks, in *Ninth IEEE International Symposium on Personal, Indoor and Mobile Radio Communications*, vol. 1, pp. 116–120, 8–11 September 1998.

14. Lan, T. and Tew, A. H., Power optimized mode selection for H.263 video coding and wireless communications, in *Proceedings of International Conference on Image Processing*, vol. 2, pp. 113–117, 1998.

15. Chakraborty, S. and Yau D. K. Y., Predicting energy consumption of MPEG video playback on handhelds, in *IEEE International Conference on Multimedia and Expo*, vol. 1, pp. 317–320, 26–29 August 2002.

16. Lu, X., Fernaine, T., and Wang, Y., Modeling power consumption of a H.263 video encoder, in *Proceedings of the 2004 International Symposium on Circuits and Systems ISCAS'04*, vol. 2, pp. 77–80, 23–26 May 2004.

17. He, Z., Liang, Y., Chen, L., Ahmad, I., and Wu, D., Power-rate-distortion analysis for wireless video communication under energy constraint. *IEEE Trans Circuits Syst Video Technol*, 2005; 14(5): 645–658.

18. Lin, J. and Ortega, A., Bit-rate control using piecewise approximation rate-distortion characteristics, *IEEE Trans. CSVT*, 1998; 8: 446–459.

19. Chiang, T. and Zhang, Y.Q., A new rate control scheme using quadratic rate distortion model. *IEEE Trans Circuits Syst Video Technol*, 1997; 7: 246–250.

20. Corbera, J. R. and Lei, S., Rate control in DCT video coding for low-delay communications. *IEEE Trans Circuits Syst Video Technol*, 1999; 9: 172–185.

21. Ding, W. and Liu, B., Rate control of MPEG video coding and recording by rate-quantization modeling. *IEEE Trans Circuits Syst Video Technol*, 1996; 6: 12–20.

22. Tian, X., Efficient transmission power allocation for wireless video communications, in IEEE Wireless Communications and Networking Conference, WCNC, 2004, vol. 4, 21–25 March 2004, pp. 2058–2063.

23. Chan, Y. S. and Modestino, J. W., A joint source coding-power control approach for video transmission over CDMA networks. *IEEE J Select Areas Commun*, 2003; 21(10): 1516–1525.

24. Kim, I. M., Kim, H. M., and Sachs, D. G., Power-distortion optimized mode selection for transmission of VBR videos in CDMA systems. *IEEE Trans Commun*, 2003; 51(4): 525–529.

25. Zhai, F., Eisenberg, Y., Luna, C. E., Pappas, T. N., Berry, R., Katsaggelos, A. K., Joint source-channel coding and power allocation for energy efficient wireless video communications, in *Proceedings 41st Allerton Conference Communication, Control, and Computing*, October 2003.

26. Eisenberg, Y., Pappas, T. N., Berry, R., and Katsaggelos, A. K., Minimizing transmission energy in wireless video communications, in *International Conference on Image Processing*, vol. 1, pp. 958–961, 7–10 October 2001.

27. Rodriguez, V., Optimal coding rate and power allocation for the streaming of scalably encoded video over a wireless link, in *IEEE ICASSP*, May 2004.

28. Zhang, Q., Zhu, W., Ji, Z., and Zhang, Y., A power-optimized joint source channel coding for scalable video streaming over wireless channel, in *IEEE International Symposium on Circuits and Systems*, vol. 5, pp. 137–140, 6–9 May 2001.

29. Ji, Z., Zhang, Q., Zhu, W., and Zhang, Y., End-to-end power-optimized video communication over wireless channels, in *IEEE Fourth Workshop on Multimedia Signal Processing*, pp. 447–452, 3–5 October 2001.

30. Zhang, Q., Ji, Z., Zhu, W., and Zhang, Y., Power-minimized bit allocation for video communication over wireless channels. *IEEE Trans Circuits Syst Video Technol*, 2002; 12(6): 398–410.

31. Lu, X., Erkip, E., Wang, Y., and Goodman, D., Power efficient multimedia communication over wireless channels. *IEEE J Select Areas Commun*, 2003; 21(10): 1738–1751.

32. Sikora, T., The MPEG-4 video standard verification model. *IEEE Trans Circuits Syst Video Technol*, 1997; 7: 19–31.

33. Berger, T., *Rate Distortion Theory*, Englewood Cliffs, NJ: Prentice Hall, 1984.

34. He, Z. and Mitra, S.K., A unified rate-distortion analysis framework for transform coding. *IEEE Trans Circuits Syst Video Technol*, 2001; 11: 1221–1236.

35. Apteker, R. T., Fisher, J. A., Kisimov, V. S., and Neishlos, H., Video acceptability and frame rate. *IEEE Multimedia*, 1995; 2(3): 32–40.

36. Benini, L., Bogliolo, A., and De Micheli, G., A survey of design techniques for system-level dynamic power management, in *IEEE Transactions on very Large Scale Integration (VLSI) Systems*, vol. 8, pp. 299–316, June 2000.

37. Akramullah, S. M., Ahmad, I., and Liou, M. L., Optimization of H.263 video encoding using a single processor computer: performance tradeoffs and benchmarking. *IEEE Trans Circuits Syst Video Technol*, 2001; 11: 901–915.

38. Deb, K., *Multi-Objective Optimization using Evolutionary Algorithms*. Chichester, UK; Wiley, 2001.

39. Rao, R., Vrudhula, S., and Rakhmatov, D. N., Battery modeling for energy aware system design. *IEEE Comput*, 2003; 36(12): 77–87.

40. Akyildiz, I. F., Su, W., Sankarasubramaniam, Y., and Cyirci, E., Wireless sensor networks: A survey. *Comput Netw*, 2002; 38(4): 393–422.

41. August, N. and Ha, D., On the low-power design of DCT and IDCT for low bit-rate video codecs, in *14th Annual IEEE International ASIC/SOC Conference*, Arlington, VA, USA, 12–15 September 2001.

42. Brooks, D., Tiwari, V., and Martonosi, M., Wattch: A framework for architectural-level power analysis and optimizations, in *Proceedings 27th Annual International Symposium on Computer Architecture*, 2000, pp. 83–94.

43. Cover, T.M. and Thomas, J. A., *Elements of Information Theory*. New York; Wiley, 1991.

44. Do, V. L. and Yun, K. Y., A low-power VLSI architecture for full-search block-matching motion estimation. *IEEE Trans Circuits Syst Video Technol*, 1998; 8(4): 393–398.

45. Duhamel, P. and Guillemot, C., Polynomial transform computation of the 2D DCT, in *Proceedings International Conference on Acoustics, Speech, and Signal Processing*, pp. 1515–1518, 1990.

46. Ebrahimi, T. and Pereira, F., *The MPEG-4 Book*. Upper Saddle River, NJ; Prentice Hall PTR, July 2002, ISBN: 0-130-61621-4.

47. Elgamel, M. A., Shams, A. M., Xi, X., and Bayoumi, M. A., Enhanced low power motion estimation VLSI architectures for video compression, in *The 2001 IEEE International Symposium on Circuits and Systems, 2001. ISCAS 2001*, vol. 4, 6–9 May 2001, pp. 474–477.

48. Gormish, M. J., Source coding with channel, distortion and complexity constraints, Ph.D. Thesis, Stanford University, March 1994.

49. Intel, Using streaming SIMD extensions in a fast DCT algorithm for MPEG encoding, *Intel Application Note AP-817*, Order No: 243651-002.

50. Intel, Using streaming SIMD extensions in a motion estimation algorithm for MPEG encoding, *Intel Application Note AP-818*, Order No: 243652-002.

51. http://www.cc.gatech.edu/fce/house/.

52. Kuhn, P., *Algorithms, Complexity Analysis and VLSI Architectures for MPEG-4 Motion Estimation*. Boston; Kluwer Academic Publishers, 1999.

53. Lee, P. Z. and Huang, F. Y., An efficient prime-factor decomposition of the discrete cosine transform and its hardware realization, in *Proceedings of International Conference Acoustics, Speech, and Signal Processing*, pp. 20.5.1–20.5.4, 1985.

54. Oppenheim, A. V. and Shafer, R. W., *Digital Signal Processing*. Englewood Cliffs, NJ: Prentice Hall, 1975.

55. Rao, K. R. and Yip, P., *Discrete Cosine Transform—Algorithms, Advantages, Applications*. London: Academic Press, 1990.

56. Zheng, D. W., Ahmad, I., and Liou, M., Real-Time software based MPEG-4 video encoding, in *2nd IEEE MPEG-4 Workshop and Exhibition*, San Jose, CA, June 2001.

6

Low-Power Design for Smart Dust Networks

Zdravko
Karakehayov
Technical University of Sofia

6.1 Introduction

Distributed sensor networks (DSNs) are composed of numerous small, low-cost, randomly located nodes. The network can be scalable to thousands of nodes that cooperatively perform complex tasks such as intelligent measurement. The network must be able to self-organize, adapt to random node spacing, execute algorithms for signal processing, and operate power as efficiently as possible. The major applications of DSNs are for monitoring environmental conditions, tracking the movements of birds and small animals, monitoring product quality, and building automation and defense networks. Smart Dust is a term recently coined at the University of California, Berkeley, to describe massively DSNs consisting of cubic-millimeter-sized motes [1,2]. The small size and anticipated low cost of the motes will help to collect information cost-effectively and less intrusively.

Each mote depends on low-capacity batteries as energy sources. Practically, the chance for battery replacement is nonexistent. As a result, every aspect of the Smart Dust networks, from mote location through computing and communication, is viewed from the low-power perspective.

6.2 Location

A deployment may leave numerous motes located in different areas of a large geographical region. The location of the motes affects energy efficiency in a number of ways. Sensor readings are of interest if only bound to a known location. Interrogation of motes before a location procedure would be a loss of energy. The global positioning system (GPS) is able to locate network nodes in outdoor environments. However, cost, power consumption, and size of the currently available GPS receivers are prohibitive for Smart Dust motes. Optical communication emerges as the most efficient method if a central station may be harnessed to provide energy for location tasks. Because motes may move, some applications would demand updating the positions regularly. Also radio-frequency (RF) communication can be used by motes to locate themselves via beacon signals from reference points [3,4].

As soon as the location procedure has been completed, some nodes will be actively involved in sensing, while others will wait for events and can be turned off to save energy. An event tracking, such as following light shadow edges over a sensor field, can be organized into two ways:

- All motes deactivate all subsystems except sensors that can obtain relevant data. If the sensors provide binary readings, they can be used to awake the motes in case of events.
- A more sophisticated power reduction approach will turn off all motes, except motes in the close vicinity of the event, completely. However, in case of a dense deployment the distance alone is not sufficient as a criterion.

Liu et al. [5] have developed a method for event tracking. The method identifies motes that will not be immediately approached by the event and can be turned off to save energy. The method is based on dual space transformation [6]. Figure 6.1 shows an example for event tracking.

With no loss of generality, it can be assumed that the event is a moving light shadow edge. The edge is presented in the primal space as the E line and the motes' locations are indicated as points. The line is uniquely defined in the primal space by the p slope and the y-intercept q. The line is transformed into the e point in the dual space; in turn, the points from the primal space are transformed into lines in the dual space. As a result, the dual space is partitioned into cells. The e point is contained in the shaded cell. Because the e point cannot intersect the m2 line before it crosses one of the cell boundaries, the M2 mote can stay turned off as long as none of M1, M3, and M4 sense a transition.

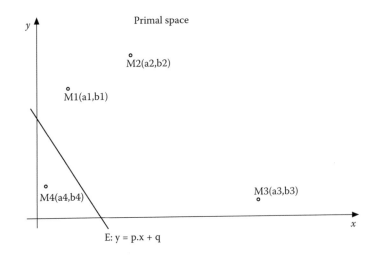

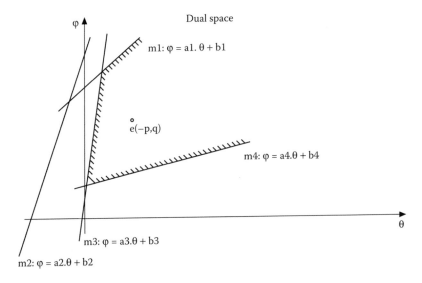

FIGURE 6.1 Primal-to-dual space transformation indicates the sequence of transitions.

6.3 Sensing

The mote's sensors vary from application to application: temperature, light, magnetic field, vibration, and acoustic. Recent advances in technology have made it possible for these sensors to be released in ultralow sizes and power versions [2,7].

Sensors convert physical variables into electrical signals. Typically, the signals are in the microvolt or millivolt range. An input signal conditioner is used to filter and amplify the signals. Energy is consumed in the sensor, amplifier, and analog-to-digital converter

(ADC). The power consumption can be reduced with appropriate power management. The ADC's resolution has a significant impact on the energy budget. For instance, if the ADC's resolution is increased from 15 to 16 bits while keeping the other parameters unchanged, the power consumption is increased from 100 to 400 mW [8].

A common method for analog-to-digital conversion is the successive approximation [9]. Because the ADC determines one bit of the result in each cycle, it would be possible to apply selective resolution. Consequently, different samples will have different numbers of bits and different energy costs. Finally, one may only want to test if the input value belongs to a certain range. In this case, a microcontroller with an on-chip analog comparator can be a power-efficient solution. Microcontrollers such as the Atmel ATmega161L are capable of turning off the comparator to reduce the power consumption [10].

6.4 Computation

Motes incorporate a processor to carry out computations locally. Functionality typically requires the processor to run in outbursts separated by idle periods. Within the idle period, the processor may enter a power reduction mode to save energy [9]. The battery lifetime is influenced by the power efficiency of a running processor and the balance between active and idle periods.

6.4.1 Asynchronous Processors

The synchronous processor's clock distribution network is characterized by significant power consumption. Moreover, synchronous systems tend to maximize supply current transients. Smart Dust motes may have analog subsystems that are influenced by the electromagnetic radiation. Asynchronous designs promise to overcome the clock-related problems. In particular, a class of asynchronous implementations, termed self-timed systems, is capable of operating as fast as circumstances allow.

6.4.2 Variable Frequency Processors

Using variable frequency processors, power consumption can be gradually controlled by scaling the clock frequency. Typically, a phase lock loop (PLL) circuitry can multiply the oscillator frequency and an adjustable prescaler can divide the oscillator frequency. Based on the current task's deadline, the clock frequency may decline as much as possible. However, if the processor completes the task ahead of the deadline and enters a power-saving mode, the energy could be minimized [11]. In this case, the task's deadline period, T_{DL}, accommodates the active period, T_{ACT}, and the power-saving period, T_{PS}, as follows:

$$T_{DL} = T_{ACT} + T_{PS} \tag{6.1}$$

Assume that the power consumption scales linearly with the clock frequency:

$$P_{ACT} = k_{ACT} \times f_{CLK} + n_{ACT} \tag{6.2}$$

$$P_{PS} = k_{PS} \times f_{CLK} + n_{PS} \tag{6.3}$$

If the task's functionality requires NC processor clocks, the energy per task,

$$E_T = P \times T_{DL} = k_{PS} \times T_{DL} \times f_{CLK} + (n_{ACT} - n_{PS})\frac{NC}{f_{CLK}} + (k_{ACT} - k_{PS})NC + n_{PS} \times T_{DL} \tag{6.4}$$

Take the first derivative and calculate the critical numbers

$$f_{CLK} = \pm\sqrt{\frac{(n_{ACT} - n_{PS})NC}{k_{PS} \times T_{DL}}} \tag{6.5}$$

Consider the following two cases for the positive value:

- Let $n_{ACT} > n_{PS}$. Based on the second-derivative test, the energy per task has a minimum for

$$f_{CLK,OPT} = \sqrt{\frac{n_{ACT} - n_{PS}}{k_{PS}}}\sqrt{\frac{NC}{T_{DL}}} \tag{6.6}$$

- If $n_{ACT} \le n_{PS}$, the clock frequency must be selected as low as possible. The power-saving mode is not used.

$$f_{CLK,OPT} = \frac{NC}{T_{DL}} \tag{6.7}$$

Equation 6.6 does not guarantee that the deadline will be met. In some cases, the calculated clock frequency must be increased to meet the deadline.

Figure 6.2 shows an example mesh plot for the clock frequency. Assume that the processor is characterized by $n_{ACT} = 1$ mW; $n_{PS} = 0.1$ mW; and $k_{PS} = 1$ mW/MHz. The example is based on 256 combinations of deadline periods and cycles per task. For two combinations, the optimal clock frequencies have been replaced by higher values.

Actual tasks, which require replacement of the optimal clock frequency, can be viewed as targets for further improvement. Optimization of the code or relaxing the timing constraints would be an appropriate course of action.

6.4.3 Variable Voltage Processors

Variable voltage processors are capable of operating over a wide voltage range. Allocating such a processor for the network nodes allows power reduction by dynamically varying the supply voltage [12–15]. The method is often termed dynamic voltage scaling (DVS).

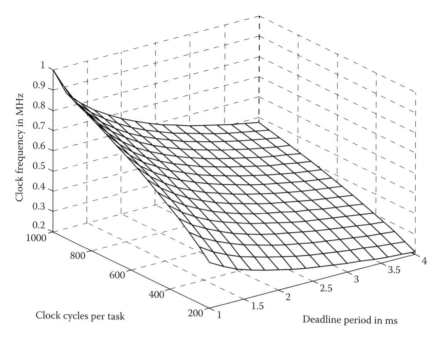

FIGURE 6.2 Mesh plot for the clock frequency.

DVS is an efficient method for power reduction; however, it imposes some limitations for the system:

- The system components must be capable of operating over a wide voltage range.
- A voltage converter loop hardware must be available.

Hong et al. [16] developed a design methodology for DVS. Figure 6.3 illustrates how to tune the voltages for extra power reduction. The tasks are specified by their arrival times, deadlines, and execution times at a nominal voltage. The schedule is viewed as a first iteration. It would be beneficial to extend the T2 task and reduce the V2 voltage. T1

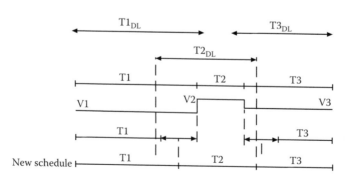

FIGURE 6.3 A schedule is modified for extra power reduction.

is scheduled for V2 to shrink the execution time. The new border between T1 and T2 is placed just in the middle of the interval indicated by an arrow; no conflict takes place with the arrival time and the change is accepted.

Similarly, T3 is scheduled for V2 to allow the extension for T2. The intention is to place the new border just in the middle of the interval marked by an arrow; however, T2 fails to meet the deadline and the border is aligned with the deadline. If the new schedule is more energy efficient, it is accepted.

6.5 Hardware–Software Interaction

A mote includes a CPU, memory, and peripherals. As a rule, peripherals possess three types of registers: data, control, and status. Data registers are employed as buffers between the CPU and peripherals, while control registers are used to adjust the I/O device functionality for a specific application. Status registers are read by the processor to check whether a specific operation is done. In spite of the huge variety of peripherals, the communication between the CPU and the I/O devices remains routine and easy to define.

Modifying one or more bits in a register, the CPU must keep the rest of the pattern unchanged. A common way to implement bit manipulation is to read a register, modify bits, and write the result back. The two memory accesses make the read–modify–write instructions power inefficient. In an attempt to improve the situation, Atmel has taken another approach with the AT91 microcontroller [17]. Instead of one control register, the microcontroller employs three registers mapped into three consecutive memory locations. The first register is used to set individual bits, the second to clear bits, and the third to obtain the current pattern. To set or reset a bit, a high bit is written to the corresponding position at the set or reset register.

In the AT91 microcontroller, a PLL circuitry and a programmable prescaler complement the ARM7TDMI core to a variable-frequency processor. The PLL circuitry multiplies the oscillator frequency; the highest multiplication factor is 64. As a result, the oscillator may run at a frequency 64 times lower than the actual clock and thus the oscillator saves energy. The programmable prescaler with a division factor of 64 allows the AT91 clock frequency to go down to 512 Hz. The CPU and embedded peripherals can be individually enabled and disabled. The ARM processor clock is enabled from the next interrupt or reset. The on-chip RAM reduces external memory accesses and allows further power reduction. Finally, the processor may switch to the 16-bit instruction set and benefit from a narrower memory.

Similar to the analog-to-digital conversion, the measurement of time intervals also falls under the accuracy–power trade-off. Figure 6.4 shows how a counter/timer determines a time interval using different clock rates. The highest possible frequency provides the highest accuracy. If a Smart Dust application is based on the AT91 microcontroller, the number of counter transitions for a 50-ms period may vary with the frequency up to 50,000.

6.6 Communication

In a wireless sensor network, communication is the major consumer of energy. Smart Dust networks have two recognized communication styles: (i) RF is characterized with

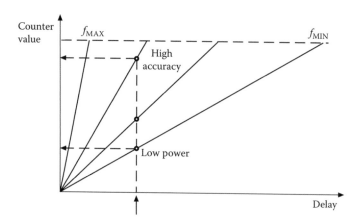

FIGURE 6.4   Using different frequencies to measure a time interval.

power consumption in the milliwatt range and (ii) optical communication is associated with a lower energy cost but requires accurate pointing. Consequently, optical communication is more suitable for interaction between network motes and a central station. The RF approach is very common for communication between motes [8,18,19].

6.6.1 Mote-to-Mote Communication

The procedures for establishing and operating a network require the motes to communicate with one another. The task of routing packets from a source to a destination can be broken down into discovering the position of the destination and the actual forwarding of packets [20]. Furthermore, channel access can be implemented by two different methods: contention and explicit organization [21]. The contention-based approach is not suitable for DSNs because of its requirement to monitor the channel for a long span of time. Because the reception and transmission have almost the same energy cost, the organized channel access is characterized with better energy efficiency. At the same time, the process of establishing time division multiple access slots or frequency bands also consumes energy. In an attempt to alleviate this problem, some protocols employ a hierarchical structure that requires partitioning the network.

The two basic schemes to limit the mote's RF transmission power are: (i) a transmitter can vary its power to cover different distances under different environmental conditions and (ii) the link can be partitioned into several short intermediate hops and use constant transmission power. Any DSN with a sufficient density of nodes can benefit from multihop communication.

The energy used to send a bit over a distance d may be written as

$$E = A \times d^n \tag{6.8}$$

where A is a proportionality constant and n depends on the environment [18,22]. The greater-than-linear relationship between energy and distance promises to reduce the energy cost when the link is partitioned.

Rewrite Equation 6.8 for *NH*, the number of hops. Also, include the energy for receiving E_R and energy for computation E_C:

$$E = A\left(\frac{d}{NH}D\right)^n NH + (E_R + E_C)NH \qquad (6.9)$$

Assume equal distances for each hop. $D > 1$ is introduced to take into account the longer path inevitably associated with multihop communication. The energy has a minimum for

$$NH_{OPT} = (d \times D)\sqrt[n]{\frac{A(n-1)}{E_R + E_C}} \qquad (6.10)$$

Figure 6.5 shows a plot for the energy per bit using different numbers of hops. The distance $d = 50$ m; $n = 4$; $A = 0.2$ fJ/m^4; $D = 1.2$; and $E_R + E_C = 30$ pJ. The energy per bit has a minimum for four hops.

A subtle effect of multihop communication is that energy consumption is distributed over the motes fairly. If the motes consume energy at about the same rate, the system lifetime is increased. Chen et al. [23] developed a coordination algorithm to increase the energy efficiency further. The algorithm is based on an assumption that when a wireless network has an ample density of nodes, only a small number of them need to be active to forward traffic. A distinctive feature of the method, named SPAN, is that the motes make a decision whether to sleep or to be active based not only on the topology of the network, but also on the amount of energy available in the battery.

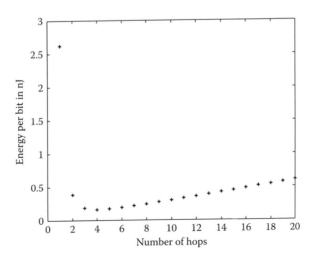

FIGURE 6.5 Energy-per-bit scales with the number of hops.

All motes of the network are dynamically split into two sets: motes that sleep and motes that stay awake to participate in the forwarding backbone topology. According to SPAN's terminology, the active motes are named coordinators. Each mote of the network makes periodic, local decisions on whether to sleep or become a coordinator. Coordinators are elected to achieve two goals: improved connectivity of the network and equal levels of energy remaining at each mote. All noncoordinator motes periodically participate in an election procedure to become coordinators; in parallel, all coordinators periodically pass through a withdrawal procedure to switch back to a sleep state. Figure 6.6 shows this election–withdrawal cycle. A mote becomes a coordinator to link two neighbor motes that cannot communicate directly or via one or two coordinators. Because several motes can run an election procedure simultaneously, there might be an overlap in the connectivity they introduce. The method attempts to minimize the number of coordinators to save energy.

To resolve contention, the election procedure is extended with a variable delay. As soon as the delay period is over, a coordinator announcement is sent out. If, at the end of

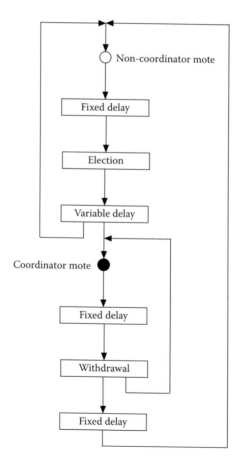

FIGURE 6.6   Span's election withdrawal cycle.

the delay, the mote receives other announcements for new coordinators, it reconsiders the need to become a coordinator. The election procedure distinguishes between the two cases:

- All applicants for coordinators have equal energy left in their batteries. In this case, the more pairs of motes the applicant connects, the shorter is the delay. Also, to rotate coordinators with time, a random value influences the delay.
- The participating motes have unequal energy available in their batteries. In this case, the delay period is calculated on the basis of the connection improvement and the amount of energy scaled to the maximum amount of energy the mote can have. The random factor is still included.

Each coordinator periodically runs a withdrawal procedure. A coordinator can go back to sleep if every pair of its neighbors can reach each other directly or via some other coordinators. Initially, the mote will stay as a coordinator if its withdrawal affects the network connectivity. However, after some time, it will switch to noncoordinator state to give other neighbors a chance to become coordinators. As shown in Figure 6.6, a mote continues to serve as a coordinator for a fixed period of time after its withdrawal announcement is sent out. Thus, the routing protocol can use the old coordinator until a new coordinator is elected.

6.6.2 Mote-to-Central Station Communication

When one or more central stations communicate with a field of dust motes, optical systems are characterized by the lowest energy budget. Two methods can be used to apply optical communication for Smart Dust: passive reflective systems and active-steered laser systems [2]. Figure 6.7 shows an example of a passive reflective device, a corner-cube retroreflector (CCR). A CCR reflects the light via three mutually orthogonal mirrors. When a light beam enters the CCR, it bounces off the mirrors and is reflected back parallel to the direction from which it entered. Because one of the mirrors is mounted on

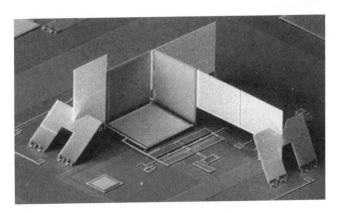

FIGURE 6.7 Microfabricated CCR. (From Hsu, V., Kahn, J. M., Pister, K. S. J., www-ee.stanford. edu/~jmk/pubs/hsu.ms.11.99.pdf. With permission.)

a spring at an angle slightly askew from perpendicularity to the other mirrors, in this state little light returns to the remote receiver. No reflection of the light is considered a low logic level. To return the light to its source, high logic level, the mirror is shifted to a position perpendicular to other mirrors. The low-to-high transition consumes less than a nanojoule [2]. The high-to-low transition requires almost no energy.

Active-steered laser systems are suitable for mote-to-mote and mote-to-central station communications. The device consists of a semiconductor diode laser, collimating lens, and a two-degree-of-freedom micromirror [2]. Central stations can use imaging receivers to process transmissions from different angles. This approach of separating transmissions according to their originating location is termed a space division multiple access (SDMA).

6.7 Orientation

Many applications will deploy motes in random orientation. Consequently, it will not be possible for all CCRs to return light to the central station. A CCR quadruplet is a solution that improves the accessibility of the motes. At the same time, some directions may be characterized with noise emissions and should be avoided. Furthermore, applications may require the motes to be invisible from a certain area.

It is proposed that the motes be magnetized and the CCR oriented to a predefined direction. When the motes fall through the air after being deployed, they will orient themselves. If the network has a sufficient density of motes, it may not need the motes, which change the orientation upon landing. This approach for zero-power orientation is even more efficient for motes floating on the water. They could freely rotate to orient themselves. Figure 6.8 shows a deployment of two types of motes that differ in their CCR orientation; two central stations interrogate the motes. The DSNs community is growing and projects that simultaneously employ a single field can benefit from SDMA.

6.8 Conclusion

The low-power design of Smart Dust networks has a lot in common with many other computer applications. By allocating variable frequency processors for the Smart Dust motes, clock frequency scaling can be applied to decrease power consumption. It is necessary to distinguish between two types of processors in order to decide whether it is more power efficient to operate quickly and then wait quietly, or just operate at the minimum speed possible. For the first case, the optimal clock frequency is calculated based on the required number of clock cycles and a deadline period. This approach also allows identifying tasks that require replacement of the optimal clock frequency. Thus, a set of tasks emerges as a target for further improvement. Variable voltage processors could combine voltage scaling with frequency scaling if the hardware overhead is not prohibitive for a cubic-millimeter-sized mote.

Hardware–software interaction also provides ample reserve for power reduction. Scaling down the theme of variable frequency from processors to counters, motes could measure time intervals, trading accuracy against number of transitions. The

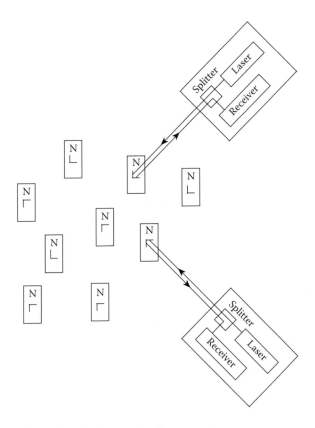

FIGURE 6.8 Each central station interrogates its own motes.

hardware–software interaction and the sensing show that a redundant accuracy wastes energy in the same way as redundant computation speed.

The energy spent for communication is crucial for the success of wireless networks such as Smart Dust. Multihop communication can help power consumption to decline significantly and avoid obstacles for RF and optical systems. As an additional benefit, multihop transmissions distribute power consumption over the motes fairly and increase system lifetime.

The location of the motes is a process specific to the network operation. Some applications may require only relative positions. Relative positions can be used to turn off motes, especially in case of event tracking. Finally, optical communication is associated with pointing and orientation. By using the Earth's magnetic field, zero-power orientation of the motes can be implemented for SDMA.

Acknowledgment

The author is grateful to Brett Warneke for important advice and to Victor Hsu for the CCR photograph.

References

1. Kahn, J. M., Katz, R. H., and Pister, K. S. J., Next century challenges: mobile networking for "Smart Dust." *J Commun Networks* 2000; 2: 188.
2. Warneke, B., Last, M., Liebowitz, B., and Pister, K. S. J., Smart Dust: communicating with a cubic-millimeter computer. *Computer* 2001; 34: 44.
3. Bulusu, N., Heidemann, J., and Estrin, D., GPS-less low-cost outdoor localization for very small devices. *IEEE Personal Commun* 2000; 7: 28.
4. Hightower, J., and Borriello, G., Location systems for ubiquitous computing. *IEEE Comput* 2001; 34: 57.
5. Liu, J., Cheung, P., Guibas, L., and Zhao, F., A dual-space approach to tracking and sensor management in wireless sensor networks. Palo Alto Research Center Technical Report P2002-10077, 2002. http://www2. parc.com/spl/projects/cosense/pub/dualspace.pdf.
6. Berg, M., Kreveld, M., Overmars, M., and Schwarzkopf, O., *Computational Geometry: Algorithms and Applications*. Berlin: Springer, 1997.
7. Gardner, J. W., *Microsensors: Principles and Applications*. New York: John Wiley & Sons, 1994.
8. Doherty, L., Warneke, B. A., Boser, B. E., and Pister, K. S. J., Energy and performance considerations for smart dust. *Int J Parallel Distributed Syst Networks* 2001; 4: 121.
9. Karakehayov, Z., Christensen, K. S., and Winther, O., *Embedded Systems Design with 8051 Microcontrollers*. New York: Marcel Dekker, 1999.
10. Atmel Corporation, *AVR RISC Microcontroller, Data Book*, 1999.
11. Karakehayov, Z., Zero-power design for Smart Dust networks, in *Proceedings of 1st IEEE International Conference on Intelligent Systems*, Varna, p. 1, 302, 2002.
12. Macken, P., Degrauwe, M., Paemel, V., and Oguey, H., A voltage reduction technique for digital systems, in *Digest of Technical Papers, 37th IEEE International Solid-State Circuits Conference*, p. 238, 1990.
13. Mudge, T., Power: a first-class architectural design constraint. *IEEE Comput* 2001; 34: 52.
14. Burd, T., *Energy-efficient processor system design*, PhD Thesis, University of California, Berkeley, 2001.
15. Sinha, A., and Chandrakasan, A., Dynamic power management in wireless sensor networks. *IEEE Design Test Comput* 2001; 18: 62.
16. Hong, I. et al., Power optimization of variable-voltage core-based systems. *IEEE Trans Computer-Aided Design Integr Circuits Syst* 1999; 18: 1702.
17. Atmel Corporation, *AT91 ARM Thumb Microcontrollers, AT91M55800A*, 2002. http://www.atmel.com.
18. Rabaey, J. M. et al., PicoRadio supports ad hoc ultra-low power wireless networking. *Computer* 2000; 33: 42.
19. Hill, J. L., and Culler, D. E., Mica: a wireless platform for deeply embedded networks. *IEEE MICRO* 2002; 22: 12.
20. Mauve, M., and Widmer, J., A survey on position-based routing in mobile ad hoc networks. *IEEE Network* 2001; 15: 30.

21. Sohrabi, K. et al., Protocols for self-organization of a wireless sensor network. *IEEE Personal Commun* 2000; 7: 16.
22. Akyildiz, I. F. et al., A survey on sensor networks. *IEEE Commun Mag* 2002; 40: 102.
23. Chen, B. et al., Span: an energy-efficient coordination algorithm for topology maintenance in ad hoc wireless networks. *ACM Wireless Networks J* 2002; 8: 481.
24. Hsu, V., Kahn, J. M., Pister, K. S. J., www-ee.stanford.edu/~jmk/pubs/hsu.ms.11.99.pdf.

7

Security Improvement of Slotted ALOHA in the Presence of Attacking Signals in Wireless Networks

Jahangir H. Sarker
University of Ottawa

Hussein T. Mouftah
University of Ottawa

7.1 Introduction

Wireless networks are vulnerable for adversaries to conduct radio interference, or jamming, attacks that effectively cause a denial of service (DoS) of either transmission or reception functionalities. Recently, research has been conducted to assess the wireless multiple access schemes in the presence of jamming or attacking signals [1]. Among the radio resource sharing multiple access schemes, the code division multiple access (CDMA) system has a special resistance against the interference attacking signals. Thus, the CDMA scheme may be the first choice as a multiple access scheme in the presence of attacking interference signals. The attacker should spread its energy evenly over all degrees of freedom in order to minimize the average capacity [2,3]. In a simplified

CDMA transmission system, with the knowledge of spreading code, the receiver is able to detect the users' signals from the attacking signals. The channel capacity can be enhanced further if the attacker state information and the effects of fading are known [2,3]. For enhancing uplink channel capacity, the attacker state information is more important than that of the effects of fading. On the other hand, the capacity of downlink channels in cooperative schemes is more sensitive to changes in fading severity [2,3].

In a hostile attacking environment, the attacker can use the same code as the legal users and transmit. As a result, it is difficult to protect these kinds of attacks. One way of preventing these kinds of attack is the use of a specific frequency hoping speed spectrum technique wherein the channel sequence is generated using a secret word, known only to the sink and the sensor nodes, as a seed [4]. However, relying on time-frequency diversity over large spectrum and to a large number of mobile nodes is inefficient. To improve the system capacity, an innovative message-driven frequency hopping was introduced and analyzed in [5]. In the message-driven frequency hopping, part of the message stream will be acting as the pseudo-random (PN) sequence, and transmitted through hopping frequency control. As a result, system efficiency is increased significantly since additional information transmission is achieved at no extra cost on either bandwidth or power [5]. In hostile jamming or attacking environment, the mobile nodes can exploit channel diversity in order to create wormholes that lead out of the jammed region, through which an alarm can be transmitted to the network operator [6].

Usually, mobile nodes of a wireless network are powered by battery, and replacement or recharging a battery during the time of operation is very difficult and most of the time impossible. Therefore, low power consumption systems are becoming important. A simple linear transformation of conventional ALOHA access, called spread ALOHA is equivalent to CDMA with a common spreading code for all users. The slotted ALOHA is the most spectral and power efficient multiple access scheme [7,8].

The slotted ALOHA multiple access scheme is a widely used random access protocol independently [9–11] for its adequate working capability for distributed wireless nodes having bursty traffic. The slotted ALOHA is also used as a part of different multiple access protocols especially for the control channels in many new wireless technologies. For instance, it is used in the random access channels of global system for mobile communications [12] and its extension general packet radio services [13,14], wideband CDMA (WCDMA) system [15], cdma2000 [16,17], IEEE 802.16 [18], IEEE 802.11 [19], and so on. A smart power saving jammer or attacker can attack only in the signaling channels, instead of attacking whole channels [20–22]. Therefore, defending the control channels from external and internal attacks [23] are very important issue and is discussed in this chapter.

A special type of DoS attack, called random packet destruction that works by transmitting short periods of noise signals, is considered as attacking signals. This random packet destruction DoS packets can effectively shut down slotted ALOHA-based networks [9–11] and the networks uses the slotted ALOHA-based signaling channels easily [12–22]. One of the main drawbacks of slotted ALOHA is its excessive collisions at higher traffic load condition. The current anti-attack measures such as encryption, authentication, and authorization [24–26] cannot prevent these types of attacks, since the random packet destruction DoS packets increase the collision and make these systems unstable.

The effects of attacking noise packet signals on the slotted ALOHA scheme have been investigated by many researchers [27–33]. The stability of slotted ALOHA can be enhanced by limiting retransmission trials in *ad hoc* networks [34]. The contributions of this chapter are outlined as follows.

1. The throughput of multichannel slotted ALOHA in the presence of random packet destruction DoS attack is presented.
2. It is shown that for any positive value of message packet arrival rate, the throughput decreases with the increase of random packet destruction DoS attacking packet rate.
3. Security improvement of slotted ALOHA using multiple channels has been investigated. Results show that a sufficient number of channels can provide the throughput of slotted ALOHA in the presence of random packet destruction DoS attack. The approximate value of that number of channels has been derived analytically.
4. Security of multichannel slotted ALOHA can be enhanced significantly by limiting the number of retransmission trials. The analytical results and graphical representation of the secured and unsecured region of slotted ALOHA by limiting the number of retransmission trials have been presented.
5. The new packet generation rate from all mobile nodes can be very high due to the false information from the network. In that situation, security of multichannel slotted ALOHA can be improved significantly by new packet rejection. The analytical results and graphical representation of the security improvement regions using new packet rejection has been investigated in this chapter.

The chapter is organized as follows. Section 7.2 describes the system model and assumptions. The security improvement analysis using multiple channels is given in Section 7.3. Limiting the number of retransmission trials to improve the security of slotted ALOHA system in the presence of random packet destruction DoS attacking signal is presented in Section 7.4. Section 7.5 shows the security improvements of slotted ALOHA system by new packet rejection. Conclusions are provided in Section 7.6.

7.2 System Model and Assumptions

Let us consider a system where a base station is located in the middle of a very large number of users having mobile units (nodes). Assume that the average value of the *new message* packet arrival rate from all active mobile nodes per time slot is λ packet per time slot. In slotted ALOHA, the throughput initially increases with the increase in the new packet generation rate, λ. The throughput reaches its maximum value for a certain value of the new packet generation rate from all active nodes. The throughput collapse and reaches to zero, if the new packet generation rate increases further. The throughput collapse is known as the security or stability problem in slotted ALOHA. The reason of throughput collapse is excessive collision. The throughput collapse can be prevented by reducing the new packet arrival rate per slot. The packet rejection can provide one of the solutions and is considered in this chapter.

Assuming that the new packet rejection probability is α. The new packet transmission rate per time slot is $\lambda(1 - \alpha)$. Let there are L parallel slotted ALOHA-based channels. The mobile nodes can transmit their packets selecting any of the L channels by random selection, without the knowledge of other mobile units' activeness. During the transmission of packets, each mobile node adjusts their packet size to fit into the time slots. Since the average new message packet transmission rate from all active mobile nodes per time slot is $\lambda(1 - \alpha)$ packet per time slot and the channel selection is *random*, the new packet transmission rate from all mobile nodes is $(\lambda/L)(1 - \alpha)$ packets per time slot.

The traditional current anti-attack measures such as encryption, authentication, and authorization cannot prevent these types of attacks [35]. The main intension of transmitting the random packet destruction DoS attacking packets is to increase the collision. One solution resides to defend the message packets from the attacking packets is to reduce the collisions. We have used three different techniques to reduce the collisions not only between among the message packets also the message packets and the attacking noise packets.

It is well known that the slotted ALOHA is widely used random access protocol, because of its distribution nature and easy to implement. But its performance is degraded due to excessive collision. Therefore, attacking signals are made to produce dummy packets/noise packets of same size to increase the collision farther [35]. In addition, assume that the attacking signals are not producing noise packets in each slot for two reasons. First, it will be detected immediately and will be removed. Secondly, it will dissipate more energy and will die soon. Let the attacking packet arrival to the base station is also Poisson with an average rate of J packet per time slot. The probability that n packets are transmitted to the same slot from the attacking node (or nodes) is

$$A_n = \frac{J^n}{n!} e^{-J} \qquad (7.1)$$

In the first collision reducing technique, we have used multiple parallel slotted ALOHA slotted channels instead of single-channel slotted ALOHA channel. For doing that the message packets can be transmit in a multiple L-channel slotted ALOHA system. Then we have the possibility of reducing collisions. Next, we intend to find the capture possibility of a message packet, in the presence of attacking packets as well as other interfering packets transmitted to the same slot.

In multiple L-channel slotted ALOHA system, the attacker packets need to transmit all L channels separately. The attacker should spread its energy evenly over all degrees of freedom in order to minimize the average capacity [2,3]. Let us assume that the attacking packets also transmitted at L parallel slotted ALOHA channels to increase the collision. The effect of receiver noise has not been consideration in this analysis, since it is very small compared to the collision.

The probability that j attacking noise packets out of n attacking noise packets will be transmitted at the same slot of an L-channel slotted ALOHA system is

$$A_{n\,|\,j} = \binom{n}{j}\left(\frac{1}{L}\right)^j\left(1 - \frac{1}{L}\right)^{n-j} \qquad (7.2)$$

Now form total probability theory, the probability that j attacking noise packets are transmitted to the same slot is

$$A_j = \sum_{n=j}^{\infty} \frac{J^n}{n!} e^{-J} \binom{n}{j}\left(\frac{1}{L}\right)^j\left(1-\frac{1}{L}\right)^{n-j} = \frac{(J/L)^j}{j!} e^{-(J/L)} \tag{7.3}$$

If the base station can receive only one message packet per time slot in the presence of attacking noise packets, then the slot is considered as successful. Let a maximum of r retransmissions are allowed. Assume the retransmitted packets are also Poisson arrival [36–38]. Thus, the aggregate message packet arrival rate is G packet per time slot. If any message packet also selects L channels by random selection, the aggregate message packet arrival rate per time slot is G/L. The system model and assumptions is presented in Figure 7.1.

A message packet is successfully received in a time slot, if three conditions are fulfills. First, the receiver will select a message packet in the presence of message packets and attacking noise packets. Secondly, the probability that no other attacking packet is transmitted at the same slot. Thirdly, the probability that no other interfering message packet is transmitted at the same slot. Therefore, the probability that a message packet is successfully transmitted can be written as

$P(\text{Su}) = P$ (the selected packet will be a message packet)

$\quad \times P$ (no other attacking packet is transmitted at the same slot)

$\quad \times P$ (no other interfering message packet is transmitted at the same slot)

$= P_M \times P_{CA} \times P_{CM}$ (7.4)

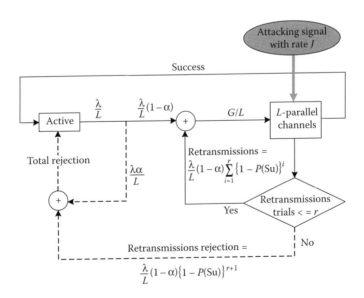

FIGURE 7.1 System model and assumptions.

The probability that a test message packet will be selected from the message packets is the ratio of the total number of message packets per time slot and the total number of message packets per time slot plus the total number of attacking noise packets per time slot. Therefore, the probability that a selected test packet is a message packet is

$$
\begin{aligned}
P_M &= P\big(\text{the selected packet is a message packet}\big) \\
&= \frac{\text{Total number of message packets per time slot}}{\underset{\substack{\text{packets per time slot}}}{\text{Total number of message}} + \underset{\substack{\text{packets per time slot}}}{\text{Total number of attacking}}}
\end{aligned}
$$

$$
\frac{\displaystyle\sum_{x=0}^{\infty} x\,\frac{(G/L)^x}{x!}\,e^{-(G/L)}}{\displaystyle\sum_{x=0}^{\infty} x\,\frac{(G/L)^x}{x!}\,e^{-(G/L)} + \sum_{y=0}^{\infty} y\,\frac{(J/L)^y}{y!}\,e^{-(J/L)}} = \frac{G/L}{G/L + J/L} = \frac{G}{G+J} \qquad (7.5)
$$

Now we have to derive the probability that no other attacking packet is transmitted at the same slot. The probability that j other attacking packets are transmitted at the same slot is derived in Equation 7.3. Therefore, the probability that no other attacking packet is transmitted at the same slot is

$$
P_{CA} = A_{j=0} = \frac{(J/L)^0}{0!}\,e^{-(J/L)} = e^{-(J/L)} \qquad (7.6)
$$

The probability that no other interfering packet is transmitted at the same slot is

$$
P_{CM} = \frac{(G/L)^0}{0!}\,e^{-(G/L)} = e^{-(G/L)} \qquad (7.7)
$$

Finally, the probability of success of a message packet in the presence of attacking signals is

$$
P(Su) = P_M * P_{CA} * P_{CM} = \frac{G}{G+J}\,e^{-(J/L)}e^{-(G/L)} = \frac{G}{G+J}\,e^{-((G+J)/L)} \qquad (7.8)
$$

The probability of failure of any message packet is $1 - P(Su)$. This unsuccessful part $(\lambda/L)(1-\alpha)\{1 - P(Su)\}$ will be transmitted during first retransmission time. The probability of two successive failures is $\{1 - P(Su)\}^2$. So, the second time retransmission part is $(\lambda/L)(1-\alpha)\{1 - P(Su)\}^2$, and so on. In general, kth time retransmission part is $(\lambda/L)(1-\alpha)\{1 - P(Su)\}^k$. Let the total number of retransmissions of a packet be r (one transmission followed r retransmission trials). The total mean offered traffic from all active users is then given by

$$
\frac{G}{L} = \sum_{k=0}^{r} \frac{\lambda}{L}(1-\alpha)\{1 - P(Su)\}^k \qquad (7.9)
$$

Simplifying Equation 7.9 and combining with Equation 7.8, we can write

$$\frac{G}{L}\frac{G}{G+J}e^{-\left(\frac{G+J}{L}\right)}=\frac{\lambda}{L}(1-\alpha)\left[1-\left\{1-\frac{G}{G+J}e^{-\left(\frac{G+J}{L}\right)}\right\}^{r+1}\right] \qquad (7.10)$$

Equation 7.10 is the basic equation of retransmission cut-off and new packet rejection algorithm of multiple L-channels slotted ALOHA in the presence of attacking signals.

7.3 Security Improvement Using Multiple Channels

The probability of success of L-channel slotted ALOHA system in the presence of attacking signal is derived in Equation 7.8. The throughput per slot of L-channel slotted ALOHA system is defined as the multiplication of average traffic arrival rate per time slot and the probability of success in the presence of attacking signal. Thus, the throughput is

$$S=\frac{G}{L}P(Su)=\frac{G^2}{L(G+J)}e^{-\left(\frac{G+J}{L}\right)}=\frac{\lambda}{L}(1-\alpha)\left[1-\left\{1-\frac{G}{G+J}e^{-\left(\frac{G+J}{L}\right)}\right\}^{r+1}\right] \qquad (7.11)$$

The above equation is the basic equation for the throughput of a message packet. Articulately, the new packet generation rate λ, number of channels L, new packet rejection probability α, attacking signal generation rate J, and number of retransmission trials r play important role in this equation.

But in this section, we will limit our discussion only the effect of L-channels in the presence of attacking channel. Therefore, we will consider in this section only the first part of the throughput result in Equation 7.11.

Figure 7.2 shows the throughput per slot, S with the variation in aggregate message packet arrival rate, G for different values of attacking packets rates of J. From Figure 7.2 we can make the following conclusions:

1. The throughput per slot S of one-channel slotted ALOHA system is very low in the presence of attacking noise packet signal (Figure 7.2a). Because of that the current one-channel slotted ALOHA-based networks [9–19] can be shut down very easily. If five-channels are used instead of one-channel, then the throughput per slot increases significantly, even with the same attacking noise packet rate, J (Figure 7.2b).
2. Since the throughput per slot, S, does not collapse with a high message packet arrival rate, G, even with a high attacking noise packet generation rate, J, with a higher number of channels, L, the security of slotted ALOHA system can be enhanced using multiple L-slotted ALOHA channels.
3. There exist an optimum point where throughput per time slot, S, is maximum for given values of message packet generation rate, G, and attacking packet generation rate, J.

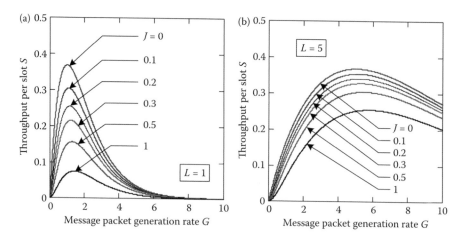

FIGURE 7.2 Throughput per slot with the variation of message packet arrival rate. (a) 1-channel system. (b) 5-channel system.

Differentiating Equation 7.11 with respect to attacking noise packet generation rate, J, we obtain

$$\frac{dS}{dJ} = -\left(\frac{1}{L}\right)\frac{G^2}{L(G+J)}\exp\left(-\frac{G}{L}\right)\exp\left(-\frac{J}{L}\right) - \frac{G^2(G+1)}{L(G+J)^2}\exp\left(-\frac{G+J}{L}\right) \qquad (7.12)$$

It is clear from the above equation that for any positive value of message packet generation rate, G, the throughput, S, decreases with the increase of attacking noise packet generation rate, J. To obtain the maximum throughput of L-channel slotted ALOHA system, differentiating Equation 7.11 with respect to message packet generation rate, G, we get

$$\frac{dS}{dG} = \frac{G^2}{L(G+L)}e^{-((G+J)/L)}\left(-\frac{1}{L}\right) + \frac{1}{L}\frac{(G+J)(2G)-G^2}{(G+J)^2}e^{-((G+J)/L)}$$

$$= \frac{e^{-((G+J)/L)}}{L(G+L)}\left\{-\frac{G^2}{L} + \frac{2GJ+G^2}{G+J}\right\} \qquad (7.13)$$

Now, putting the differentiation result Equation 7.13 equal to zero, we obtain the optimum value of the message packet arrival rate from all active mobile nodes,

$$G_{\text{opt}} = \frac{(L-J)\pm\sqrt{(L-J)^2+8LJ}}{2} \qquad (7.14)$$

For positive values of L and J, only the positive sign is applicable in Equation 7.14. The numerical results of Equation 7.14 are depicted in Figure 7.3. It can be said from the

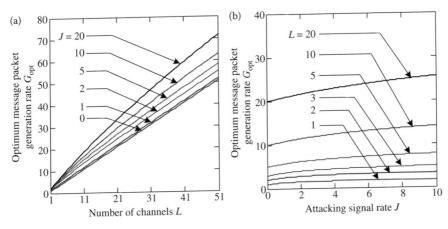

FIGURE 7.3 The optimum message packet generation rate, G_{opt}. (a) G_{opt} vs. L. (b) G_{opt} vs. J.

numerical results of Equation 7.14 and Figure 7.3a that for a given attacking signal rate J, the optimum message traffic rate, G_{opt}, increases linearly with the increase of number of channels, L. On the other hand, the optimum message traffic rate, G_{opt}, increases very flat linearly with the increase of attacking signal rate J, for a given number of channels, L (Figure 7.3b).

Using the value of optimum message packet arrival rate, G_{opt}, in Equation 7.11, we can obtain the optimum throughput per time slot as

$$S_{opt} = \frac{\left\{ L - J + \sqrt{(L-J)^2 + 8LJ} \right\}^2}{2L \left\{ L + J + \sqrt{(L-J)^2 + 8LJ} \right\}} e^{-(1/2L)\left\{ L + J + \sqrt{(L-J)^2 + 8LJ} \right\}} \qquad (7.15)$$

The optimum throughput per time slot is shown Figure 7.4, using numerical results of Equation 7.15. Figure 7.4 shows that the optimum throughput can be increased significantly using multiple channels.

Now question may arises that what is the optimum number of channels, L, that provides maximum throughput. To answer this question, the optimum L can be obtained by setting Equation 7.13 is equals to zero. Therefore, the optimum number of channels is

$$L_{opt} = \frac{G(G+J)}{2J+G} \qquad (7.16)$$

The above equation shows that the optimum number of channels, L_{opt}, increases linearly with the increase in aggregate message traffic arrival rate, G. However, the same decreases with the increase of attacking packet arrival rate, J.

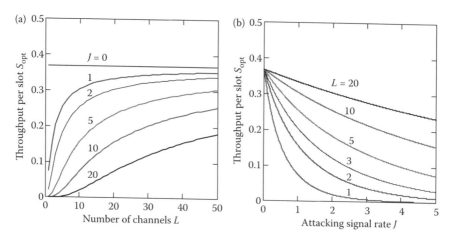

FIGURE 7.4 Maximum throughput per time slot in the presence of attacking noise packets. (a) S_{opt} vs. L. (b) S_{opt} vs. J.

7.4 Security Improvement by Limiting the Number of Retransmission Trials

In a normal data transmission system, every packet must be transmitted successfully. On the other hand, in the case of contention-based access protocol or for real-time data transmission, we can cut the retransmission number, which will avoid the undesirable security problem of slotted ALOHA [39]. Over a long time period, the total offered traffic load G will have an optimum value G_{opt} depending on the other parameters like new packet generation rate per time slot and the number of retransmission trials. In the case of *access* or real-time traffic, transmission packets are identical in nature for each user and the access procedure is limited by time. For a secured operation of L-channels slotted ALOHA-type system, with a higher value of new packet generation rate per time slot, the retransmission trials should be controlled. The purpose of the retransmission trial control is to get the optimum value of offered traffic load from all users G_{opt}, which will make the system secured or stable. Here, in this chapter a simplified assumption is considered: if the traffic generation rate from all active users in a given time slot is less than or equal to the optimum packet arrival per time slot, the system is secured. This assumption is reasonable for slotted ALOHA system [36–38].

The throughput per slot of L-channels slotted ALOHA system by limiting the number of retransmission trials can be obtained from Equation 7.11 as

$$S = \frac{\lambda}{L}\left[1-\left\{1-\frac{G}{G+J}e^{-((G+J)/L)}\right\}^{r+1}\right] \qquad (7.17)$$

Where the relationship between λ and G can be obtained by solving Equation 7.10. The main purpose of our system model is to maximize the throughput per slot, S, by adjusting the transmission trials, r, and the new packet generation rate per slot, λ/L, for a given value of attacking packet arrival rate J. We have already derived the maximum throughput of L-channels slotted ALOHA system S_{opt} in Equation 7.15. And it occurs when the aggregate traffic generation rate, G_{opt}, which is shown in Equation 7.14. Now putting G_{opt} into Equation 7.10 and after simplification we get

$$\frac{\lambda_{opt}}{L} = \frac{\left(\dfrac{\left\{ L - J + \sqrt{(L-J)^2 + 8LJ} \right\}^2}{2L\left\{ L + J + \sqrt{(L-J)^2 + 8LJ} \right\}} \right) e^{-(1/2L)\left(L + J + \sqrt{(L-J)^2 + 8LJ} \right)}}{\left\{ 1 - \left(\dfrac{\left(L - J + \sqrt{(L-J)^2 + 8LJ} \right)}{\left(L + J + \sqrt{(L-J)^2 + 8LJ} \right)} \right) e^{-(1/2L)\left(L + J + \sqrt{(L-J)^2 + 8LJ} \right)} \right\}^{r+1}} \tag{7.18}$$

The above equation is the basic equation for the secured transmission method. The secured transmission method can be stated as follows: for a call establishment system design or for a real-time traffic transmission, the time out is the most important parameter. This time out is the time to transmit the access information from mobile to base station plus the switching time. From the value of the time to transmit the access information or real-time transmission plus the propagation delay, we can find the maximum allowable retransmission trials, r, that is, how many retransmission trials are possible for a given time. From this value of r, and the value of J, we can find the optimum new packet generation rate per time slot, λ_{opt}/L per time slot using Equation 7.18.

Figure 7.5 shows the variation in optimum new packet generation rate per time slot, λ_{opt}/L, with the variation in the number of channels, L, using Equation 7.18. Similarly, Figure 7.6 shows the variation in optimum new packet generation rate per time slot, λ_{opt}/L, with the variation in attacking signal rate, J, using Equation 7.18.

Without any retransmission attempts ($r \rightarrow 0$), the optimum new packet generation rate per time slot can be obtained using Equation 7.18 as

$$\left(\frac{\lambda_{opt}}{L} \right)_{r \rightarrow 0} =$$

$$\frac{\left(\left\{ L - J + \sqrt{(L-J)^2 + 8LJ} \right\}^2 \right) \Big/ \left(2L\left\{ L + J + \sqrt{(L-J)^2 + 8LJ} \right\} \right) e^{-(1/2L)\left(L + J + \sqrt{(L-J)^2 + 8LJ} \right)}}{1 - 1 + \left(\left(L - J + \sqrt{(L-J)^2 + 8LJ} \right) \Big/ \left(L + J + \sqrt{(L-J)^2 + 8LJ} \right) \right) e^{-(1/2L)\left(L + J + \sqrt{(L-J)^2 + 8LJ} \right)}}$$

$$= \frac{L - J + \sqrt{(L-J)^2 + 8LJ}}{2L} \tag{7.19}$$

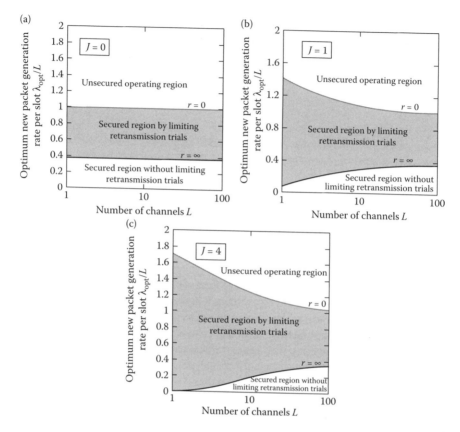

FIGURE 7.5 Security improvement by limiting the retransmission trials. (a) Attacking signal rate $J = 0$. (b) Attacking signal rate $J = 1$. (c) Attacking signal rate $J = 4$.

In the other extreme, without any retransmission cut-off ($r \to \infty$), the optimum new packet generation rate per time slot can be obtained using Equation (7.18) as

$$
\left(\frac{\lambda_{opt}}{L}\right)_{r \to \infty} = \frac{\left\{ L - J + \sqrt{(L-J)^2 + 8LJ} \right\}^2}{2L\left\{ L + J + \sqrt{(L-J)^2 + 8LJ} \right\}} e^{-\frac{1}{2L}\left(L + J + \sqrt{(L-J)^2 + 8LJ} \right)} \tag{7.20}
$$

Therefore, the security improvement area by limiting the number of retransmission trials, r, is

$$
Ar = \left(\frac{\lambda_{opt}}{L}\right)_{r \to 0} - \left(\frac{\lambda_{opt}}{L}\right)_{r \to \infty}
$$

$$
= \frac{L - J + \sqrt{(L-J)^2 + 8LJ}}{2L} - \frac{\left\{ L - J + \sqrt{(L-J)^2 + 8LJ} \right\}^2}{2L\left\{ L + J + \sqrt{(L-J)^2 + 8LJ} \right\}} e^{-(1/2L)\left(L + J + \sqrt{(L-J)^2 + 8LJ} \right)}
$$

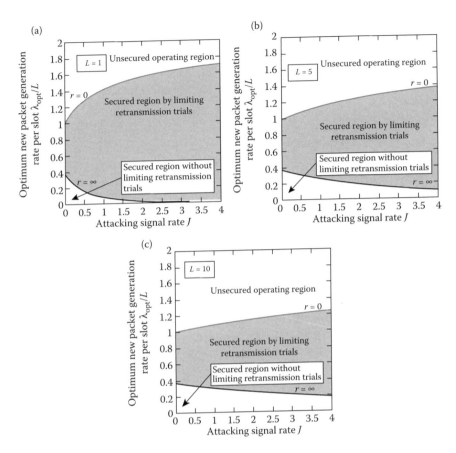

FIGURE 7.6 Security improvement by limiting the retransmission trials. (a) 1-channel system. (b) 5-channel system. (c) 10-channel system.

$$= \frac{L-J+\sqrt{(L-J)^2+8LJ}}{2L}\left[1-\frac{L-J+\sqrt{(L-J)^2+8LJ}}{L+J+\sqrt{(L-J)^2+8LJ}}e^{-(1/2L)\left(L+J+\sqrt{(L-J)^2+8LJ}\right)}\right]$$

(7.21)

The gray parts indicated in Figures 7.5 and 7.6 shows the numerical results of Equation 7.21.

7.5 Security Improvement by New Packet Rejection

The main purpose of this chapter is to obtain the secured transmission of L-channel slotted ALOHA system. It is already shown that if L-channel slotted ALOHA system provides maximum throughput, then the system is secured. To obtain a secured stabilized the L-channel slotted ALOHA system; if limiting the retransmission trials is not sufficient, then it can be achieved by the expense of newly generated packet rejection.

The maximum throughput per slot of a L-channel slotted ALOHA is S_{opt} is derived in Equation 7.15, and it occurs when the aggregate traffic generation rate, G_{opt}, which is shown in Equation 7.14. The aggregate message packet generation rate per time slot G/L, by limiting the number of retransmission trials and new packet rejection is shown in Equation 7.10. Combining Equations 7.10 and 7.14 and after simplification we can write

$$\frac{\lambda_{opt}}{L} =$$

$$\frac{1}{(1-\alpha)} \frac{\left(\left(\left\{L-J+\sqrt{(L-J)^2+8LJ}\right\}^2\right) \middle/ \left(2L\left\{L+J+\sqrt{(L-J)^2+8LJ}\right\}\right)\right) e^{-(1/2L)\left(L+J+\sqrt{(L-J)^2+8LJ}\right)}}{1-\left\{1-\left(\left(L-J+\sqrt{(L-J)^2+8LJ}\right) \middle/ \left(L+J+\sqrt{(L-J)^2+8LJ}\right)\right) e^{-(1/2L)\left(L+J+\sqrt{(L-J)^2+8LJ}\right)}\right\}^{r+1}}$$

(7.22)

The secured operating region of L-channel slotted ALOHA system with and without limiting the retransmission trials is shown in Figure 7.7. Please note that here the y-axis should be multiplied by

$$\frac{L-J+\sqrt{(L-J)^2+8LJ}}{2L(1-\alpha)}.$$

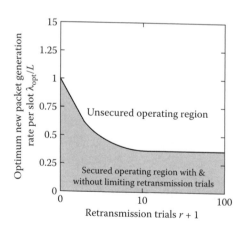

FIGURE 7.7 Secured and unsecured operating regions of multichannel L-slotted ALOHA in the presence of attacking noise packets. The y-axis should be multiplied by $(L-J+\sqrt{(L-J)^2+8LJ})/2L(1-\alpha)$.

From Figure 7.7, it can be said that the maximum value of new packet generation rate per slot, λ_{opt}/L with new packet rejection is

$$\text{UR} = \frac{L - J + \sqrt{(L-J)^2 + 8LJ}}{2L(1-\alpha)} \qquad (7.23)$$

On the other hand, lower limit of the new packet generation rate per slot, λ_{opt}/L with new packet rejection is

$$\text{LR} = \frac{\left\{ L - J + \sqrt{(L-J)^2 + 8LJ} \right\}^2}{2L(1-\alpha)\left\{ L + J + \sqrt{(L-J)^2 + 8LJ} \right\}} e^{-(1/2L)\left(L + J + \sqrt{(L-J)^2 + 8LJ} \right)} \qquad (7.24)$$

From Figure 7.7, we can conclude that the aggregate message packet generation rate, G, never reaches its optimum point, if the new packet generation rate per slot, λ_{opt}/L, is less than LR packet per time slot. The reason is that, we started to get the result of Equation 7.24 with the aggregate message packet generation rate, G_{opt}. So, it is unnecessary to control the retransmission attempt for a secured operation of L-channels slotted ALOHA, if the new packet generation rate per slot, λ_{opt}/L, is less than LR packet per time slot, where α is the newly generated packet rejection probability.

The optimum new packet generation rate per slot, λ_{opt}/L, with new packet rejection is shown in Equation 7.22. At the same way, the optimum new packet generation rate per slot, λ_{opt}/L, without new packet rejection is shown in Equation 7.18. Therefore, the secured area with new packet rejection can be obtained by subtracting Equation 7.18 from Equation 7.22. Therefore, the security improvement area by new packet rejection and limiting the number of retransmission trials, r, is

$$\text{AN} = $$

$$\frac{\alpha}{(1-\alpha)} \frac{\left(\left\{ L - J + \sqrt{(L-J)^2 + 8LJ} \right\}^2 \right) \Big/ \left(2L\left\{ L + J + \sqrt{(L-J)^2 + 8LJ} \right\} \right) e^{-(1/2L)\left(L + J + \sqrt{(L-J)^2 + 8LJ} \right)}}{1 - \left\{ \left(L - J + \sqrt{(L-J)^2 + 8LJ} \right) \Big/ \left(L + J + \sqrt{(L-J)^2 + 8LJ} \right) \right\}^{r+1} - e^{-(1/2L)\left(L + J + \sqrt{(L-J)^2 + 8LJ} \right)}}$$

$$\qquad (7.25)$$

The numerical results of Equation 7.25 are shown in Figure 7.8. Figure 7.8 shows clearly that the secured operating can be increased significantly by increasing the new packet rejection probability, α.

7.6 Conclusions

The security improvement of L-channels slotted ALOHA in the presence of random attacking noise packets signals is studied in this chapter. We have used three different

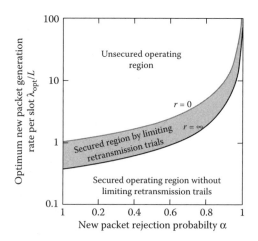

FIGURE 7.8 Secured and unsecured operating regions with new packet rejection. The *y*-axis should be multiplied by $\left(L - J + \sqrt{(L-J)^2 + 8LJ}\right)/2L(1-\alpha)$.

techniques to improve the secured packet transmission of the slotted ALOHA system by reducing the collisions. Since the random packet destruction DoS attacking noise packet increase the collision, we intend to use multiple channels in the slotted ALOHA protocol to reduce the collisions in the first technique. The use of multiple channels in the slotted ALOHA protocol reduces the packets collisions between not only the message packets themselves, but also between a message packet and other attacking noise packets. In the second technique, we have used retransmissions cut-off by limiting the number of retransmission trials. The retransmissions cut-off technique can limit the aggregate packet flow and form the optimum message packet flow in the presence of attacking noise packet. It is possible that the second technique retransmissions cut-off technique is not enough to control the flow of message packets. Because of that the third technique called new packet rejection probability is introduced. Using the third technique, the system is secured or stable with almost any high value of new packet generation rate per slot.

References

1. Xu, W., Ma, K., Trappe, W., and Zhang, Y., Jamming sensor networks: Attack and defense strategies. *IEEE Netw* 2006; 20(3): 41–47.
2. Nikjah, R., and Beaulieu N.C., On antijamming in general CDMA systems—Part I: Multiuser capacity analysis, *IEEE Trans Wireless Commun* 2008; 7(5, Part 1): 1646–1655.
3. Nikjah, R., and Beaulieu N.C., On antijamming in general CDMA systems—Part II: Antijamming performance of coded multicarrier frequency-hopping spread spectrum systems, *IEEE Trans Wireless Commun* 2008; 7(3): 888–897.
4. Mpitziopoulos, A., Gavalas, D., Pantziou, G., and Konstantopoulos, C., Defending wireless sensor networks from jamming attacks, in *IEEE Personal, Indoor and Mobile Radio Communications (PIMRC), September 2007*, Athens, Greece, pp. 1–5.

5. Ling, Q., and Li, T., Message-driven frequency hopping: Design and analysis. *IEEE Trans Wireless Commun* 2009; 8(4): 1773–1782.

6. Cagalj, M., Capkun, S., and Hubaux, J-P., Wormhole-based antijamming techniques in sensor networks. *IEEE Trans Mobile Comput* 2007; 6(1): 100–114.

7. Abramson, N., Fundamentals of packet multiple access for satellite networks. *IEEE J Select Areas Commun* 1992; 10(2): 309–316.

8. Abramson, N., The throughput of packet broadcasting channels. *IEEE Trans Commun* 1977; 25(1): 117–128.

9. Su, W., Alchazidis, N., and Ha, T. T., Multiple RFID tags access algorithm. *IEEE Trans Mobile Comput* 2010; 9(2): 174–187.

10. Chen, W-T., An accurate tag estimate method for improving the performance of an RFID anticollision algorithm based on dynamic frame length ALOHA. *IEEE Trans Automat Sci Eng* 2009; 6(1): 9–15.

11. Shin, W. J., and Kim, J. G., A capture-aware access control method for enhanced RFID anti-collision performance, *IEEE Commun Lett* 2009; 13(5): 354–356.

12. Mouly, M., and Pautet, M-B., *The GSM System for Mobile Communications*. Authors, 1992.

13. ETSI EN 300 940, GSM 04.08 Version 7.1.2, Digital cellular telecommunications system (Phase 2+); Mobile radio interface layer 3 specification. 1998.

14. Premkumar, K., and Chockalingam, A., Performance analysis of RLC/MAC and LLC Layers in a GPRS protocol stack, *IEEE Trans Veh Technol* 2004; 53(5): 1531–1546.

15. Cho, S. H., and Park, S. H., Stabilised random access protocol for voice/data integrated WCDMA system. *Electron Lett* 2001; 37(19): 1197–1199.

16. 3GPP, TSG-RAN: MAC protocol specification, 3G TS 25.321 v3.4.0, June 2000.

17. 3GPP2, TSG-C: MAC standard for cdma2000 spread spectrum systems, release A — addendum I, 3GPP2 C.S003-A-1 v.1.0, September 2000.

18. Choi, Y-J., Park, S., and Bahk, S., Multichannel random access in OFDMA wireless networks, *IEEE J Select Areas Commun* 2006; 24(3): 603–613.

19. Richard, T. B., Vishal Misra, M., and Rubenstein, D., An analysis of generalized slotted-ALOHA protocols. *IEEE/ACM Trans Netwo* 2009; 17(3): 936–949.

20. Tague, P., Li, M., and Poovendran, R., Probabilistic mitigation of control channel jamming via random key distribution, in *IEEE Personal, Indoor and Mobile Radio Communications, (PIMRC), September 2007*, Athens, Greece, pp. 1–5.

21. Chan, A., Liu, X., Noubir, G., and Thapa, B., Broadcast control channel jamming: Resilience and identification of traitors, in *IEEE International Symposium on Information Theory 2007*, June 2007, pp. 2496–2500.

22. Ma, L., and Shen, C-C., Security-enhanced virtual channel rendezvous algorithm for dynamic spectrum access wireless networks, in *3rd IEEE Symposium on New Frontiers in Dynamic Spectrum Access Networks (DySPAN)*, October 2008, pp. 1–9.

23. Tague, P., Li, M., and Poovendran, R., Mitigation of control channel jamming under node capture attacks. *IEEE Trans Mobile Comput* 2009; 8(9): 1221–1234.

24. Matoba, O., Nomura, T., Perez-Cabre, E., Millan, M. S., and Javidi, B., Optical techniques for information security. *Proc IEEE* 2009; 97(6): 1128–1148.

25. Langheinrich, M., and Marti, R., Practical minimalist cryptography for RFID privacy. *IEEE Syst J* 2007; 1(2): 115–128.

26. Cao, X., Zeng, X., Kou, W., and Hu, L., Identity-based anonymous remote authentication for value-added services in mobile networks. *IEEE Trans Veh Technol* 2009; 58(7): 3508–3517.

27. Bhadra, S., Bodas, S., Shakkottai, S., and Vishwanath, S., Communication through jamming over a slotted ALOHA channel. *IEEE Trans Inform Theory*, 2008; 54(11): 5257–5262.

28. Zander, J., Jamming in slotted ALOHA multihop packet radio networks. *IEEE Trans Commun* 1991; 39(10): 1525–1531.

29. Clare L. P., and Baker, J. E., The effects of jamming on control policies for frequency-hopped slotted, ALOH., in *IEEE Global Communications Conference (GLOBECOM)*, 1990, pp. 1132–1138.

30. Pronios, N. B., and Polydoros, A., Slotted ALOHA-type fully connected networks in jamming—Part I: Unspread, ALOH., in *IEEE MILCOM 1987*, Vol. 3, 19–22 October 1987, pp. 910–914.

31. Yang, K., and Stuber, G. L., Throughput analysis of a slotted frequency-hop multiple-access network, *IEEE J Select Areas Commun* 1990; 8(4): 588–602.

32. Ma, R. T., Misra, V., and Rubenstein, D., An analysis of generalized slotted-aloha protocols. *IEEE/ACM Trans Netw* 2009; 17(3): 936–949.

33. Xue, Y., and Kaiser, T., Pursuing multiuser diversity in an OFDM system with decentralized channel state information, in *IEEE International Conference in Communications (ICC)*, Vol. 6, 20–24 June 2004, pp. 3299–3303.

34. Sarker, J. H., and Mouftah, H. T., Effect of jamming signals on wireless ad hoc and sensor networks, in *IEEE Global Communications Conference (GLOBECOM)*, Honolulu, Hawaii, USA, November–December, 2009, pp. 1–6 (to be appear).

35. Zhou, B., Marshall, A., Zhou, W., and Yang, K., A random packet destruction dos attack for wireless networks, in *IEEE International Conference in Communications (ICC)*, May 2008, Beijing, China, pp. 1658–1662.

36. Zander, J., and Kim, S-K., *Radio Resource Management for Wireless Networks*. Artech House, 2001.

37. Wong, E. W., and Yum, T-S. P., The optimal multicopy ALOHA. *IEEE Trans Autom Control* 1994; 39(6): 1233–1236.

38. Sarker, J., Stable and unstable operating regions of slotted ALOHA with number of retransmission trials and number of power levels, in *IEE Proceedings, Communications*, Vol. 153, no. 3, June 2006, pp. 355–365.

39. Sakakibara, K., Muta, H., and Yuba, Y., The effect of limiting the number of retransmission trials on the stability of slotted ALOHA systems, in *IEEE Transactions on Vehicular Technology*, Vol. 49, Issue 4, July 2000, pp. 1449–1453.

8

Analysis and Classification of Reliable Protocols for Pervasive Wireless Sensor Networks

Ahmed Badi
Florida Atlantic University

Imad Mahgoub
Florida Atlantic University

Michael Slavik
Florida Atlantic University

8.1 Introduction

Wireless sensors are one of the fastest developing pervasive communication technologies [1–4]. The availability of small, cheap low-power embedded processors, radio transceivers and sensors, often integrated on a single chip leads to the use of sensing, computing and wireless communication for monitoring and interacting with the physical world. These wireless sensor devices are assembled with the hardware components mentioned above, an energy source, in most cases battery together with networking and application firmware and software. Depending on the size of the network and the complexity required of each sensor, the cost of sensor devices could vary from hundreds of dollars to a few dollars. The size of a single sensor node can also vary. Sensors can be deployed in large numbers to form networks that are used to collect data or to pervasively monitor the physical environment.

A wireless sensor network (WSN) is a telecommunication network consisting of spatially distributed sensors and a base station. These sensors monitor physical or environmental conditions in a cooperative manner [5,6]. Sensors collect data from their surrounding environment, and use their networking infrastructure to aggregate and send the collected data to the base station. These networks can be more accurately described as distributed systems where participants agree to receive and forward data messages sent by other network participants. The sensors self-organize to form distributed systems that can be used for a variety of purposes. Military applications such as monitoring of troop movement and target tracking originally motivated the development of WSNs. However, currently, WSNs are found in many civilian applications as well. They can be useful in applications such as security and surveillance, smart spaces, monitoring of natural habitats and eco-systems, healthcare applications, home automation, traffic control, industrial process control, and structural health monitoring.

8.1.1 Design Requirements of Wireless Sensor Networks

The sensing capabilities of the wireless sensors cover physical measurements of phenomena such as temperature, sound, vibration, pressure, moisture, light intensity, magnetism, motion, radiation, or pollutants among many other physical and environmental quantities. The price, size, and self-organization features of the sensors make them cost-effective solutions for many problems. Wireless *ad hoc* sensor network design requirements include the following:

8.1.1.1 Energy Optimization

In many applications, the sensors are battery powered as shown in Figure 8.1. They are usually placed in remote areas where manual service of sensor nodes may not be possible.

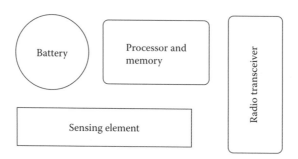

FIGURE 8.1 Components of sensor device.

In this case, the node's lifetime will be dependent on the battery's lifetime, thereby requiring the optimization of energy consumption.

8.1.1.2 Self-Configuration

With a large number of nodes in a sensor network and their potential placement in hostile locations, individual node configuration is not possible. Therefore, it is essential that the network be self-configuring. In addition, nodes may fail due to energy exhaustion, malfunction, or destruction and new nodes may be added to the network. For these reasons, the network must be able to periodically reconfigure itself so that it can continue to function. Also, depending on the nature of the application the network needs to maintain some degree of connectivity.

8.1.1.3 Scalability

WSNs are assumed to have a large number of mostly stationary sensors. Networks of 10,000 or even 100,000 nodes are envisioned and network scalability is a major issue.

8.1.1.4 In-Network Signal Processing

To improve the quality of data collected, it is often useful to fuse data from multiple sources. This requires the transmission of data and control messages to some master node before sending it to the base station. This will impose some requirements on the network's architecture.

8.1.1.5 In-Network Query Processing

The sensor network may collect a large amount of data. This may overwhelm the user who may not be able to process all of the collected information. Instead, selected nodes within the network will collect the data from their neighbors and create a representative message.

8.2 Reliability in Wireless Sensor Networks

This section deals with the survey and classification of WSNs' reliable protocols. The classification follows the transmission control protocol (TCP/IP) network communication stack model. A summary overview of this stack model is given in Section 2.2.1.

WSNs reliability research is still in its infancy. There is no unified definition for the reliability of the wireless sensor and each work has defined reliability differently and in line with their approach. Table 8.1 shows the classification done by Cinque et al. [7] in numerous IEEE and ACM publications and proceedings on WSNs over the past few years. It shows that while routing and MAC have attracted their share of attention, research in WSNs reliability acquired only 5% of those publications. This clearly shows the need for reliability studies in WSNs field.

Designing WSNs is challenging since they need to be treated as embedded systems, wireless communication systems, and distributed systems, simultaneously. The wireless communication medium is inherently unpredictable and unreliable. This places extra requirements on WSNs communication protocols to provide robustness and resilience to communication failures.

TABLE 8.1 Classification of IEEE/ACM WSNs Related Research Work in WSNs

WSNs Research Area	IEEE/ACM Publications Percentage
Routing	29
MAC	14
Localization	13
Energy-efficiency	11
Topology	07
Placing	5
Reliability	5
Security	4
Technology	3
OS	3
Performance	2
Modeling	2
Synchronization	2

8.2.1 Importance of Wireless Sensor Networks Reliability

Currently, industrial involvement in WSN technology is lagging. This can be attributed in part to the unanswered reliability concerns and issues. System and technology reliability issues are major concerns in practical and industrial settings. The WSNs' technology will not gain wide adoption and reach its full potential until and unless reliability and other industrial and practical concerns have been addressed. Without effective industrial involvement, none of the applications that are enabled by the WSNs' technology can be realized.

As the WSNs research matures, it needs to move beyond studies that are focused on energy conservation and resource constraints. To build trust in using these systems, more emphasis should be placed on studying and analyzing the reliability and dependability of WSNs. So far, WSNs energy efficiency research has not taken reliability into consideration as a performance parameter or as a design constraint.

8.2.2 Classification of WSNs Reliable Protocols Using the TCP/IP Stack Model

The TCP/IP stack was instrumental in classifying and organizing studies for WSNs energy conservation research and protocols [8]. A similar organizing framework does not exist for WSNs reliability protocols and WSNs reliability evaluation research. Reliability can be studied as a coverage problem or as a message delivery problem [9]. The information data transport reliability is affected by the message delivery reliability. The networking protocol control messages reliability, which affects protocol's correctness, is also dependent on the message delivery reliability.

A previous survey work [9] discussed WSNs reliability challenges and introduced some of the transport protocols that addressed WSNs reliability. The protocols presented are classified according to their message data stream type as single packet vs. block of packets vs. periodic stream of packets. The limitation of this classification approach is that it leaves out a considerable number of WSN reliability studies that approached the problem from a more abstract level and did not necessary fall under any of the categories in the above classification. We present more general classification terms that attempt to organize the WSNs reliable protocols under categories that follow the TCP/IP stack layers.

One way to measure reliability is to specify a "data-delivery probability" [9]. Higher data-delivery probability requirements imply higher energy cost. This is true regardless of the definition used for reliability. Different types of data streams within the same network may require different reliability measures [9], for example, single packet delivery reliability as in the case of delivering aggregated data to the sink, block of data delivery reliability as in the case of code update, and periodic reporting data reliability.

In this section, we attempt to classify the reliable protocols proposed for WSNs using the TCP/IP stack layer as a classification platform.

8.2.2.1　TCP/IP and Open Systems Interconnection Network Stack Models

The open systems interconnection (OSI) model is a standard developed by the International Organization of Standards (ISO) as how to transmit messages between any two telecommunicating points in a network [10,11]. The standard defines seven layers of functions that take place at each end of a communication. Each layer is responsible for a number of logical steps that it implements. Several performance parameters of the OSI stack for wired networks have been optimized. These parameters include latency, fairness, and throughput.

In the OSI model, the communication process between two points in a network is divided into seven layers: Application, Presentation, Session, Transport, Network, Medium Access Control, and Physical layers [10]. An advantage of this view is that the complexity of the communication process is also divided among the different layers making the implementation of such systems manageable. The programming and hardware that furnishes the seven layers, also known as the network stack, is usually found partly in the computer operating system, in several stand-alone applications such as web browsers, and in the network firmware and hardware interfaces that are common parts of any computer system.

The above discussion presented the well-known OSI stack. It describes a fixed, seven-layer stack for networking communication protocols. Similarly, there is another layered stack protocol, which is the simpler five-layer stack model, also known as the TCP/IP protocol stack shown in Figure 8.2. There are lots of similarities between the two protocols since they attempt to define the same communication process. WSN stack has more in common with the TCP/IP stack. These five layers are summarized below:

Application Layer: The application layer sits at the top of the communication stack. It generates the data that will be sent out or it will be the entity that ultimately receive and decodes the data. At this layer, the communicating partners are identified, quality of service is defined and identified, data encryption and decryption is performed, and user authentication and privacy issues are considered. In the seven-layer protocol stack model, this layer is further divided into the presentation and session layers.

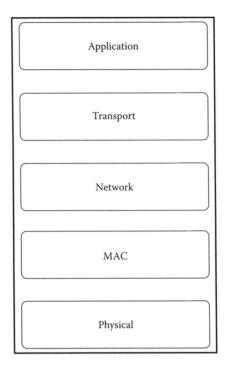

FIGURE 8.2   TCP/IP five layer protocol stack model.

Transport Layer: The transport layer provides transparent data transfer between hosts. It is responsible for end-to-end error recovery and flow control. It is also responsible for providing a reliable, error-free communication over an unreliable communication medium and ensuring complete data transfer. The well-known TCP and user datagram protocol (UDP) are implementations of the transport layer functionality.

Network Layer: The network layer performs error control, source to destination routing by ensuring the sending of data messages in the right direction to the right destination on outgoing transmissions, and receiving incoming packet transmissions. This layer is also responsible for flow control, data segmentation, and desegmentation. IPv4, IPv6, and X.25 are the most commonly used implementations for this layer.

Medium Access Control (MAC) Layer: The MAC layer regulates the usage of the shared communication medium. Before transmitting frames, a station must first gain access to the medium. For a local area network (LAN) this can be the token in a token ring network. In a wireless network scenario the medium is the radio channel.

The IEEE 802.11® [12] is a wireless communication MAC standard that is widely adopted. In this standard, as a condition to access the medium, the MAC layer checks the value of its network allocation vector (NAV). This is a counter resident at each node and it represents the amount of time for which the communication medium will be busy. Prior to transmitting a frame, a node calculates the amount of time necessary to send the frame based on the frame's length and the channel's data rate. The node places a value representing this time in the header's duration field of the frame. When other nodes receive the frame, they examine this duration field value and use it as the basis for setting their corresponding NAVs. Before a station can attempt to send a frame, its NAV must be zero. This process reserves the medium for the sending station.

In general, contention-based medium access is implemented by the distributed coordination function (DCF), which is a random back off timer that stations use if they detect a busy medium. If the channel is in use, then the station must wait a random period of time before attempting to access the medium again. This ensures that multiple stations wanting to send data do not transmit at the same time. The random delay causes stations to wait for different periods of time and prevents them from sensing the medium at exactly the same time, finding the channel idle, transmitting, and colliding with each other. The back off timer significantly reduces the number of collisions and corresponding retransmissions especially when the number of active users increases.

For radio-based LANs, a transmitting station cannot listen for collisions while sending data, mainly because the station cannot have its receiver on while transmitting the frame. As a result, the receiving station needs to send an acknowledgment (ACK) if no errors were detected in the received frame. If the sending station does not receive an ACK within a specified period of time, the sending station will assume that there was a collision or radio frequency (RF) interference and retransmits the same frame again.

Physical Layer: The physical layer is the bottom layer of the OSI and the TCP/IP stack models. It provides the hardware means of sending and receiving data on a carrier and performs services requested by the MAC layer.

The physical layer is the most basic network layer, providing only the means of transmitting raw bits rather than packets over a physical data link. This layer transmits the bit stream through the network as an electrical or electromagnetic signal. It provides bit-by-bit node-to-node delivery, signal modulation and demodulation, equalization filtering, training sequences, pulse shaping, and other signal processing of physical signals. The physical layer determines the bit rate in bits per second (bits/s), also known as channel capacity, digital bandwidth, maximum throughput, or connection speed. The physical layer also defines half duplex or full duplex transmission mode.

8.2.2.2 TCP/IP and OSI Stacks in Wireless *Ad Hoc* and Wireless Sensor Networks

Since the inception of the ISO OSI and TCP/IP layered communication stack models for the wired LAN and wide area network (WAN), the goals have been set to achieve compatibility and simplification of functional description of separate units. The optimizations of the stack have also been in the direction of improving the latency, quality of service (QoS), reliability, and throughput matrices.

With the emergence of *ad hoc* and wireless networks, the OSI and TCP/IP stack models were ported as is to this new technology. Research has been active in the study of ways to enhance the stack to optimize it and bring it up to face the new challenges found in the wireless communication field. In the medium access control layer, token ring protocols have been replaced by a new set of protocols suitable for wireless communication, for example, IEEE 802.11 [12], IEEE 802.15.4® (*ZigBee*) [13], and S-MAC [14] protocols. In the networking and routing layer, several new editions have been introduced, for example, *ad hoc* on-demand distance vector (AODV) [15], dynamic source routing (DSR) [16], zone routing protocol (ZRP) [17], LEACH [18], TEEN [19], APTEEN [20], PEGASIS [21], and directed diffusion [22] to mention a few.

8.2.3 Transport Layer Reliable Protocols

In [23], the authors argue that while node redundancy, inherent in WSNs increase the fault tolerance, no guarantee on the reliability levels can be assured. Further, the frequent communication failures within WSNs impact the network's reliability over time and make it more challenging to achieve a desired reliability. Another issue is that flooding of *bursty* raw data causes broadcast storms, which can result in collisions, contentions, and power wastage. In-network processing is an optimization that reduces redundancy resulting in fewer collisions, contentions, and enhances responsiveness. In order to model the reliability of information transport, all the operations on data starting from its generation to dissemination, aggregation, and transmission need to be considered.

A framework for modeling reliability in WSNs is currently missing [23]. Such a framework will be useful in classifying existing transport protocols and in comparing their reliability performance. Next, we present examples of reliable transport layer protocols proposed for WSNs.

8.2.3.1 Event-to-Sink Reliable Transport (ESRT) Protocol

The first reliable transport protocol proposed for WSNs is the event-to-sink reliable transport (ESRT) protocol [24]. The concept of event-to-sink reliability is introduced in [24] as an alternative to the classical source to destination reliability. This is more applicable to WSNs since data flows from nodes to sink are loss-tolerant. Event detection and the reliable relaying of information to the sink is what matters in determining the reliability and successful operation of the network. This is regardless of how many sensors did successfully deliver information about the detected event. The event radius defines a circle around the event within which all the enclosed nodes will be able to sense (detect) the event.

ESRT is a centralized protocol that runs only on the sink and thus leveraging its abundance of computing and power resources and relieving the resource constrained nodes. In ESRT, congestion control is identified as an important factor for reliable data flow since packet loss due to congestion can impair event detection at the sink. The reliability requirements are determined by the application layer. The transport layer reliability is defined by a reliability factor, that is, the ratio between the number of received packets at the sink to the optimal number of packets that is required for reliable event detection. Ideally, the ratio should be maintained as close as possible to one. The reporting frequency is defined as the number of reports that the nodes need to generate per unit time to achieve required event detection reliability. This factor is calculated and broadcasted to all nodes by the sink. The protocol operation also relies on congestion detection. To do this the sink relies on the nodes setting a congestion flag bit on their reply messages. A node will monitor its buffer fullness and the rate at which the buffer is getting filled. The congestion flag bit will be set if the node predicts that it will experience a buffer overflow during the next reporting period.

The ESRT protocol identifies five network states as shown below, these states are:

1. (NC, LR) not congested, low reliability.
2. (NC, HR) not congested, high reliability.
3. (C, HR) congested, high reliability.
4. (C, LR) congested, low reliability.
5. (OOR) optimal operating region.

The protocol will use the reporting frequency to control the transition of any of the states to the OOR state.

8.2.3.2 Reliable Multi-Segment Transport Protocol

The reliable multi-segment transport (RMST) protocol, a transport layer protocol designed to be used with directed diffusion [22] is presented in [25]. The authors point out that the emphasis on energy conservation in sensor networks implies that poor paths should not be selected during route discovery. Nor should they be artificially bolstered via MAC automatic repeat request (ARQ) mechanisms. Given the high wireless link loss rate (between 2% and 30%), the authors address the tradeoffs between implanting reliability in the MAC layer (i.e., hop-to-hop) vs. in the transport layer (end-to-end). Several design choices were explored. MAC layer ARQ refers to the hop-to-hop recovery of missing or erroneous packets. This includes using RTS/CTS and ACK control messages. If the design is using no ARQ, then the transmissions do not employ MAC layer reliability mechanisms. In this mode, reliability is deferred to the transport or application layer. There are several benefits to this approach including:

- Significant amount of overhead associated with the RTS/CTS and ACK exchange can be avoided.
- Routing protocols attempt to select high quality paths for data transmission.

The performance of caching and non-caching designs was also discussed in [25]. In caching, each node in the path caches the fragments that make up a larger data entity. When a node senses a missing fragment, a repair request is sent to the next hop. If the requested fragment is in the neighbor's local cache a response is sent. Otherwise, the request is forwarded towards the data source. In the non-caching design, only the sinks can detect missing fragments and data errors.

8.2.3.3 Analysis and Classification of WSNs Reliable Transport Protocols

Shaikh et al. [23] present a definition for data transport reliability as the sink that detects the phenomenon of interest within an application-specific time bound. Existing WSNs data transport protocols were classified into two categories: end-to-end (e2e) and event-to-sink. The RMST protocol [25] was used as an example for e2e. It relies on the selective NACK to detect message loss at the sink. The ESRT protocol [24] was used as an example for event-to-sink category. From the RMST reliability analysis, the equation for the protocol's reliability is derived as given by Equation 8.1. Similarly, for the ESRT transport protocol, the reliability can be derived as shown in Equation 8.2. Using Equation 8.1, the positive impact of increasing the number of retransmissions on RMST performance can be studied.

$$R_{RMST} = 1(1 - R_R) * (R_R * R_{MLD}))^r, \tag{8.1}$$

where r is the number of retransmissions, R_R the routing reliability, and R_{MLD} the reliability of message detection.

$$R_{ESRT} = 1 - (1 - R_R)^n, \tag{8.2}$$

where R_R is the routing reliability and n the number of sources reporting the phenomenon to the sink.

As another example for ESRT protocols, the RBC [26] protocol was presented. Following similar analysis as for ESRT and RMST, the reliability equation for the RBC protocol can be derived as given in Equation 8.3. From Equations 8.2 and 8.3, the performance of these two event-to-sink protocols can be analyzed.

$$R_{RBC} = 1 - (1 - [1 - (1 - R_R) * (1 - (R_R * R_{MLD}))^r])^n, \tag{8.3}$$

where r is the number of retransmissions, R_R the routing reliability, and R_{MLD} is the reliability of message detection.

8.2.3.4 Improving Transport Reliability by Using MAC Layer ARQ

The authors in [23] present an analysis for the effect of MAC retry limit on reliability. If p is defined as the probability of success for a single attempt across one hop, and R is the number of MAC attempts, then the probability of success (P_h) with MAC ARQ in R retries is given by the following equation:

$$P_h = \sum_{i=0}^{R-1} p.(1 - p)^i. \tag{8.4}$$

This can be simplified as

$$P_h = 1 - (1 - p)^R. \tag{8.5}$$

For H hops, the end-to-end reliability (P_e) is

$$P_e = (P_h)^H. \tag{8.6}$$

Table 8.2 shows the probability of message arrival vs. MAC retry limit for 40 hops and error rate of 0.10 per hop. It is very interesting to observe that for such a high link error rate, only few retransmissions are effective in achieving a high probability of message arrival. Table 8.3 shows the probability of message arrival at the sink over six hops when using MAC ARQ (RTS/CTS and ACK) and when not using ARQ. The results in Table 8.3 show clearly that the probability of arrival drops sharply while increasing error rate for non-ARQ, but stays high when using MAC ARQ. This shows the effectiveness of MAC layer ARQ in combating the unreliable nature of the wireless link.

TABLE 8.2 Probability of Arrival Across 40 Hops with an Average Error Rate of 0.10 per Hop, Given Fixed Number of Retries per Hop

Number of MAC Retries	Probability of Delivery
1	0.20
2	0.73
3	0.95
4	0.99

8.2.4 Network Layer Reliable Protocols

The networking and routing algorithms have an immense impact on the overall network reliability and other QoS measurements. The layer uses several mechanisms to improve the end-to-end network communication reliability. Several reliable protocols that use these mechanisms have been proposed. In this section, we present our survey of some of these protocols and discuss the effectiveness of the algorithms used.

8.2.4.1 Reliable Routing Using Graph Theory Analysis

Graph theory architectural models are used intensively in computer networks to construct mathematical representations for these networks. In [27], a reliability analysis using graph enumeration is presented. Reliability is defined as the probability that the network will function. The network is modeled as a graph $G = (V, E)$ composed of elements that can fail statistically independent of one another and their failure probabilities are known. Failure states enumeration, minimum paths enumeration, and cuts enumeration between source and destination are used to calculate data-delivery probability. From the analysis it is clear that this is a source-to-sink communication scenario.

The problem of computing a measure for reliability and the maximum message delay in WSNs is addressed in [28]. Reliability is defined as the probability that there exists an operational communication path between the sink and at least one operational sensor in the target cluster (target area). The failure of one or more nodes may not cause the data sources to be disconnected from the data sinks. This can be considered an event-to-sink

TABLE 8.3 Probability of Arrival Across 6 Hops

Wireless Channel Error Rate	Probability of Arrival	
	with ARQ	without ARQ
0.01	0.99	0.96
0.05	0.99	0.75
0.10	0.98	0.54
0.15	0.98	0.39
0.20	0.98	0.27
0.25	0.98	0.19
0.30	0.97	0.11

reliability scenario. The network is modeled by a probabilistic graph $G = (V, E)$, where every node in the network is represented by a node in V. Each node v in V has an associated operational probability P_v. An edge exists between two nodes if they are within communication range from each other.

8.2.4.2 Multiple Routes and Erasure Codes Reliable Protocols

A simple wireless communication network graph was used to derive the reliability definition used by AboElFotoh et al. [28]. The component reliabilities are as given below:

$$\text{Rel}(G) = P_B(P_{H1}(1 - (1 - P_{S1})(1 - P_{S2})) + (1 - P_{H1})P_G \cdot P_{H2} \cdot P_{S2}). \tag{8.7}$$

In the reliability analysis given, many networking factors were eliminated from playing a role in the reliability calculation. An example is the effect of MAC layer retransmissions on link layer reliability.

Some of the surveyed work introduced mechanisms for improving WSNs reliability while observing their resource constraint. Kim et al. [29] start by introducing

$$\text{Number of packets received} = P_{\text{success}}{}^*\text{Number of packets sent}. \tag{8.8}$$

The per packet delivery success probability (P_{success}) can be obtained through empirical studies. This value is fixed and nothing more can be done to improve it. The technique used here is to increase the number of packets sent sufficiently enough so that all the data are received. The following mechanisms for improving the network's reliability were suggested:

- Link layer retransmissions
- Use of erasure codes [30–34], which add redundancy to each message thus allowing the construction of m original messages from any received n messages (provided that $n > m$)
- Providing alternative routes (multiple routes) to replace failing links

Desirable properties for erasure code algorithms used in WSNs are pointed out in [35] as:

- The ratio $\frac{m}{n}$ needs to be dynamic since it is dependent on the current channel error conditions.
- There should be no restriction on the packet length.
- Encoding and decoding should be inexpensive in their memory and processing requirements.

The authors point out limitations with traditional erasure codes [30,34] used in [29] such as Reed-Solomon codes since they fail to satisfy the above properties. A suggested solution is to use the new classes of erasure codes such as Luby transform [32], Raptor [31], and Online codes [33] that have better performance for WSNs.

Similar to the technique above, Dulman et al. [36] suggest improving reliability by sending the data message through multiple paths. The algorithm tries to find k disjoint paths between source and destination and adding redundancy to each packet so that only E ($E < k$) packets need to reach the destination for the original message to be constructed. The algorithm assigns a "reputation coefficient" to each path based on its past performance. The degree of redundancy added to each message depends on the number of paths (k), and on the reputation coefficients of the available paths. The algorithm and the associated analysis are built around the notion of source to destination reliability.

Limitations of Erasure Codes and Multiple Routes Reliable Protocols: The limitations with using erasure codes and multiple routes to improve reliability in WSNs are the following:

- They depend on the existence of multiple routes between source and destination. This may not be applicable to hierarchal topologies that are the typical architectures for large-scale WSNs.
- The surveyed erasure code protocols failed to study the energy consumption overhead of adding redundant bits to each packet.
- There is an energy overhead introduced by transmitting more packets than needed to construct the original message.

8.2.4.3 Reliable Routing Using Link Connectivity Statistics

Woo et al. [37] state that the dynamic and lossy nature of the wireless communication medium posses a major challenge to reliable self-organizing multi-hop networks. This is more problematic with the simple, low-power transceivers commonly used in sensor networks. To improve communication reliability, the authors propose capturing link connectivity statistics dynamically by using efficient and adaptive link estimators. Routing decisions should exploit such connectivity statistics.

Each node maintains a neighborhood table. The table stores link status, quality, and routing information. The link quality estimators can be very efficient in implementing cost-based routing. A challenge arises that in dense networks, a node may receive packets from more nodes than it can represent in its neighborhood table. The challenge is for a node to decide in which nodes it should invest its limited neighborhood table resources to maintain link statistics. The problem is that if a node is not in the table, there is no place to record its link quality information. The authors then attempt to develop an algorithm for neighborhood management that will keep a sufficient number of good neighbors in the table regardless of the network density. The problem addressed here has aspects in common with cache management and with database statistical estimation techniques.

8.2.4.4 Reliable Routing Using Link Rating Parameter

The neighborhood table concept is extended in [38]. A reliable and energy-efficient network routing protocol is introduced for hierarchical WSNs. The protocol defines and uses different message reliability requirement levels that are dependent on the message type and importance. The protocol introduces the link rating parameter for each of the nodes one hop links. This can be defined mathematically as follows:

$$\text{Link Rating} = \frac{\sum R_{\text{L}}/P_{\text{L}}}{N}, \tag{8.9}$$

where R_L is the measured reliability for level L, P_L the power value needed to achieve level L's reliability, N the total number of message reliability levels defined.

The link rating parameter, as defined in Equation 8.9, can be used for any number of message reliability levels (N). The nodes calculate and store a link rating value for each of their one hop neighbors. At the beginning of each network reclustering cycle (*round*), new set of nodes elect themselves as cluster heads. The new cluster heads will generate cluster head advertisement messages that are *broadcasted* to nodes within their coverage region. When a noncluster head node receives the cluster head advertisement messages, the node will join the cluster head with the best link rating parameter value resulting in the optimum reliability, and node to cluster head link power setting.

8.2.5 MAC Layer Reliable Protocols

The MAC layer uses few mechanisms to improve the per-hop communication reliability. Several reliable protocols that rely on these mechanisms have been proposed. In this section, we present some of these protocols classified by the MAC mechanisms used.

8.2.5.1 Reliable Protocols Using MAC Layer Retransmission

Several studies have attempted to analyze the various WSNs parameters and their impact on the overall network reliability and performance. As an example, MAC retransmission is considered to be very effective in improving reliability. Nevertheless, it is also an expensive operation and wastes valuable resources [39]. Challenges to achieving reliability in WSNs are summarized as being due to:

- Nature of the wireless medium and burst errors
- Resource constraints
- Algorithms to achieve reliability cannot be computationally extensive

As stated in Vuran et al. [40], one of the major factors affecting the reliability in multi-hop networks is the local retransmission reliability mechanism implemented in the MAC layer. The performance of this mechanism depends mainly on the maximum number of retransmissions for packet success. The effect of this mechanism on the overall network performance is investigated and the results show that although there is a significant difference between the maximum number of retransmissions, $Rt_{max} = 4$ and $Rt_{max} = 7$, further increase in the retransmission limit to 10 does not have a significant effect on the overall network reliability.

8.2.5.2 Reliable Protocols Using MAC Layer Contention Window Size

Contention is one of the major sources of packet drop. For this reason, contention resolution mechanisms are needed at the MAC layer. Contention resolution is performed via contention window adjustment. A node selects its random back-off time between (0,cw), where cw is the contention window size. This window size is initially set to minimum and as the contention level increases in the vicinity of the node, the size of the contention window is increased. Hence, the current value of the cw is representative of the local contention level.

The interaction between the contention window size and retransmission techniques for improving end-to-end data-delivery reliability is studied in [41]. The authors point out that the radio link exhibits varying reliability over time, space and from node to node. Similar to [24], where the primary source for data loss is due to noise and environmental effects rather than congestion.

The observation that not all messages in a WSN need the same reliability guarantees made in [42]. A classification of message delivery reliability requirements based on the message type is presented, along with a reliable MAC algorithm. The algorithm proposed uses different contention window sizes, combined with different number for the retransmission attempts, where both depends on the message type. This causes important messages to experience short delays, and host nodes to attempt to retransmit them sooner. The important messages stay alive in the node buffers longer due to their higher number of retransmission attempts, giving them better reliability in the form of a better chance of being received by the destination.

The important findings in [43] are the following:

- Link layer retransmissions are necessary for improving reliability.
- A small number of retransmissions are sufficient for a satisfactory reliability improvement.
- The cost of higher reliability is higher network overhead due to longer path lengths and excessive retransmissions.

8.2.5.3 Reliable Protocols Using MAC RTS/CTS Messages

In Volgyesi et al. [44], the focus is on the reliable multihop bulk transfer of data. The authors argue that the vast majority of research is focused on reliable and power-efficient transfer of small amount of data. However, in a few WSN applications, reliable transfer of mass data is essential. The authors present bulk transfer service that achieves reliability by employing a simplified version of IEEE 802.11. This is achieved by RTS/CTS handshake as shown in Figure 8.3 to protect long packet bursts and provide a simple, efficient flow control by allowing only one data stream to communicate at any given time.

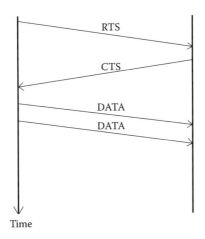

FIGURE 8.3   RTS/CTS handshake to protect longer packets bursts [44].

8.2.5.4 Reliable Protocols Using MAC ACK Messages

The concept of event to sink reliability is used in [43]. The sink sends its query to the network and indicates in the query whether reliable delivery of the response is required, by setting a flag in the query. The authors address the issue of a parent node failing after receiving a message from its child. In this case, the child will not receive an ACK and a dynamic route switching algorithm will be executed. This will select a new parent through which the data will be sent.

8.2.6 Radio and Physical Layer Reliable Protocols

The relationship between radio range and reliability was discussed in [28]. An increase in the radio range results in an increase in the one hop reliability. The relationship between radio range and power is also discussed, as radio range is proportional to at least the square of the power. Therefore, increasing the radio range is effective in improving the one-hop reliability, but extremely taxing on energy.

The work in [45] presents a cost-based reliable algorithm. The network tries to set the one hop retry limit and the transmission power to a setting that minimizes the power consumption metrics while meeting some reliability requirements constraint. Here again the probability of one hop successful transmission is shown to be proportional to the transmission power as given by the equation below:

$$P_{(\text{successfulTx})} = f\left(\frac{\text{Gain} * \text{Txpower}}{\text{Noisepower}}\right) \tag{8.10}$$

The main theme in [46] is to study reliability as a coverage and connectivity problem. Necessary and sufficient condition for a random grid network to cover a unit square region was derived. Similarly, it provided sufficient conditions for connectivity. The results can be used to determine the tradeoff between node diameter, reliability, and power consumption since the radio propagation radius is directly related to the transmission power. For a network with n nodes, if $r(n)$ is the transmission or sensing radius and $D_{ij}(n)$ is the number of transmissions required to traverse from node i to node j. The diameter of the grid is then defined as

$$D(n) = \max_{i,j} D_{ij}(n). \tag{8.11}$$

From which the authors derive the upper and lower bounds on the diameter of the grid as:

$$\sqrt{2} < r(n)D(n) < \frac{2}{1 - \frac{2}{\sqrt{\pi c}}}, \tag{8.12}$$

where c is a parameter that decides how much power each node uses.

Two hardware empirical studies are conducted in [37] using Berkeley Mica motes running TinyOS [47]. The goal is to test the reliability of the wireless communication

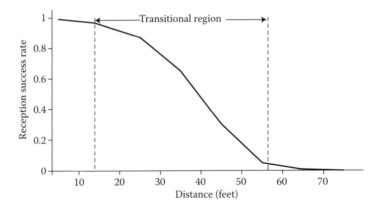

FIGURE 8.4 Reception probability of all links in a network with a line topology [37].

link. For the first experiment, a group of sensors are placed linearly with a 2 feet spacing between each pair. One node is chosen as a transmitter sending periodic packets at a given power level. The rest of the nodes are acting as receivers counting the number of received packets. Figure 8.4 shows the results obtained. As expected, there is a region within which all nodes have good reception. The size of this region depends on the transmission power. There is also a distance beyond which all nodes have a poor connectivity. Between the two points there is a transitional region within which the overall connectivity drops smoothly. The graph also highlights the fact that this transitional region is dominant compared to the other good connectivity and poor connectivity region.

To test whether the link quality is stable over time, the second experiment is conducted in an indoor environment using a pair of nodes. One node is configured as a data source, sending eight packets per second and the other node configured as a receiver. The test is carried out for a period of 20 min at a distance of 15 feet. The nodes are then brought to a distance of 8 feet and the experiment is continued for a total time of 4 h. The results obtained in the experiment are summarized here in Table 8.4. At each distance the mean link quality is relatively stable, but there is significant variation in the instantaneous link

TABLE 8.4 Reception Probability Variation Over Time Across a Single Hop

Time (Minutes)	Reception Probability		
	Max	Min	Mean
0–50	0.80	0.55	0.675
50–100	0.83	0.54	0.685
100–150	0.80	0.55	0.675
150–200	0.75	0.50	0.625
200–250	0.80	0.52	0.660
250–300	0.75	0.45	0.600

quality. The authors in [37] also state the observation that some distant pairs have better reception than relatively closer ones. The significance of these experiments is that they highlight the challenges that wireless communication introduces and its effect on the overall network reliability.

8.2.7 Cross-Layer Reliability

In [48], the authors point out the fact that routing techniques so far did not differentiate between data with high reliability requirements and data with low reliability require- ments. Thus, the network will undergo the same overhead and cost regardless of the importance of the data. To improve data-delivery reliability for critical data, the pro- posed approach is to create multiple routes between source and destination. The algo- rithm begins by assigning different reliability levels based on the message type. The number of routes through which the message is sent is a function of the message reli- ability requirement level. To improve the data-delivery reliability the authors present the ETX Metric algorithm [48]. This algorithm attempts to find paths with the smallest number of expected transmissions (including retransmissions) required to deliver a packet. It predicts the number of retransmissions required using per link measurements of packet loss ratio in both directions of each wirelesslink. This packet loss ratio is cal- culated using the number of probes received in w seconds over the link and the actual number of probes that should have been received. The protocol has these characteristics: it is based on delivery ratio, which affects throughput; can use precise link loss ratio to make fine-grained decisions between routes; can penalize routes with more hops, which tend to have lower throughput due to interference.

From the algorithm description, the data-delivery reliability is translated to a throughput and latency problem. The AODV [15], DSR [16], or other ad hoc protocols are given as possible routing strategies to work on top of the ETX algorithm. Drawback: this solution may be fit for source-to-destination communication style common to *ad hoc*-type networks. This may not be the case for the typically hierarchal WSNs.

8.2.7.1 Cross-Layer Reliable Protocol Using Embedded Message Reliability Flag

To improve network lifetime, nodes can buffer messages until their buffers are filled and then send all messages to the upper level nodes. The impact of this in-network buffering on reliability is studied in [49]. A one-bit reliability flag in the message is proposed. If the flag is not set, then the network will do its best effort to deliver the message while con- serving energy. If the flag is set, the network will transform itself into a reliable data- delivery system using MAC layer ACKs.

8.2.8 Multi-Constraints Reliable Protocols

The ESRT protocol discussed earlier is introduced in [24]. This protocol was extended in [50] to address the pressing WSNs energy constraint. It optimizes the network's energy performance by requiring only a subset of the node within the event detection range to participate in communicating the event to the base station. The subset is chosen such that the smallest number of nodes that are needed to maintain a satisfactory level of reliability will be included.

A reliability model that simultaneously addresses coverage and connectivity is presented in [51]. The authors define K-Reliability as the weighted average value of normalized lifetime and normalized invulnerability that is restricted by K-Coverage ratio. Since coverage and connectivity are dependent on node density, an analysis into the effect of the network node density is given. The K-reliability is defined as given in

$$K\text{-Reliability} = w * R_{\mathrm{L}} + (1 - w) * R_{\mathrm{I}}, \tag{8.13}$$

where R_{L} is the normalized lifetime, and R_{I} is the normalized invulnerability.

A multiconstrained QoS multipath distributed routing algorithm for WSNs is introduced in [52]. Reliability is defined here as the ratio of the number of unique packets successfully received by the sink to the total number of packets sent by the different node sources. The algorithm selects the multihop routing subset that satisfies the required QoS. The QoS measurement parameters are defined as being reliability, energy-efficiency, and delay-bounded message delivery. The protocol in [38] that was presented earlier in Section 3 is also considered multi-constraint since it provides a networking solution that is reliable, and simultaneously optimizes the network energy performance.

8.3 Wireless Sensor Networks Reliability Techniques, Challenges, and Open Issues

In this section, we summarize WSNs reliability techniques, challenges, and point out some open issues.

8.3.1 Wireless Sensor Networks Reliability Techniques

Table 8.5 is a summary of the WSNs reliable protocols covered in this work. From the surveyed protocols literature, some of the proposed techniques for improving WSNs network reliability can be recognized as follows:

- Link layer retransmissions.
- Increasing the one hop transmission power. This leads to longer transmission range, but very taxing on the power requirements.
- Use of ACKs and NACKs (at the link or transport levels).
- Use of multiple disjoint paths to send the same message and adding redundancy to each packet (erasure codes).

8.3.2 Wireless Sensor Networks Reliability Challenges

Some of the surveyed work point out certain reliability challenges specific to WSNs. These are summarized below as follows:

- Common cause failure (CCF). This is defined as node failure that can be due to special type of events that have a low probability of occurrence, but will disable a

TABLE 8.5   Classification of Reliable Protocols for Pervasive Wireless Sensor Networks

Protocol	OSI Layer	Failure Cause	Technique(s)	Cross-Layer?	Multi-Constraint?	Remarks
ESRT [24]	Transport	Network congestion	Congestion control	No	No	Event, rather than message reliability
RMST [26]	Transport	Wireless medium	Multiple routes	No	No	Limited scalability due to multihop routing
Reliable Routing using State Enumeration [28]	Network	Node hardware failure	Minimum path enumeration	No	No	Reliability is defined as the probability that the network will function
Multiple Routes, Erasure Codes Reliable Protocols [30,32,37]	Network	Wireless medium	Multiple routes	No	No	Sending multiple copies of the same message can be very taxing on energy
Reliable Routing using Link Connectivity Statistics [38]	Network	Wireless medium	Use best route	Yes	No	May place buffering demands on the nodes in dense networks
Reliable Routing using Link Rating Parameter [39]	Network	Wireless medium	Use best route	Yes	Yes	Reliability to energy tradeoff
Reliable Protocols using MAC Retransmissions [39,41]	MAC	Wireless medium	Message retransmission	No	No	MAC retransmission is effective, but may waste valuable energy and bandwidth
Reliable MAC Protocols using Contention Window Size [24,43,44]	MAC	Wireless medium	Contention window resizing	No	No	Requires careful consideration of the interaction between window size and number of retransmission attempts

Reliable Protocol using MAC RTS/CTS [45]	MAC	Wireless medium	RTS/CTS MAC messages	No	No	Energy efficient for bulk data transfers
Reliable Protocol using MAC ACK [44]	MAC	Wireless medium	MAC ACKs	No	No	More effective if combined with dynamic route switching
Cost-based Reliable Routing [46]	Radio	Wireless medium	MAC retry limit, transmit power	Yes	Yes	Reliability is defined in terms of network coverage and connectivity
Reliable Routing using Message Reliability Flag [50]	MAC	Wireless medium	MAC ACKs	No	Yes	Guaranteed message delivery for important data
Extended ESRT [51]	MAC	Collisions	Multiple reports	No	Yes	Reducing the number of event reporting nodes will reduce collisions and thus improves energy efficiency
Multi-constraint Reliable Model [52]	Network	Sensing range	Topology control algorithm	No	Yes	A reliability model based on coverage and connectivity constraints
Multi-constraint QoS Multipath Routing[53]	Network	Wireless medium	Multiple routes	No	Yes	Reliability is defined as the ratio of the number of packets received to total number of packets sent

large number of nodes when they do occur [53]. Shrestha et al. [54] argue and discuss the need for incorporating CCF in the reliability calculation of WSNs since communication failure can be due to link failure or node failure, which in turn can be due to CCF. Examples of CCF failure are flooding, fire, and mud slides.

- Silent failure of nodes due to energy depletion. This is another example of reliability challenges specific to WSNs. Marcello et al. [7] point out and analyze the challenges introduced by this type of failure.
- Event detection reliability in unattended WSNs. Detection failure can be attributed to hardware malfunction, software bugs, or due to environmental noise. Depending on the failure cause, the failure can last for a short period, or be present for a considerable length of time. Analyses and mathematical modeling of detection reliability are presented in [55].
- Data quality deterioration due to aggregation. This is a reliability challenge in multihop and hierarchical WSNs [56].
- Coverage reliability due to stochastic node placement. Ishizuka et al. [57] discuss this type of coverage reliability, along with analysis and mathematical modeling.

8.3.3 Wireless Sensor Networks Reliability Open Issues

As summarized in Willig and Karl [9], wireless senor networks data-delivery reliability has room for research in the following areas:

- Design and evaluation of new mechanisms for improving reliability taking into consideration, the complex behavior of the wireless channel and the interaction between the different networking stack layers.
- Ways to adaptively control the mixture of mechanisms used in the network according to the current data-delivery reliability measurements and the target reliability.
- Consideration of timing aspects and the effect of reliability demands on real-time requirements and on the network's ability to meet deadlines.

8.4 Conclusions

In this chapter, we surveyed and classified several reliable protocols for WSNs. The WSNs reliability research is still in its early stages and found to be very diverse. There is also no unified definition for Wireless Sensor's reliability and each work has defined reliability differently and in line with their approach. The TCP/IP network stack model is instrumental in classifying WSNs energy conservation research and protocols. Following similar approach, our effort here is to classify the reliability research following the TCP/IP stack model. Arguably, the main design constraint confronted by WSNs protocols and solutions is the limited energy resources. The reliable protocols surveyed in this chapter indirectly highlighted the fact that past research in reliability for WSNs is in isolation from energy conservation research. In order for this field to gain wide acceptance, specially in industry, WSNs research needs to move beyond studies that are concerned only with energy conservation and resource constraints. Future research and studies in WSNs need to start considering reliability as a design constraint and as a performance parameter in their proposed protocols and solutions.

References

1. Boulis, A. *Models for Programmability in Sensor Networks*, chapter 7, CRC Press, Boca Raton, FL, 2004.
2. Mahgoub, I. and Ilyas, M. *Sensor Network Protocols*. CRC Press, Boca Raton, FL, 2005.
3. Mahgoub, I. and Ilyas, M. SMART DUST: Sensor Network Applications, Ar-chitecture, and Design, CRC Press, Boca Raton, FL, 2006.
4. Su, W., Cayirci, E., and Akan, O. B. *Overview of Communication Protocols for Sensor Networks*, chapter 4. CRC Press LLC, Boca Raton, FL, 2006.
5. Ilyas, M. and Mahgoub, I. *Handbook of Sensor Networks: Compact Wireless and Wired Sensing Systems*, CRC Press, Boca Raton, FL, 2005.
6. Shah, R. C., Petrovic, D., and Rabaey, J. M. *Energy-Aware Routing and Data Funneling in Sensor Networks*, chapter 9, Taylor and Francis Group, CRC Press LLC, Boca Raton, FL, 2006.
7. Cinque, M., Cotroneo, D., De Caro, G., and Pelella, M. Reliability requirements of wireless sensor networks for dynamic structural monitoring, in *Proceedings of The First Workshop on Applied Software Reliability*, Philadelphia, PA, USA,2006.
8. Ahmed, B., and Imad, M. *Wireless Sensor Networks (WSNs): Optimization of the OSI Network Stack for Energy Eciency*, vol. 3, pp. 1523–1533. London, Taylor and Francis Group, Auerbach Publishing Inc, 2008.
9. Willig, A. and Karl, H. Data transport reliability in wireless sensor networks—a survey of issues and solutions. *Praxis der Informationsverarbeitung und Kommunikation*, 2005; 28(2): 86–92.
10. Day, J. D. and Zimmermann, H. The osi reference model, in *Proc IEEE* 1983; 71: 1334–1340.
11. Neumann, J. The reality of osi, in *Proceedings of the 13th Conference on Local Computer Networks*, pp. 157–161, Minneapolis, MN, USA, October 1988.
12. IEEE 802.11 standard documentations. http://www.ieee802.org/11/.
13. IEEE 802.15.4 standard documentations. http://www.ieee802.org/15/.
14. Ye, W., Heidemann and J., Estrin, D. An energy-efficient mac protocol for wireless sensor networks. *IEEE INFOCOM* 2002; 3: 1567–1576.
15. Perkins, C. E. and Royer, E. M. *Ad hoc* on-demand distance vector routing, in *IEEE 2nd Workshop on Mobile Computer Systems and Applications* (WMCSA 99), pp. 90–100, 1999.
16. Johnson, D. B. Routing in *ad hoc* networks of mobile hosts, in *Workshop on Mobile Computing Systems and Applications*, pp. 158–163, 1994.
17. Haas, Z. J. A new routing protocol for the reconfigurable wireless networks, in *6th IEEE International Conference on Universal Personal Communications (ICUPC97)*, vol. 2, pp. 562–566, 1994.

18. Heinzelman, W. R. Chandrakasan, A., and Balakrishnan, H. Energy-effcient communication protocol for wireless microsensor networks. *Proceedings of the 33rd Hawaii International Conference on System Sciences*, vol. 2, p. 10, January 2000.

19. Manjeshwar, A. and Agrawal, D. Teen: A routing protocol for enhanced efficiency in wireless sensor networks, in *Proceedings of the 15th International Parallel and Distributed Processing Symposium*, p. 2009–2015, April 2001.

20. Manjeshwar, A. and Agrawal, D. Apteen: A hybrid protocol for efficient routing and comprehensive information retrieval in wireless sensor networks, *International Parallel and Distributed Processing Symposium IPDPS*, 2002.

21. Lindsey, S. and Raghavendra, C. Pegasis: Power-efficient gathering in sensor information systems. *IEEE Aerospace Conf Proc* 2002; 3: 1125–1130.

22. Intanagonwiwat, C., Govindan, R., Estrin, D., Heidemann, F., and Silva, J. Directed diffusion for wireless sensor networking, in *IEEE/ACM Transactions on Networking*, vol. 11, pp. 1–16, Feburay 2003.

23. Shaikh, F., Khelil, A., and Suri, N. On modeling the reliability of data transport in wireless sensor networks, in *IEEE 15th Euromicro International Conference on Parallel, Distributed and Network-Based Processing*, pp. 395–402, 2007.

24. Sankarasubramaniam, Y., Akan, O., and Akyildiz, I. Esrt: Event-to-sink reliable transport in wireless sensor networks. *IEEE/ACM Trans Network* (TON), 2005; 13(5): 1003–1016.

25. Stann, F. and Hheidemann, J. RMST: Reliable data transport in sensor networks, in *IEEE International Conference on Sensor Net Protocols and Applications (SNPA)*, pp. 102–112, 2003.

26. Zhang, H., Arora, A., Choi, Y., and Gouda, M. G. Reliable bursty convergecast in wireless sensor networks, in *Proceedings of the 6th ACM International Symposium on Mobile Ad Hoc Networking and Computing (MobiHoc)*, pp. 266–276, 2005.

27. Lucet, C. and Manouvrier, J. F. Exact methods to compute network reliability, in *Proceedings of the of 1st International Conference on Mathematics Methods in Reliability*, September 1997.

28. AboElFotoh, H., Iyengar, S., and Chakarbarty, K. Computing reliability and message delay for cooperative wireless distributed sensor networks subject to random failures. *IEEE Trans Reliab* 2005; 54(1): 145–155.

29. Kim, S. Fonseca, R., and Culler, D. Reliable transfer on wireless sensor networks, in First *Annual IEEE Communications Society Conference on Sensor and Ad Hoc Communications and Networks*, pp. 449–459, October 2004.

30. Blömer, J., Kalfane, M., Karpinski, M., Karp, R., Luby, M., and Zuckerman, D. An xor-based erasure-resilient coding scheme. Technical Report TR-95-048, International Computer Science Institute, August 1995.

31. Byers, J. W., Luby, M., Mitzenmacher, M., and Rege, A. A digital fountain approach to reliable distribution of bulk data. *ACM SIGCOMM Comput Commun Rev* 1998; 28(4): 56–67.

32. Luby, M. Lt codes, in *IEEE 43rd Symposium on Foundations of Computer Science (FOCS)*, 2002.

33. Maymounkov, P. Online codes, Technical Report TR2002-833, New York University, 2002.

34. Rizzo, L. Effective erasure codes for reliable computer communication protocols. *ACM SIGCOMM Comput Commun Review* 1997; 27(2): 24–36.
35. Kumar, R., Paul, A., Ramachandran U., and Kotz D. On improving wireless broadcast reliability of sensor networks using erasure codes, in *2nd International Conference on Mobile Ad hoc and Sensor Networks (MSN 2006)*, pp. 155–170, 2006.
36. Dulman, S., Nieberg T., Wu, J., and Havinga, P. Trade-off between traffic overhead and reliability in multipath routing for wireless sensor networks, in *Proceedings of the IEEE Wireless Communications and Networking (WCNC 2003)*, vol. 3, pp. 1918–1922, IEEE, Silver Spring, MD, 2003.
37. Woo, A., Tong, T., and Culler, D. Taming the underlying challenges of reliable multihop routing in sensor networks, in *Procedings of the ACM Sensys*, 2003, pp. 14–27, 2003.
38. Ahmed, B., Imad, M., and Mohammad, I. Using individualized link power settings for energy optimization in hierarchical wireless sensor networks (WSNS), in *IEEE 6th International Symposium on High Capacity Optical Networks and Enabling Technologies HONET* 2009, pp. 172–178, December 2009.
39. Rachit, A., Emanuel, P., and Brendan, F. Adaptive wireless sensor networks: A system design approach to adaptive reliability, in *IEEE 2nd International Conference on Wireless Communication and Sensor Networks (WCSN 2006)*, 2006.
40. Vuran, M., Gungor, V., and Akan, O. On the interdependence of congestion and contention in wireless sensor networks, in *ACM 3rd International Workshop on Measurement, Modeling, and Performance Analysis of Wireless Sensor Networks (SenMetrics 2005)*, pp. 897–909, 2005.
41. Gnawali, O., Yarvis, M., Heidemann, J., and Govindan R. Interaction of retransmission, blacklisting, and routing metrics for reliability in sensor network routing, in *First Annual IEEE Communications Society Conference on Sensor and Ad Hoc Communications and Networks (IEEE SECON 2004)*, pp. 34–43, IEEE, Silver Spring, MD, 2004.
42. Ahmed, B., Imad, M., and Mohammad, I. Mac layer dynamic backoff scheme for message delivery reliability in wireless sensor networks, in *IEEE 5th International Symposium on High Capacity Optical Networks and Enabling Technologies HONET 2008*, pp. 86–90, November 2008.
43. Baydere, S., Safkan, Y., and Durmaz, O. Lifetime analysis of reliable wireless sensor networks. *IEICE Trans Commun* 2005; E88-B(6): 2465–2472.
44. Volgyesi, P. Nadas, A., Ledeczi, A., and Molnar, K. Reliable multihop bulk transfer service for wireless sensor networks, in *Proceedings of the IEEE 13th Annual International Symposium and Workshop on Engineering of Computer Based Systems (ECBS 2006)*, pp. 112–122, IEEE, Silver Spring, MD, 2006.
45. Kwon, H., Kim, T. H., Choi, S., and Lee, B. G. A cross layer strategy for energy-efficient reliable delivery in wireless sensor networks, *IEEE Trans Wireless Commun* 2006; 5(12): 3689–3699, December 2006.
46. Shakkottai, S., Srikant, R., and Shro, N. Unreliable sensor grid: Coverage, connectivity and diameter, in *IEEE Societies Twenty-Second Annual Joint Conference of the IEEE Computer and Communications (INFOCOM 2003)*, vol. 2, pp. 1073–1083, 2003.

47. Levis, P. Tinyos online documentation. http://www.tinyos.net.
48. Joseph, L. and Uma, G. Reliability based routing in wireless sensor networks. *Int J Comput Sci Network Security* 2006; 6(12): 331–338.
49. Durmaz, O. and Baydere, S. Impact of in-network buffering on the reliability of sensor networks, in *Proceedings of Second International Workshop on Sensor and Actor Network Protocols and Applications (SANPA 2004)*, 2004.
50. Othman, F., Bouabdallah, N. and Boutaba, R. *Energy Conservation in Reliable Wireless Sensor Networks.* pp. 2404–2408, May 2008.
51. Cai, W., Jin, X., Zhang, Y., Chen, K., and Tang J. *Research on Reliability Model of Large-Scale Wireless Sensor Networks*, pp. 1–4, 2006.
52. Huang, X. and Fang, Y. Multiconstrained QoS multipath routing in wireless sensor networks. *Wireless Networks* 2008; 14: 465–478. 10.1007/s11276-006-0731-9.
53. Page, L. and Perry, J. E. A model for system reliability with common-cause failures. *IEEE Trans Reliab* 1989; 38(4): 406–410.
54. Shrestha, A., Xing, L., and Liu, H. Infrastructure communication reliability of wireless sensor networks, in *Proceedings of the 2nd IEEE International symposium on dependable, automatic and secure computing* (DASC06), pp. 250–257, October 2006.
55. Hsu, M. T., Lin, F., Chang, Y. S., and Juang, T-Y. The reliability of detection in wireless sensor networks: Modeling and analyzing, in Kuo, T. W., Sha, E., Guo, M., Yang, L., Shao, Z. editors, *Embedded and Ubiquitous Computing, volume 4808 of Lecture Notes in Computer Science*, pp. 432–443. Springer, Berlin, 2007.
56. Benson, J. P., O'Donovan, T., and Sreenan, J. *Reliability Control for Aggregation in Wireless Sensor Networks*, pp. 833–840, 2007.
57. Ishizuka, M. and Aida, M. The reliability performance of wireless sensor networks configured by power-law and other forms of stochastic node placement. *IEICE Trans. Commun.* 2004; E87-B(9): 2511–2520.

9

Positioning and Location Tracking in Wireless Sensor Networks

Yu-Chee Tseng
National Chiao-Tung University

Chi-Fu Huang
National Chiao-Tung University

Sheng-Po Kuo
National Chiao-Tung University

9.1 Introduction

Locations of devices or objects are important information in many applications. This is particularly true for wireless sensor networks, which usually need to determine devices' context. For outdoor environments, the most well-known positioning system is the global positioning system (GPS) [1]. This positioning system uses 24 satellites set up by the U.S. Department of Defense to enable global three-dimensional positioning services; it has two levels of accuracy: stand positioning service and precise positioning service. The accuracy provided by GPS is around 20–50 m.

In addition to the GPS system, positioning can also be done using some wireless networking infrastructures. Taking the PCS cellular networks as an example, the E911 emergency service requires determining the location of a phone call via the base stations of the cellular system. Several location estimation models, such as angle of arrival (AoA);

time of arrival (ToA); received signal strength (RSS); phase of arrival (PoA); and assisted GPS (A-GPS), are widely used in cellular networks and wireless sensor networks.

Much work has been dedicated recently to positioning and location tracking in the area of wireless ad hoc and sensor networks. The purpose of this chapter is to review the recent progress in this direction. GPS is not suitable for wireless sensor networks for several reasons:

- It is not available in an indoor environment because satellite signals cannot penetrate buildings.
- For more fine-grained applications, higher accuracy is usually necessary in the positioning result.
- Sensor networks have their own battery constraint, which requires special design.

Location information can be used to improve the performance of wireless networks and provide new types of services. For example, it can facilitate routing in a wireless ad hoc network to reduce routing overhead. This is known as geographic routing [2,3]. Through location-aware network protocols, the number of control packets can be reduced. Service providers can also use location information to provide some novel location-aware or follow-me services. The navigation system based on GPS is an example. An user can tell the system for his destination and the system will guide him there. Phone systems in an enterprise can exploit locations of people to provide follow-me services. Other types of location-based services include *geocast* [4,5], by which an user can request to send a message to a specific area, and *temporal geocast*, by which an user can request to send a message to a specific area at a specific time. In contrast to traditional multicast, such messages are not targeted at a fixed group of members, but rather at members located in a specific physical area.

In this chapter, Section 9.2 introduces some fundamental distance estimation models; Section 9.3 discusses some positioning and location tracking algorithms. In Section 9.4, some experimental systems are reviewed and Section 9.5 gives a summary.

9.2 Fundamentals

To position an object or a device, the basic step is to use a reference point to determine the distance and angle between the device and the reference point. This has been exploited in the radar systems widely used in military applications. This section describes several such basic approaches. The next subsection discusses how to use multiple reference points jointly to estimate the location of a device.

9.2.1 ToA, TDoA, and AoA

In the ToA approach, the signal traveling time is used to estimate the distance between a device and the reference point. Such systems typically use signals that move at a slower speed, such as ultrasound, to measure the time of signal arrival. Figure 9.1a illustrates

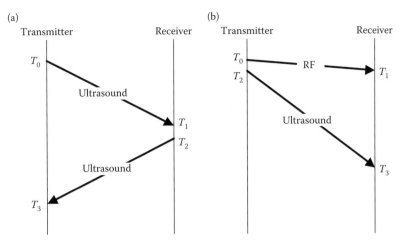

FIGURE 9.1   (a) ToA measurement; (b) TDoA measurement.

this idea. An ultrasound signal is sent from the transmitter to the receiver; in turn, the receiver sends a signal back to the transmitter. After this two-way handshake, the transmitter can infer the distance from the round-trip delay of the signals:

$$\frac{((T_3 - T_0) - (T_2 - T_1)) \times V}{2},$$

where V is the velocity of the ultrasound signals. The error of such measurement may come from the processing time of signals (such as computing latency and the unknown delay $T_2 - T_1$ at the receiver's side).

Another distance estimation technique is the time difference of arrival (TDoA). Although similar to the ToA scheme, this method uses two signals that travel at different speeds, such as the radio frequency (RF) and ultrasound. Figure 9.1b shows how TDoA works; transmission in one direction is sufficient. At T_0, the transmitter sends an RF signal, followed by an ultrasound signal at time T_2. The receiver can then determine its distance to the transmitter by

$$((T_3 - T_1) - (T_2 - T_0)) \times \left(\frac{V_{RF} \times V_{US}}{V_{RF} - V_{US}} \right),$$

where V_{RF} and V_{US} are the traveling speeds of RF and ultrasound signals, respectively. For TDoA, in addition to errors caused by processing time, the receiver must also know the precise value of $(T_2 - T_0)$ to determine the distance.

The AoA approach is another commonly used method for positioning [6,7]. Such approaches require an antenna array or an array of ultrasound receivers, which can determine the angle and orientation of the received signals.

9.2.2 Positioning by Signal Strength

Besides using the signal traveling time, another distance estimation technique is to use the property of signal degradation while traveling in a space to determine the mutual distance. Because signals traveling in a space typically reduce in strength with respect to the distance that they travel, the RSS can be measured at the receiver's side. A mathematical propagation model can be derived to estimate the distance d between a transmitter and a receiver as follows [8]:

$$PL(d) = PL(d_0) + 10n\log\left(\frac{d}{d_0}\right) \propto \left(\frac{d}{d_0}\right)^n,$$

where $PL(\cdot)$ is the path loss function with respect to distance measured in decibels; n is a loss exponent that indicates the rate at which the loss increases with distance; and d_0 is the reference distance determined from a measurement close to the transmitter. The path loss exponent n usually ranges from 2 to 4.

Using path loss may incur significant errors. For example, Figure 9.2 shows an experimental result based on IEEE 802.11b. As can be seen, a trend for the relation between distance and signal strength does exist; however, the curve is unstable in small ranges. The true signal strength model is complex and many uncontrollable environmental factors (such as shadows and terrain) are present.

To solve the preceding problem, it is necessary to model the error for signal attenuation. One possibility is to include a random variable in the preceding path loss function as follows:

$$PL(d) = PL(d_0) + 10n\log\left(\frac{d}{d_0}\right) + X_\rho,$$

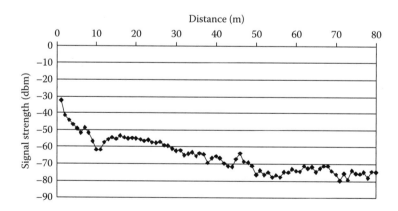

FIGURE 9.2   Signal strength versus distance in IEEE 802.11b.

where $X\rho$ is a zero-mean Gaussian random variable with a standard deviation ρ. Due to the existence of such errors, errors will occur as well when positioning a device based on the signal strength. Assuming the similar error model in measuring distances, Slijepcevic et al. [9] further analyzed the location errors in a wireless sensor network and proved that the distribution of location error can be approximated by a family of Weibull distributions.

9.3 Positioning and Location Tracking Algorithms

The previous section discussed how to estimate the distance between two devices. If an object knows its distances to multiple devices at known locations, one may estimate its location. Several such methods are discussed here.

9.3.1 Trilateration

Trilateration is a well-known technique in which the positioning system has a number of *beacons* at known locations. These beacons can transmit signals so that other devices can determine their distances to these beacons based on received signals. If a device can hear at least three beacons, its location can be estimated. Figure 9.3a shows how trilateration works; A, B, and C are beacons with known locations. From A's signal, one can determine that the object should be located at the circle centered at A. Similarly, from B's and C's signals, it can be determined that the object should be located at the circles centered at B and C, respectively. Thus, the intersection of the three circles is the estimated location of the device.

The preceding discussion has assumed an ideal situation; however, as mentioned earlier, the distance estimation always contains errors that will, in turn, lead to location errors. Figure 9.3b illustrates an example in practice. The three circles do not intersect in a common point. In this case, the maximum likelihood method may be used to estimate the

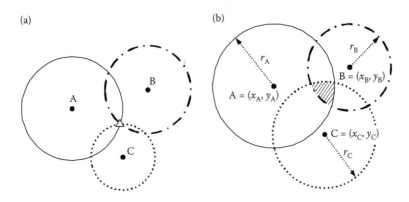

FIGURE 9.3 Trilateration method: (a) ideal situation and (b) real situation with errors.

device's location. Let the three beacons A, B, and C be located at (x_A, y_A), (x_B, y_B), and (x_C, y_C), respectively. For any point (x, y) on the plane, a difference function is computed.

$$\sigma_{x,y} = |\sqrt{(x-x_A)^2 + (y-y_A)^2} - r_A|$$
$$+ |\sqrt{(x-x_B)^2 + (y-y_B)^2} - r_B|$$
$$+ |\sqrt{(x-x_C)^2 + (y-y_C)^2} - r_C|$$

where r_A, r_B, and r_C are the estimated distances to A, B, and C, respectively. The location of the object can then be predicted as the point (x, y) among all points such that $\sigma_{x,y}$ is minimized.

In addition to using the ToA approach for positioning, the AoA approach can be used. For example, in Figure 9.4, the unknown node D measures the angle of ADB, BDC, and ADC by the received signals from beacons A, B, and C. From this information, D's location can be derived [6].

9.3.2 Multilateration

The trilateration method has its limitation in that at least three beacons are needed to determine a device's location. In a sensor network, in which nodes are randomly deployed, this may not be true. Several multilateration methods are proposed to relieve this limitation.

The *ad hoc* localization system (AHLoS) [10] is a distributed system for location discovery. In the network, some beacons have known locations and some devices have unknown locations. The AHLoS enables nodes to discover their locations by using a set of distributed iterative algorithms. The basic one is *atomic multilateration*, which can estimate the location of a device of unknown location if at least three beacons are within its sensing range. Figure 9.5 shows an example in which, initially, beacon nodes contain only nodes marked as having a GPS. Device nodes 1, 2, 3, and 4 are at unknown locations. In the first iteration, as Figure 9.5a shows, the locations of nodes 1, 2, and 3 will be determined.

The atomic multilateration is further extended to an *iterative multilateration* method. Specifically, once the location of a device is estimated, its role is changed to a beacon

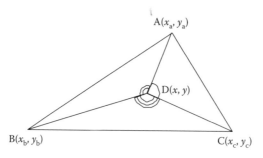

FIGURE 9.4 Angle measurement from three beacons A, B, and C.

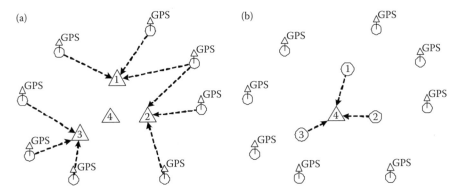

FIGURE 9.5   (a) Atomic multilateration and (b) iterative multilateration in AHLoS.

node so as to help determine other devices' locations. This is repeated until all hosts' locations are determined (if possible). As Figure 9.5b shows, in the second iteration, the location of node 4 can be determined with the help of nodes 1, 2, and 3, which are now serving as beacons.

The iterative multilateration still has its limitation. For example, as Figure 9.6 shows, it is impossible to determine the locations of nodes 2 and 4 even if the locations of nodes 1, 3, 5, and 6 are known. The *collaborative multilateration* method may relieve this problem, because it allows one to predict multiple potential locations of a node if it can hear fewer than three beacons. For example, in Figure 9.6, from beacon nodes 1 and 3, two potential locations of node 2 may be guessed (the other potential location is marked by 2'). Similarly, from beacon nodes 5 and 6, one may guess two potential locations of node 4 (the other potential location is marked by 4'). Collaborative multilateration allows the estimation of the distance between nodes 2 and 4. With this information, the locations of nodes 2 and 4 can be estimated, as the figure shows.

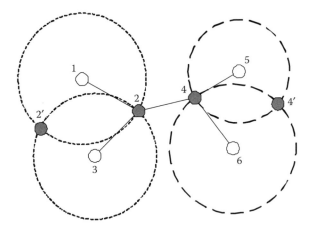

FIGURE 9.6   Collaborative multilateration in AHLoS.

9.3.3 Pattern Matching

Another type of location discovery is by pattern matching. Instead of estimating the distance between a beacon and a device, this approach tries to compare the received signal pattern against the training patterns in the database. Thus, this method is also known as the *fingerprinting* approach. The basic idea is that the RSS at a fixed location is not necessarily a constant. It typically moves up and down, so it would be better to model signal strength by a random variable. This is especially true for indoor environments.

The main idea is to compare the received signals against those in the database and determine the likelihood that the device is currently located in a position. A typical solution has two phases (refer to Figure 9.7):

- *Off-Line Phase.* The purpose of this phase is to collect signals from all base stations at each training location. The number of training locations is decided first. Then, the RSSs are recorded (for a base station that is too far away, the signal strength is indicated as zero). Each entry in the database has the format: $(x, y, \cdot ss_1, ss_2, \ldots, ss_n\grave{O})$, where (x, y) is the coordinate of the training location, and $ss_i, i = 1 \ldots n$, is the RRS at the training location from the ith base station. These entries are stored in the database. Note that, for higher accuracy, one may establish multiple entries in the database for the same training location. From the database, some positioning rules, which form the positioning model, will then be established.
- *Real-Time Phase.* With a well-trained positioning model, one can estimate a device's location given the signal strengths collected by the device from all possible base stations. The positioning model may determine a number of locations, each associated with a probability. However, the typical solution is to output only the location with the highest likelihood.

There are several similarity searching methods in the matching process; two approaches are introduced next.

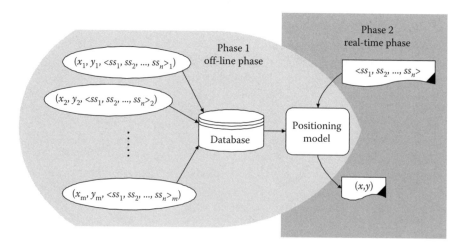

FIGURE 9.7   Pattern matching approach.

9.3.3.1 Nearest Neighbor Algorithms

The simplest approach is the *nearest neighbor in signal space* (NNSS) approach [11,12]. In the first phase, only the average signal strength of each base station at each training location is recorded. Then, in the second phase, the NNSS algorithm computes the *Euclidean distance* in the signal space between the received signal and each record in the database. Euclidean distance means the square root of the summation of square of the difference between each RSS and the corresponding average signal strength from the access point under consideration. The training location with the minimum Euclidean distance is then chosen as the estimated location of the device. Because this algorithm only picks existing locations in the database, to improve its accuracy, it is suggested that the training set be dense enough.

One variant of the basic NNSS algorithm is *NNSS-AVG*. To take the uncertainty of a device's location into consideration, this method tries to pick a small number of training locations that closely match the RSSs (such as those with smaller Euclidean distances). Then, it infers the location of the device to be a function of the coordinates of the selected training locations. For example, one may take the average of the x and y coordinates of the selected training locations as the estimated result.

9.3.3.2 Probability-Based Algorithms

The probability-based positioning approach regards signal strength as a probability distribution [13]. In NNSS, because the RSSs are averaged out, the probability distribution would disappear. So, the probability-based approach will try to maintain more complete information of signal strength distribution. The prediction result is typically more accurate.

The core of the probability-based model is the Bayes rule:

$$p(l|o) = \frac{p(o|l)\,p(l)}{p(o)} = \frac{p(o|l)\,p(l)}{\sum_{l' \in L} p(o|l')\,p(l')},$$

where $p(l|o)$ is the probability that the device is at location l given an observed signal strength pattern o. The prior probability that a device is resident at l is $p(l)$, which may be inferred from history or experience. For example, people may have a higher probability to appear in a hallway or lobby. If this is not available, $p(l)$ may be assumed to be a uniform distribution. L is the set of all training locations. The denominator $p(o)$ does not depend on the location variable l, so it can be treated as a normalized constant whenever only relative probabilities are required.

The term $p(o|l)$ is called the likelihood function; this represents the core of the positioning model and can be computed in the off-line phase. There are two ways to implement the likelihood function [13]:

- *Kernel Method.* For each observation o_i in the training data, it is assumed that the signal strength exhibits a Gaussian distribution with mean o_i and standard deviation σ, where σ is an adjustable parameter in the model. Specifically, given o_i, the probability to observe o is

$$K(o;o_i) = \frac{1}{\sqrt{2\pi}\,\sigma} \exp\left(\frac{(o-o_i)^2}{2\sigma^2} \right).$$

- Based on the kernel function, the probability $p(o\Omega l)$ can be defined as

$$p(o|l) = \frac{1}{n_1} \sum_{l_i \in L, l_i = l} K(o;o_i),$$

where n_1 is the number of training vectors in L obtained at location l. Intuitively, the probability function is a mixture of n_1 equally weighted density functions. Also note that the preceding formulas are derived, assuming that only one base station exists. With multiple base stations, the probability function will be multivariated, and the probability will become the multiplication of multiple independent probabilities, each for one base station.

- *Histogram Method.* Another method to estimate the density functions is to use histogram, which is related to *discretization* of continuous values to discrete ones. A number of *bins* can be defined as a set of nonoverlapping intervals that cover the whole random variables. The advantage of this method is in its ease in implementation and low computational cost. Another reason is that its discrete property can smooth out the instability of signal strengths.

The probability-based methods can adapt to different environments. To further reduce the computational overhead, Youssef et al. [14] proposed a method by clustering training data in the database.

9.3.4 Location Tracking

Location tracking means that a device's location can be derived based on some history traces. Because the trace of a device may indicate where it may move in the next step, this information can be used to improve the accuracy of positioning results. For example, one possibility is to consider the relative distances between consecutive moves of a device in a short period of time. These distances are typically not long. Using this information can reduce errors in tracking results.

In Bahl et al. [11], a Viterbi-like tracking algorithm is proposed for location tracking. The Viterbe algorithm is typically used in communications theory for recognizing the most likely message that is transmitted over a noisy channel. In location tracking, because various environmental factors may interfere with signals, the Viterbi algorithm is also suitable for selecting the most likely location of a device. The idea behind the Viterbi-like tracking algorithm is to take the continuity of a user's track in the past into consideration so as to come up with a better guess of the user's current location.

Figure 9.8 shows the details of the Viterbi-like tracking algorithm. Each time the mobile device receives signals from the access points, it computes a set of k most likely locations. This may be obtained from the NNSS-AVG algorithm described earlier. After receiving continuous h samples, the Viterbi-like algorithm can generate an $h * k$ map, which is an h-stage graph in which each stage contains k possible locations of the device at that stage. The possible locations are modeled by vertices. Edges are established between continuous stages and a weight is assigned to each edge equal to the Euclidean distance of the two incident vertices.

Under the assumption that a user may not move too far away from his current location in a short period of time, the Viterbi-like tracking algorithm computes a shortest path in the $k * h$ map. This shortest path can be viewed as the most likely trajectory of the mobile user. Then, the user's current location can be guessed to be the head of this shortest path. Note that, for this reason, the Viterbi-like algorithm may have $h - 1$ periods of delay.

The variances of environments may also complicate the problem. The radio channel condition in working hours may significantly differ from that during off hours. The positioning model may need to adapt to such factors. Recalibration is sometimes inevitable, but laborious. An environmental profile may need to be established to conquer this problem.

9.3.5 Network-Based Tracking

Special concerns—power saving, bandwidth conservation, and fault tolerance—arise when a solution is designed for a wireless sensor network. At the network level, location tracking may be done via the cooperation of sensors. Tseng et al. [15] addressed these issues using an agent-based paradigm. Once a new object is detected by the network, a mobile agent will be initiated to track the roaming path of the object. The agent is mobile because it will choose the sensor closest to the object to stay. The agent may invite some

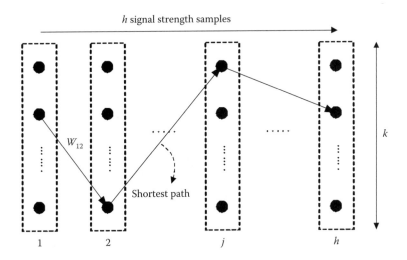

FIGURE 9.8   Viterbi-like location tracking algorithm.

nearby slave sensors to cooperatively position the object and inhibit other irrelevant (i.e., farther) sensors from tracking the object. More precisely, only three agents will be used for the tracking purpose at any time and they will move as the object moves. The trilateration method is used for positioning. As a result, the communication and sensing overheads are greatly reduced. Because data transmission may consume a lot of energy, this agent-based approach tries to merge the positioning results locally before sending them to the data center. These authors also address how to conduct data fusion.

Figure 9.9 shows an example. The sensor network is deployed in a regular manner and it is assumed that each sensor's sensing distance equals the distance between two neighboring sensors. Initially, each sensor is in the *idle* state, searching for new objects. Once detecting a target, a sensor will transit to the *election* state, trying to serve as the *master agent*. The nearest sensor will win. The master agent will then dispatch two neighboring sensors as the *slave agents*; master and slave agents will cooperate to position the object. In the figure, the object is first tracked by sensors $\{S_0, S_1, S_2\}$ when resident in A_0, then by $\{S_0, S_1, S_6\}$ when in A_1, by $\{S_0, S_5, S_6\}$ when in A_2, and so on.

The master agent is responsible for collecting all sensing data and performing the trilateration algorithm. It also conducts data fusion by keeping the tracking results while

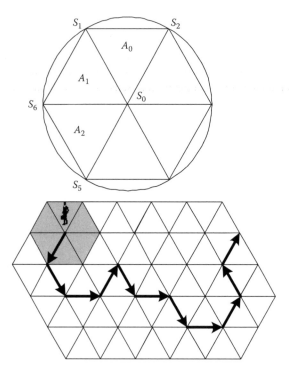

FIGURE 9.9 Roaming path of an object (dashed line) and the migration path of the corresponding master agent (arrow). Sensors that ever host a slave agent are marked by black. (From Tseng et al. *In International Workshop Information Processing Sensor Networks (ISPN)*, vol. 2634, pp. 625–641, 2003. Also to be published in *The Computer Journal*. With permission.)

it moves around. At a proper time, the master agent will forward the tracking result to the data center. Two strategies are proposed for this purpose: *threshold-based* (TB) strategy, which will forward the result when the amount of data reaches a predefined threshold value *T*, and *distance-based* (DB) strategy, which will make a decision based on the routing distance from the agent's current location to the data center and the direction in which the agent is moving.

9.4 Experimental Location Systems

In this section, several location systems are introduced. Although they may not be specially designed for wireless sensor networks, these design concepts and experiences will benefit future implementations of positioning systems in wireless sensor networks.

9.4.1 Active Badge and Bat

The *Active Badge* system [16] is a cell-based location system in which objects are each attached with a badge that periodically emits infrared signals with a unique ID. Infrared receivers mounted at known positions collect these signals and relay them over a wired network. As a result, the system knows in which infrared cell a badge currently stays. The disadvantage of this badge system is that it is hard to deploy in a large-scale environment and that infrared is sensitive to external light, such as sunlight.

A successor of the Active Badge system is the *Bat* system [17], which consists of a collection of wireless transmitters, a matrix of receiver elements, and a central RF base station. The wireless transmitters, called bats, can be carried by a tagged object and/or attached to equipment. The sensor system measures the time of flight of the ultrasonic pulses emitted from a bat to receivers installed in known and fixed positions. It uses the time difference to estimate the position of each bat by trilateration.

The RF base station coordinates the activity of bats by periodically broadcasting messages to them. Upon hearing a message, a bat sends out an ultrasonic pulse. A receiver that receives the initial RF signal from the base station determines the time interval between receipt of the RF signal and receipt of the corresponding ultrasonic signal. It then estimates its distance from the bat. These distances are sent to the computer, which performs data analysis. By collecting enough distance readings, it can determine the location of the bat within 3 cm of error in a three-dimensional space at 95% accuracy. This accuracy is quite enough for most location-aware services; however, the deployment cost is high.

9.4.2 Cricket

Cricket is a system that can provide location-dependent applications [18]. Rather than explicitly tracking users' locations, Cricket helps devices learn their locations and lets them decide whether to advertise them, for preservation of privacy. Cricket does not rely on any centralized management or control and no explicit coordination occurs between beacons. To obtain information about a space, every object is attached to a *listener*, a small device that listens to messages from beacons mounted on ceilings and walls.

Similar to the Bat system, Cricket uses a combination of an RF signal and ultrasound to evaluate the distances between beacons and listeners (i.e., TDoA). A beacon sends the space information over an RF and an ultrasonic pulse at the same time. When the listener hears the RF signal, it uses the first few bits as training information and then turns on its ultrasonic receiver. It then listens for the ultrasonic pulse, which will usually arrive in a short time. The listener uses the time difference between the receipt of the first bit of RF information and the ultrasonic signal to determine its distance from the beacon.

9.4.3 RADAR and Nibble

The *RADAR* location system [12] tries to take advantage of the already existing RF data network formed by IEEE 802.11 access points. IEEE 802.11 networks are now becoming more prevalent in many office and public areas, so no extra hardware cost is incurred. In addition, users can enjoy data communications. RADAR uses the nearest neighbor technology of pattern matching discussed in Section 9.3 to infer objects' locations.

The *Nibble* [19] also adopts the IEEE 802.11 infrastructure for positioning purposes. Nibble uses the probability-based approach in Section 9.3.3.2. It relies on a fusion service to infer the location of an object from measured signal strengths. Data are characterized probabilistically and input into the fusion service. The output of the fusion service is a probability distribution over a random variable that represents some context.

9.4.4 CSIE/NCTU Indoor Tour Guide

The authors have also developed a prototype indoor tour guide system at the Department of Computer Science and Information Engineering, National Chiao Tung University (CSIE/NCTU), Taiwan. The hardware platforms of this project include several Compaq iPAQ PDAs and laptops. Each mobile station is equipped with a Lucent Orinoco Gold wireless card. Signal strengths are used for indoor positioning. The probability-based pattern-matching algorithm in Section 9.3.3.2 is used. Figure 9.10 shows the system architecture. The concept of logical areas is used to identify offices, rooms, lobbies, and so on. The manager is the control center responsible for monitoring each user's movements, configuring the system, and planning logical areas and events. The location server takes care of the location discovery job and the service server is in charge of message delivery. The database can record users' profiles; the gateway can conduct location-based access control to the Internet.

One of the innovations in this project is that an event-driven messaging system has been designed. A short message can be delivered to a user when he enters or leaves a logical area. The event-driven message can also be triggered by a combination of time, location, and property of location (such as who is in the location and when the location is reserved for meetings). A user can set up a message and a corresponding event to trigger the delivery of the message. The manager will check the event list periodically and initiate messages, when necessary, with the service server. Messages can be unicast or broadcast. The expectation is that streaming multimedia can be delivered in the next stage. The system can also be applied to support a smart library. Another innovation is to provide location-based access control. In certain rooms, such as classrooms and meeting

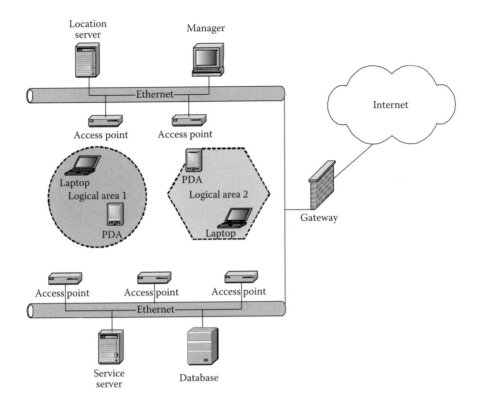

FIGURE 9.10   System architecture of the CSIE/NCTU tour guide system.

rooms, users may be prohibited from accessing certain sensitive web pages. These rules can be organized through the manager and set up at the gateway.

9.5 Conclusions

In this chapter, some fundamental techniques in positioning and location tracking have been discussed and several experimental systems reviewed. Location information may enable new types of services. Accuracy and deployment costs are two factors that may contradict each other, but both are important factors for the success of location-based services.

References

1. Enge, P. and Misra, P. Special issue on GPS: the global positioning system. *Proc. IEEE*, 1999; 87: 3–15.
2. Ko, Y. and Vaidya, N. GeoTORA: a protocol for geocasting in mobile ad hoc networks, in *8th International Conference Network Protocols (ICNP)*, pp. 240–250, 2000.

3. Navas, J. C. and Imielinski, T. GeoCast—geographic addressing and routing, in *ACM/IEEE MOBICOM*, pp. 66–76, 1997.

4. Ko, Y. and Vaidya, N. Geocasting in mobile ad hoc networks: location-based multicast algorithms, in *IEEE Workshop Mobile Computing System Applications (WMCSA)*, pp. 101–110, 1999.

5. Liao, W-H., Tseng, Y-C., Lo, K-L., and Sheu, J-P. GeoGRID: a geocasting protocol for mobile ad hoc networks based on GRID. *J. Internet Technol.*, 2000; 1(2): 23–32.

6. Niculescu, D. and Nath, B. Ad hoc positioning system (APS) using AoA, in *IEEE INFOCOM*, San Francisco, pp. 1734–1743, 2003.

7. Priyantha, N. B., Miu, A. K. L., Balakrishnan, H., and Teller, S. J. The Cricket compass for context-aware mobile applications, in *ACM/IEEE MOBICOM*, pp. 1–14, Rome, Italy, 2001.

8. Rappaport, T. S., *Wireless Communications. Principles and Practice*. IEEE Press, Piscataway, NJ, 1996.

9. Slijepcevic, S., Megerian, S., and Potkonjak, M. Characterization of location error in wireless sensor networks: analysis and applications, in *International Workshop Information Processing Sensor Networks (IPSN)*, vol. 2634, pp. 593–608, 2003.

10. Savvides, A., Han, C. C., and Srivastava, M. B. Dynamic fine-grained localization in ad hoc networks of sensors, in *ACM/IEEE MOBICOM*, Rome, pp. 166–179, 2001.

11. Bahl, P., Balachandran, A., and Padmanabhan, V. Enhancements to the RADAR user location and tracking system. Technical report MSR-TR-00-12, Microsoft Research, 2000.

12. Bahl, P. and Padmanabhan, V. N. RADAR: an in-building RF-based user location and tracking system, in *IEEE INFOCOM*, vol. 2, pp. 775–784, 2000.

13. Roos, T., Myllymaki, P., Tirri, H., Misikangas, P., and Sievanen, J. A probabilistic approach to WLAN user location estimation. *Int J Wireless Inf Netw*, 2002; 9(3): 155–164.

14. Youssef, M., Agrawala, A., and Shankar, U. WLAN location determination via clustering and probability distributions, in *IEEE PerCom*, 2003.

15. Tseng, Y-C., Kuo, S-P., Lee, H-W., and Huang, C-F. Location tracking in a wireless sensor network by mobile agents and its data fusion strategies, in *International Workshop Information Processing Sensor Networks (IPSN)*, vol. 2634, pp. 625–641, 2003.

16. Want, R., Hopper, A., Falcao, V., and Gibbons, J. The active badge location system. Technical report 92.1, Olivetti Research Ltd. (ORL), 1992.

17. Addlesee, M., Durwen, R., Hodges, S., Newman, J., Steggles, P., Ward, A., and Hopper, A. Implementing a sentient computing system. *Computer*, 2001; 34(8): 50–56.

18. Priyantha, N. B., Chakraborty, A., and Balakrishnan, H. The Cricket location-support system, in *ACM/IEEE MOBICOM*, Boston, MA, USA, pp. 32–43, 2000.

19. Castro, P., Chiu, P., Kremenek, T., and Muntz, R. R. A probabilistic room location service for wireless networked environments, in *Ubicomp*, 2001, pp. 18–34.

10

Wireless Location Technology in Location-Based Services

Junhui Zhao
Beijing Jiaotong University

Xuexue Zhang
Beijing Jiaotong University

10.1 Introduction

Over the last decade, wireless communications has expanded significantly, with an annual increase of cellular subscribers averaging about 40% worldwide. Currently, it is estimated that there are between 36 and 46 million cellular users in the United States alone, representing over 20% of the U.S. population. In the next few years, it is expected that a total of about 200 million wireless telephones will be in use worldwide, and that in the next 10 years, the demand for mobility will make wireless technology the main source for voice communication, with a total market penetration of 50–60% [1].

Meanwhile, depending on wireless positioning, geography information systems (GIS), application middleware, application software, and support, the location-based service (LBS) is in use in every aspect of our lives. In particular, the growth of mobile technology makes it possible to estimate the location of the mobile station (MS) in the LBS. In the LBS, we tend to use positioning technology to register the movement of the MS and use the generated data to extract knowledge that can be used to define a new research area that has both technological and theoretical underpinnings.

Nowadays, the subject of wireless positioning in the LBS has drawn considerable attention. While wireless service systems aim to provide support to the tasks and interactions of humans in physical space, accurate location estimation facilitates a variety of applications that include areas of personal safety, industrial monitoring and control, and a myriad of commercial applications, for example, emergency localization, intelligent transport systems, inventory tracking, intruder detection, tracking of fire-fighters and miners, and home automation. Besides applications, various methods are used for obtaining location information from a wireless link. However, although a variety of different methods may be employed for the same type of application, factors including complexity, accuracy, and environment play an important role in determining the type of distance measurement system applied for a particular use [2].

In the wireless systems in the LBS, transmitted signals are used in positioning. By using characteristics of the transmitted signal itself, the location estimation technology can estimate how far one terminal is from another or estimate where that terminal is located. In addition, location information can help optimize resource allocation and improve cooperation between wireless networks [2–4].

The remainder of the chapter is organized as follows. In Section 10.2, estimation of position-related parameters (or data collection) is studied. Section 10.3 introduces cellular network fundamentals. In Section 10.4, the cellular network, including fundamentals, cellular LBSs, and so on, will be discussed. Section 10.5 shows the location precision of the systems. Section 10.6 concludes the chapter.

10.2 Study on the Estimation of Position-Related Parameters (or Data Collection)

Positioning, as well as navigation, has a long history. As long as people move across the Earth's surface, they want to determine their current location. Seafarers, especially, need precise location information for long journeys. In the past, they determined where they were by observing the stars and lighthouses; now, they rely on electronic systems.

Thus, we can conclude that positioning, especially wireless positioning, plays an important role in the LBS. In order to realize its potential applications, an accurate estimation of position should be performed even in challenging environments that have multi-path and non-line-of-sight (NLOS) propagation. To achieve an accurate position estimation, details of the position estimation process as well as its theoretical limits should be well understood [3].

Position estimation is defined as the process of estimating the position of a node, called the "target" node, in a wireless network by exchanging signals between the target node and a number of reference nodes. The position of the target node can be estimated by the target node itself, which is called self-positioning, or it can be estimated by a central unit that obtains information via the reference nodes, which is called remote-positioning (network-centric positioning) [3]. Another divisive condition is whether or not the position is directly estimated from the signals traveling between the nodes, on which the positioning can be separated into direct positioning and two-step positioning, which are shown in Figure 10.1.

As shown in Figure 10.1, direct positioning refers to the case in which the position estimation is performed directly from the signals traveling between the nodes, while two-step positioning obtains certain information from the signals first, and then estimates the position based on an analysis of those signal parameters. In the first step of a two-step positioning algorithm, signal parameters, such as time of arrival (TOA), received signal strength (RSS), and so on, are obtained. Then, in the second step, using the signal parameters obtained in the first step, the position of the target node is estimated. Additionally, in the second step of position estimation, techniques such as fingerprinting approaches, geometric or statistical, can be used because of the accuracy requirements and system constraints [3].

In addition, in considering how to determine the location of a mobile user, the system can also be divided into two categories: tracking and positioning.

If a sensor network determines the location, we talk about tracking, while if the wireless system determines the location itself, we talk about positioning. When using tracking, users have to wear a specific tag that allows the sensor network to track the user's position. The location information is first available in the sensor network; and in the mobile system, the location information is directly available and does not have to be transferred wirelessly when using positioning. In addition, the positioning system does not have to consider privacy problems because the location information is not readable by other users.

Systems using tracking as well as positioning are based on the following basic techniques, or a combination of these techniques.

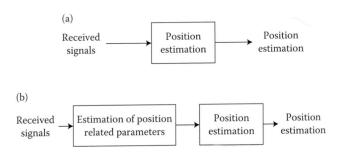

FIGURE 10.1 (a) Direct positioning and (b) two-step positioning. (From Gezici, S. *Wirel Pers Commun Int J*, 2008; 44(3): 263–282. With permission.)

10.2.1 Cell of Origin

Cell of origin (COO) is a mobile-positioning measurement used for finding the position of the terminal, which is the basic geographical coverage unit of a cellular system, when the system has a cellular structure [5]. Wireless transmitting technologies have a restricted range: if the cell has a certain identification, it can be used to determine a location. Additionally, it may be used by emergency services or for some commercial uses. COO is the only positioning technique that is widely used in wireless networks [5].

Most commercially used systems rely on "enhanced" COO. The global system for mobile communications (GSM) relies on the MSs constantly obtaining information on the signal strength from the closest six base stations (BS) and locking on to the strongest signal (the reality is a little more complex than this, encompassing parameters that can be optimized by each individual network, including the signal quality and variability. Most networks try to reduce power consumption, but the overall effect approximates to each phone locking onto the strongest signal). So-called "splash maps," which are generated by the networks, can be employed to predict signal coverage when we plan and manage our networks. These maps can be processed to analyze the area that will be dominated by each BS and to approximate each area by a circle [6].

Although COO positioning is not as precise as other measurements, it offers other unique advantages: it can quickly identify the location (generally in about 3 s) and does not need equipment or network upgrades, making it easy to deploy to existing customer bases. The American National Standards Institute (ANSI) and the European Telecommunications Standards Institute (ETSI) recently formed the T1P1 subcommittee, which is dedicated to creating standardization for positioning systems using TOA, assisted global positioning system (AGPS), and enhanced observed time difference besides COO [5].

10.2.2 Time of Arrival

TOA means the travel time of a radio signal from a single transmitter to a remote single receiver. Electromagnetic signals move at light speed, thus the communication runtimes are very short owing to its high speed. If the signal speed is assumed as a nearly constant light speed, we can use the time difference between sending and receiving the signal to calculate the spatial distance between the transmitter and receiver. The TOA positioning technology uses the absolute TOA at a certain BS and the required distance can be directly calculated from the TOA when the velocity of the signals is known. TOA data from two BSs will narrow the position of the MS into two circles and the data from a third BS is required to solve the precision problem with the third circle matching in a single point [6].

In TOA, location estimates are found by determining the points of intersection of circles or spheres whose centers are located at the fixed stations and the radii are the estimated distances to the target. Figure 10.2 shows a simple geometric arrangement for determining the location of a target MS. In this figure, the MS is located on the same plane as BS1, BS2, and BS3 [2].

In Figure 10.2, three BSs are in use, two of which are located on the *x*-axis with BS1 at the origin in order to simplify the calculations. The coordinates of BS1, BS2, and BS3 are

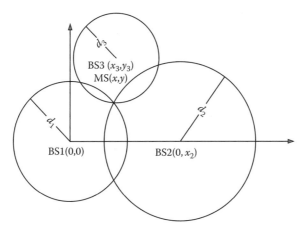

FIGURE 10.2   Determine the location of a target MS using TOA.

known in advance, and distances d_1, d_2, and d_3 are calculated by multiplying the measured signal propagation time between each BS and the target node by the speed of light [2].

The equations for the three intersecting circles whose centers are at the fix stations and radii equal to distances from the target are

$$d_1 = x^2 + y^2,$$ (10.1)

$$d_2 = (x - x_2)^2 + y^2,$$ (10.2)

$$d_3 = (x - x_3)^2 + (y - y_3)^2.$$ (10.3)

These equations can be solved directly for x, y, which are the coordinates of the MS:

$$x = \frac{d_1^2 - d_2^2 + x_2^2}{2 \cdot x_2},$$ (10.4)

$$y = \frac{x_3^2 + y_3^2 + d_1^2 - d_3^2 - 2 \cdot x \cdot x_3}{2 \cdot y_3},$$ (10.5)

We see that the coordinates of the target can be accurately estimated because, as seen in Figure 10.2, the position determined is the only one where all three circles intersect.

10.2.3 Time Difference of Arrival

Similar to the TOA technique, time difference of arrival (TDOA) technology is the measured time difference between departing from one station and arriving at the other station. Unlike the TOA method, which uses the transit time between transmitter and receiver directly to find distance, the TDOA method calculates location from the differences of the arrival times measured on pairs of transmission paths between the target and fixed terminals. Both TOA and TDOA are based on the time of flight (TOF) principle of distance measurement, where the sensed parameter, time interval, is converted to distance by multiplication by the speed of propagation, but TDOA locates the target at intersections of hyperbolas or hyperboloids that are generated with foci at each fixed station of a pair [2].

Even in the absence of synchronization between the target node and the reference nodes, the TDOA estimation can be performed well, if there is synchronization among the reference nodes [3]. In this measurement, the difference between the arrival times of two signals traveling from the target node to the two reference nodes is estimated. In this case, we can determine the position of the target node on a hyperbola, with foci at the two reference nodes, as shown in Figure 10.3 [3].

In Figure 10.3, d_1 and d_2 are the estimations of TOA for each signal traveling between the target node and a BS. We can then obtain the difference between the two distances. Since the target node and the reference nodes are not synchronized, the TOA estimates include a timing offset, which is the same in all estimates as the reference nodes are synchronized, in addition to the TOF. Therefore, the parameters of the estimated TDOA can be obtained as

$$\tau_{TDOA} = \tau_1 - \tau_2 \tag{10.6}$$

where τ_i for $i = 1, 2$, shows the estimated TOA for the signal traveling between the target node and the ith fix stations.

Although the cross-correlation-based TDOA estimation works well for single path channels and white noise models, its performance can degrade considerably over multi-path channels and colored noise.

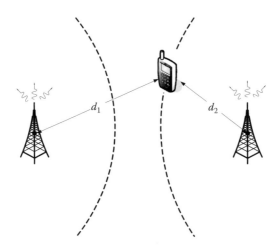

FIGURE 10.3   A TDOA measurement defines a hyperbola passing through the target node with foci at the reference nodes.

10.2.4 Angle of Arrival

By calculating the line-of-sight (LOS) path from the transmitter to receiver, the angle of arrival (AOA) determines the location of the MS in areas of sparse cell site density, or where cell sites are linearly arranged. This distance measurement and location positioning may be the oldest approach and easiest to understand and carry out.

The AOA approach is introduced briefly as:

- AOA uses multiple receivers (two or more) to locate a phone
- AOA yield is 99%
- Accuracy varies, but can get sub-100 m
- Speed and direction of travel is available
- AOA functions for any phone [network 4]

In a wireless system, AOA is a principle component. Using radar, only one fixed station is required in two or three dimensions to determine the location of an MS. There are two methods of AOA and TOF in use. When using AOA alone, at least two fixed terminals are required, or at least two separate measurement parameters by a single terminal in motion [3].

If antennas with direction characteristics are used, arrive direction of a certain signal can be found out. Obtaining two or more direction parameters from fixed positions to the MSs, we can calculate the location of the terminal in motion. Because of the difficulty in constantly turning an antenna for measuring, receivers use a kind of antenna that lines up in all directions with a certain angle difference.

Location and distance are estimated by triangulation in an AOA system. An example is shown in Figure 10.4. To simplify calculations, two BSs are located on the x-axis in a global coordinate system, separated by a distance D. The AOA of the two BSs are

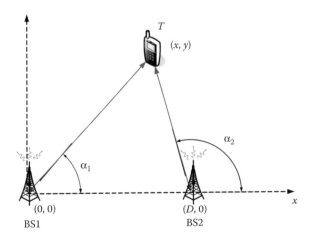

FIGURE 10.4 Triangulation in two dimensions.

α_1 and α_2. From trigonometry, we can determine the coordinates of the target station (x, y) to be

$$x = \frac{D\tan(\alpha_2)}{\tan(\alpha_2) - \tan(\alpha_1)},$$ (10.7)

$$y = \frac{D\tan(\alpha_1)\tan(\alpha_2)}{\tan(\alpha_2) - \tan(\alpha_1)}.$$ (10.8)

The signal-using angle of the arriving measurement cannot be measured exactly, as shown in Figure 10.5. The respective uncertainty of α_1 and α_2 in the measurement is $\Delta\alpha_1$ and $\Delta\alpha_2$. The estimated coordinates of the target stations are then contained within the superposed region in Figure 10.5. The size of this region, which indicates the possible error of target location, is a factor of the AOA measurement accuracy, the angles themselves, and the distance from the target station to the two BSs. The positioning error is represented by the distance from the estimated location at point T whose coordinates are $(\perp x, \perp y)$ to the true location (x, y) [2]:

$$\text{error} = \sqrt{(x - \hat{x})^2 + (y - \hat{y})^2}.$$ (10.9)

10.2.5 Received Signal Strength

RSS is a well-known location method that uses a known mathematical model describing signal path loss with distance. The RSS measurement-based location systems are

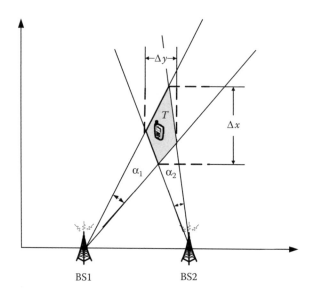

FIGURE 10.5   Position uncertainty due to antenna beam width.

potential candidates to enable indoor location-aware services due to pervasively available wireless local area networks and hand-held devices. On average, the intensity of electromagnetic signals decreases even in a vacuum with the square of the distance from their source. Given a specific signal strength, we can compute the distance to the sender. If the relationship between signal strength and distance is known, analytically or empirically, the distance between two terminals can be determined. When several BSs and a target are involved, triangularization can be applied to determine the target's location [2,7].

Compared with the TOF measurement, the RSS has several advantages. It can work on an existing wireless communications system that has little or no hardware changes. Actually, it only needs the ability to read an RSS indicator (RSSI) output, which is provided on nearly all receivers, and is used to interpret the reading by using dedicated location estimation software. In this RSS measurement, the modulation method, data rate, and system timing precision are not relevant. In addition, coordination or synchronization for distance measurement between the transmitter and the receiver are not required [2].

Unfortunately, this method is inaccurate because obstacles such as walls or clays can reduce the signal strength. In addition, due to the interference and multi-path on the radio channel, the variations in signal strength are quite large, thus the positioning accuracy is generally less than that when using the TOF measurement. In order to achieve the required accuracy in a location system, many more fixed or reference terminals are needed than the minimum number required for triangulation [2].

Two basic classes of the systems are used for positioning estimation: those that are implemented based on known analytic relationships of the radio propagation, and those that are involved in searching a database, which in a location-specific survey includes the measured signal strengths. A third class can be defined as a combination of the two—a database formed from the use of analytic equations or derived from ray tracing software [2].

10.3 Infrastructure of Positioning in Cellular Network

Positioning is a process of obtaining the spatial position of a mobile target station. There are several methods for doing this, each differing from the other in a number of parameters, such as quality, overhead, and so on. In general, positioning is determined by the following elements:

- One or several parameters observed by measurement methods
- A positioning method for position calculation
- A descriptive or spatial reference system
- An infrastructure and protocols for coordinating the positioning process [8]

Location capability was added to cellular communication for the physical security of the holders of handsets, at least in some countries where cellular providers are obliged by telecommunication regulations to provide positioning as a non-subscription service. Once it became available for the infrastructure and/or handset models to provide location, it was natural that the services' range based on location would begin to enlarge. In

Europe and other regions in the world, it is these commercial services that are generating location inclusion capability in the cellular networks [2].

10.3.1 Cellular Network Fundamentals

In fact, all cellular systems are quite similar, except their air interfaces differ significantly. In addition, it is the air interface that absolutely affects the performance of the positioning function. For the air interfaces of second-generation GSM and CDMA IS-95 and third-generation WCDMA (UMTS) as well as CDMA2000, a comparison of several parameters is given in Table 10.1 [2].

The transmission direction between MSs and BSs is employed in two ways. The forward channel on which data promulgate from the BS to the MS is a communication link when the BS is considered the origin. On a reverse channel, the direction of data promulgation is from MS to BS. While considering the MS as the origin, the downlink direction is from BS to MS, and the uplink is from MS to BS. A handset-based location system measures performance on downlink data while a network-based system measures characteristics of the uplink signal [2].

Between the MS and BS, data are arranged in a hierarchy of frames and time slots. The process of communication is carried out on physical channels that are divided into traffic channels and control channels. Traffic channels are composed of the information, speech, or data that, after a set-up call, is transferred between an MS in the network and a terminal in any other fixed station or other cellular network. Control channels, on the

TABLE 10.1 Comparison of Several Parameters in Different Cellular Systems

Feature	GSM		CDMA IS-95	
Major frequency band	Uplink	Downlink	Uplink	Downlink
	890–915 MHz	935–960 MHz	824–849 MHz	869–894 MHz
	1710–1785 MHz	1805–1880 MHz		
	1850–1910 MHz	1930–1990 MHz		
Symbol/chip rate	270.8 kb/s		1288 kb/s	
Channel width	200 kHz		1250 kHz	
Multiple access	Time division (TDMA)		Code division (CDMA)	
Modulation	GMSK (Gaussian Minimum Shift Keying)		Phase shift keying	
Power control	Yes		Yes	
Feature	WCDMA (UMTS)		CDMA2000	
Major frequency bands	Uplink	Downlink	Uplink	Downlink
	920–1980 MHz	2110–2170 MHz	821–835 MHz	866–880 MHz
Symbol/chip rate	4096 kb/s		3686.4 kb/s	
Channel width	5000 kHz		4500 kHz	
Multiple access	Code division (CDMA)		Code division (CDMA)	
Modulation	Phase shift keying		Phase shift keying	
Power control	Yes		Yes	

other hand, are mainly to set up and terminate calls, to synchronize slot time and frequency assignments, and to facilitate handover between mobile and adjacent cells between an MS and a BS [2].

10.3.2 Classification of Positioning Infrastructures

With respect to different criteria, positioning and positioning infrastructures can be classified into several kinds. In all these kinds, integrated and stand-alone positioning infrastructures, terminal and network-based positioning, as well as satellite, cellular, and indoor infrastructures are the most common distinctions [8].

10.3.2.1 Integrated and Stand-Alone Infrastructures

An integrated infrastructure is a wireless network that is used for both communication and positioning. Originally, these networks were designed for communication only, now are experiencing for other application as localizing their users from standard mobile devices, which is especially adapted to cellular networks. The components of the cellular networks can be reused BSs and mobile devices as well as protocols of location and mobility management. An integrated approach has the advantage that the network does not need to be built from scratch and that roll-out and operating costs are manageable, while a stand-alone infrastructure works independently of the communication network the user is attached to. In an integrated approach, measurements in most cases must be done on the existing air interface, whose design has not been optimized for positioning but for communication, and hence the resulting implementations seem to be somewhat complicated and cumbersome in some cases. In addition, in contrast to an integrated infrastructure, the infrastructure and the air interface in a stand-alone infrastructure are intended exclusively for positioning and are very straightforward in their designs [8].

10.3.2.2 Network- and Terminal-Based Positioning

There are some differences between network- and terminal-based positioning, including the site that works on the measurements and calculation of the position of the fix stations. All this is done by the network in the network-based positioning system, while in the terminal-based positioning system, it is the terminal that carries out the function discussed below [8].

In mobile-based location systems, the MS estimates its location from the received signals from some BSs or from the GPS. In GPS-based estimations, the MS receives the signal from at least four satellites that are in the current network of 24 GPS satellites and measures its parameters. The parameter measured by the MS for each satellite is the time that the satellite signal takes to reach the MS. A high degree of accuracy is characteristic of the GPS systems, which also provides global location information. In addition, there is a hybrid technique that uses both in the GPS technology and in the cellular infrastructure. In this case, the cellular network is used to aid the GPS receiver, which is embedded in the mobile handset so that it can improve accuracy and/or acquisition time.

Network-based location technology, on the other hand, is based on some existing networks (either cellular or WLAN) to determine the position of an MS by measuring its signal parameters when received from the network BSs. In this technology, the BSs receive the signals transmitted from an MS and then send them to a central site for further processing and data fusion, in which case, an estimate of the MS location can be provided. A significant advantage of network-based techniques is that the MS is not involved in the location-finding process, thus the technology to modify the existing handsets is not required. However, unlike GPS location systems, many aspects of network-based location have not yet been fully studied [9].

10.3.2.3 Satellites, Cellular, and Indoor Infrastructures

Another criterion to classify positioning is to consider the type of network in which it is implemented and operated [8].

10.4 Cellular Networks

A cellular network is a wireless network composed of several cells, each made up of at least one transceiver of fixed-location called a cell site or BS. In order to provide radio coverage over an area that is wider than that of one cell, these cells cover different areas, in which case, a variable number of terminal in motion can be used in any cell as well as moved from one cell to another during transmission.

Cellular networks offer a number of advantages as follows:

- Increased capacity
- Reduced power usage
- Larger coverage area
- Reduced interference from other signals [network 6]

10.4.1 Global Positioning System Solution

Global navigation satellite systems (GNSS) are the standard generic term for satellite navigation systems that provide autonomous geo-spatial positioning with global coverage. It provides reliable positioning, navigation, and timing services to worldwide users on a continuous basis in all weather, day and night, and anywhere on or near the Earth. GNSS include GPS, GLONASS, Galileo, BeiDou (COMPASS) Navigation Satellite System. As of 2010, the U.S. NAVSTAR GPS is the only fully operational GNSS. The application areas include aviation, surveying and mapping, public transportation, time and frequency comparisons and dissemination, space and satellite operations, law enforcement and public safety, technology and engineering, and GIS.

A GPS receiver calculates its position by precisely timing the signals sent by the GPS satellites high above the Earth. The receiver utilizes the messages it receives to determine the transit time of each message and computes the distances to each satellite. These distances along with the satellites' locations are used with the possible aid of trilateration to compute the position of the receiver [10].

10.4.2 Cell Identification

Cell identification, or so-called cell-ID, can be either handset- or network-based and is the most basic positioning technology available for cellular systems. In order to communicate, a handset connects with a separate base transceiver located in a network cell [2]. Mobile terminals with built-in GPS receivers are becoming more and more usable; therefore, the public deployment of LBS is increasingly feasible. The coming LBS technology is no longer reactive only, but more and more proactive, which enables users to subscribe for some special events and be notified when a point of interest comes within proximity. However, for mobile terminals, power consumption with continuous tracking is still the main problem. In this section, this problem and solutions proposed for energy-efficient combination of GPS and GSM are defined as the cell-ID positioning for MSs. Several approaches for extending the battery lifetime are introduced, and how to combine these strategies into existing middleware solutions is shown. Simulations based on a realistic proactive multi-user context confirm the approach [11].

10.4.3 Problems and Solutions in Cellular Network Positioning

Application of specific positioning technologies usually depends strongly on the type of cellular network involved. The bandwidth of the cellular signal, to a great extent, determines the precision reached in the measurements of the TOA, the fading degree, and the effects of the multi-path propagation.

10.4.3.1 Narrowband Networks

Both the analog advanced mobile phone system (AMPS) and the U.S. digital cellular standard (USDC) have a limited bandwidth of 30 kHz. A system based on the coverage of digital receivers that are connected to antennas of the existing BS was developed. The system mentioned above uses the TDOA measurement and changes processing for correlation of controlled channel signals. Time stamps are contained in the controlled channel messages, in order that in the vicinity of the located mobile unit, copies originating at different receivers can be connected together to produce the data on the time difference that is needed for TDOA positioning. Doppler shifts are also detected in the signals promoting MS location by estimating the speed and bearing of the MS.

To wake deep fading of the systems, which involve MSs with a narrow bandwidth, space diversity antennas are used for BSs. In addition, AOA measurement is also used to reduce multi-path effect and provide an additive method for a TDOA system to improve the location accuracy.

10.4.3.2 Code Division Multiple Access

Code division multiple access (CDMA) is a form of direct sequence spread spectrum communications. In general, spread spectrum communications is distinguished by three main aspects:

- The signal occupies a much greater bandwidth than that necessary to send the information. This has many advantages, such as immunity to interference and jamming as well as multi-user access.

- The bandwidth is determined based on a code that is independent from the data. This code independence distinguishes it from standard modulation schemes in which the data modulation determines the spectrum somewhat.
- To recover the data, the receiver synchronizes to the code. The use of an independent code and synchronous reception allows multiple users to access the same frequency band at the same time.

To protect the signal, the value of the used code is pseudo-random, which appears to be random, but is actually deterministic. In this case, the receivers can rebuild the code for synchronous detection [network 7].

10.4.3.3 Global System for Mobile Communications

GSM was first developed by the CEPT, whose services follow an integrated services digital network and are divided into electronic services and data services. The bandwidth of a GSM signal is 200 kHz, which makes it potentially more accurate than that of AMPS or time division multiple access (TDMA) in TDOA positioning.

A GSM network is a public land mobile network (PLMN), which also includes the TDMA and CDMA networks. GSM uses the following to distinguish it from the PLMN:

- Home PLMN (HPLMN)—the so-called HPLMN is the GSM network where a GSM user is a subscriber in it. All of the above implies that the subscription data of the GSM user reside in the HLR in that PLMN.
- Visited PLMN (VPLMN)—the VPLMN is the GSM network where a subscriber is currently registered. The subscriber may be registered in his/her HPLMN or in another PLMN, in which case, the subscriber is defined as outbound roaming (from HPLMN's perspective) and inbound roaming (from VPLMN's perspective). The HPLMN is the VPLMN at the same time, when the subscriber is currently registered in his/her HPLMN.
- Interrogating PLMN (IPLMN)—the IPLMN is the PLMN containing the GMSC that handles mobile terminating (MT) calls.

10.5 Precision and Accuracy

The error in the positioning accuracy is caused by the timing accuracy of base station, the cellular structure, and the antenna direction of base station and terminal. In addition, there are other important factors, including the multi-path wireless channel, the obstacle between the transmitter and receiver (NLOS), multiuser interference, and the available base station for position.

The U.S. Federal Communications Commission (FCC) announced the positioning requirement of the emergency call "911" (E-911) in 1996, which requires that all wireless cellular signals should provide the location services with an accuracy of 125 m to enable the MS to issue E-911. The systems should also provide the information at higher precision and three-dimensional position. Currently, the requirement of the positioning accuracy is: the positioning program that is based on the cellular network and does not include terminal calls for the positioning accuracy, at least 67% is not below 150 m and at least 95% is not below 300 m; the positioning program that is based on the MS and the

MS is changeable calls for the positioning accuracy, at least 67% is not below 50 m and at least 95% is not below 150 m. The announcement of the U.S. FCC clearly defined the E-911 positioning services, which will be the basic function for the cellular network, especially the 3G network.

10.5.1 Study of the Multi-Path Promulgate

The multi-path promulgate is the basic reason for the appearance of the error in the measured values of the signal character. For the TOA and TDOA positioning principle, even if the signal can LOS spread between the MS and the BS, the multi-path promulgate will still cause the measurement error. Because the performance of the delay estimator based on the technology of interrelated can be affected by the multi-path promulgate, the arrival time of the reflected wave and direct wave are in the same chip gap. Today, there are more and more way to improve the multi-path promulgate problem.

10.5.2 NLOS Promulgate

The NLOS promulgate is the necessary condition to obtain the exact measured values of the signal character. The GPS system realizes the precise location based on the LOS promulgate of the signal. However, to realize the LOS promulgate between the MS and several BSs is difficult, even without multi-path and bringing in the high-precision timing technology, the NLOS promulgate can still cause the measurement error of the TOA and TDOA. Thus, the NLOS promulgate is the main reason affecting the positioning accuracy of all kinds of cellular network, and the key to enhance the accuracy is how to reduce the interference in the process of the NLOS promulgate. Currently, there are some methods to reduce the interference in the process of the NLOS promulgate. One is to distinguish the LOS and NLOS promulgate using the standard deviation of the TOA measurement values. As we all know, the measurement value of the NLOS promulgate standard deviation is much higher than the LOS promulgate. Therefore, by using the a priori information of measurement error estimation, the measurement value of NLOS for some time can be adjusted close to that of LOS. Another is to reduce the weight of the NLOS measurement value in the nonlinear least squares algorithm, which also needs to judge, which MSs obtain the NLOS measurement value first. The final method is to optimize the algorithm to improve the positioning accuracy via adding a constraint polynomial in the least squares algorithm. This constraint polynomial is characterized the measurement value under the condition of NLOS promulgate being higher than the actual distance.

10.5.3 CDMA Multi-Address Access Interference

Multi-address interference significantly reduces the performance of the CDMA system. The CDMA system is a time-varying system, in which the background channel noise and the relative position between BSs and users are continuous; in addition, the joining and leaving of users are stochastic. All of these factors result in the changes in the received signal's properties continuously. Additionally, in recent years, various types of multi-carrier CDMA systems have been employed. Under appropriate conditions, the signals of multi-carrier CDMA will propagate through multi-path channels with little loss. The system using only a few subcarriers to deal with the intersymbol interference and the interchip interference is introduced in. In a channel of a typical indoor environment, this system is more optimal than the Rake receiver [12].

10.5.4 Other Sources of Positioning Error

In addition, the relative position between each BS involved in the positioning, the difference in the geometric dilution of precision caused by the diversity of the relative position between MSs and BSs can also affect the performance of the positioning algorithm and cause the difference in positioning accuracy.

10.6 Conclusions

In this chapter, we presented the basic principle, techniques, and systems of wireless location technology in LBSs.

GNSS is widely used to determine the current location in many LBS. GNSS receivers are cheap, and the corresponding location result is accurate. However, the location only works if a direct LOS between the satellites and the receivers is given. Cellular location is often viewed as the most promising technology for LBS, as it can cover a large geographic area and have a high number of mobile subscriber. Different location technologies are proposed in the corresponding industry association, for example 3GPP and 3GPP2. Indoor location is based on radio, infrared, or ultrasound technologies with a small coverage, such as in a single building. This chapter will serve as foundation for understanding the implementation of LBS in subsequent chapters.

References

1. Caffery, J. J. *Wireless Location in CDMA Cellular Radio Systems*. Norwell, MA: Kluwer Academic, 1999.
2. Bensky, A. *Wireless Positioning Technologies and Ap-plications*, vol. 9, Norwood, MA: Artech House, pp. 223–241, 2007.
3. Gezici, S. A survey on wireless position estimation. *Wirel Pers Commun Int J*, 2008; 44(3): 263–282.
4. Barton. R. J., Zheng, R., Gezici, S., and Veeravalli, V. V. Signal processing for location estimation and tracking in wireless environments. *EURASIP J Adv Signal Process* 2008: 1–3.
5. Cell of origin (telecommunications). http://en.wikipedia.org/wiki/Cell_of_origin_ (telecommunications), accessed 4 April 2010.
6. Time of arrival. http://en.wikipedia.org/wiki/Time_of_arrival, accessed 4 April 2010.
7. Schiller, J. *Location-Based Services*. The Morgan Kaufmann Series in Data Management Systems. San Francisco, CA: Morgan Kaufmann, 2004.

8. Küpper, A. *Location-Based Services: Fundamentals and Operation.* New York: Wiley, 2005.
9. Sayed Ali, H., Alireza, T., and Nima, K. Network-based wireless location: Challenges faced in developing techniques for accurate wireless location information. *IEEE Signal Process Mag*, 2005; 22(4): 24–40.
10. Global navigation satellite system. http://en.wikipedia.org/wiki/Global_Navigation_Satellite_System, accessed 4 April 2010.
11. Nico, D. and Peter, R. Combining GPS and GSM cell-ID positioning for proactive location-based services, in *Fourth Annual International Conference on Mobile and Ubiquitous Systems: Networking & Services*, 2007, pp. 1–7, 2007.
12. Lining, W. and Guangxin, Y. Effect of MAI on MC-CDMA's acquisition performance, in *Proceeding in IEEE International Conference on Communication Technology ICCT'98*, Vol. 2, pp. 22–24, 1998.
13. Manesis, T. and Avouris, N. Survey of position location techniques in mobile systems, in *Proceedings of the 7th Interna-tional Conference on Human Computer Interaction with Mo-bile Devices & Services ACM*, vol. 111, pp. 291–294, 2005.
14. Gustafsson, F. and Gunnarsson, F. Positioning using time-difference of arrival measurements, in *IEEE International Conference on Acoustics, Speech, and Signal Processing*, vol. 6, pp. 553–556, 2003.
15. Ming, G., Pan, Y., and Fan, P. *Advances in Wireless Networks: Performance Modelling, Analysis and Enhance-ment.* Hauppauge, NY: Nova Science, 2008.

11

Next-Generation Technologies to Enable Sensor Networks

Joel I. Goodman
MIT Lincoln Laboratory

Albert I. Reuther
MIT Lincoln Laboratory

David R. Martinez
MIT Lincoln Laboratory

11.1 Introduction

Several important technical advances make extracting more information from intelligence, surveillance, and reconnaissance (ISR) sensors very affordable and practical. As shown in Figure 11.1, for the radar application, the most significant advancement is expected to come from employing collaborative and network-centric sensor netting. One important

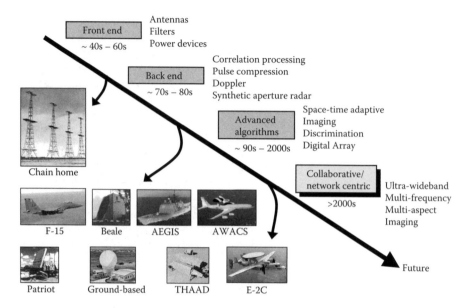

FIGURE 11.1   Radar technology evolution.

application of this capability is to achieve ultrawideband multifrequency and multiaspect imaging by fusing the data from multiple sensors. In some cases, it is highly desirable to exploit multimodalities, in addition to multifrequency and multiaspect imaging.

Key enablers to fuse data from disparate sensors are the advent of high-speed fiber and wireless networks and the leveraging of distributed computing. ISR sensors need to perform enough on-board computation to match the available bandwidth; however, after some initial preprocessing, the data will be distributed across the network to be fused with other sensor data so as to maximize the information content. For example, on an experimental basis, MIT Lincoln Laboratory has demonstrated a virtual radar with ultrawideband frequency [1]. Two radars, located at the Lincoln Space Surveillance Complex in Westford, Massachusetts, were employed; each of the two independent radars transmitted the data via a high-speed fiber network. The total bandwidth transmitted via fiber exceeded 1 Gbits/s (billion bits per second). One radar was operating at X-band with 1-MHz bandwidth, and the second was operating at Ku-band with a 2-MHz bandwidth. A synthetic radar with an instantaneous bandwidth of 8 MHz was achieved after employing advanced ultrawideband signal processing [2].

These capabilities are now being extended to include high-speed wireless and fiber networking with distributed computing. As the Internet protocol (IP) technologies continue to advance in the commercial sector, the military can begin to leverage IP-formatted sensor data to be compatible with commercial high-speed routers and switches. Sensor data from theater can be posted to high-speed networks, wireless, and fiber, to request computing services as they become available on this network. The sensor data are processed in a distributed fashion across the network, thereby providing a larger pool of resources in real time to meet stringent latency requirements. The availability of distributed processing in a grid-computing architecture offers a high degree of robustness throughout the network. One important application to benefit from these advances is the ability to geolocate and identify mobile targets accurately from multiaspect sensor data.

11.1.1 Geolocation and Identification of Mobile Targets

Accurately geolocating and identifying mobile targets depends on the extraction of information from different sensor data. Typically, data from a single sensor are not sufficient to achieve a high probability of correct classification and still maintain a low probability of false alarm. This goal is challenging because mobile targets typically move at a wide range of speeds, tend to move and stop often, and can be easily mistaken for a civilian target. While the target is moving the sensor of choice is the ground moving target indication (GMTI). If the target stops, the same sensor or a different sensor working cooperatively must employ synthetic aperture radar (SAR). Before it can be declared foe, the target must often be confirmed with electro-optical or infrared (EO/IR) images. The goal of future networked systems is to have multiple sensors providing the necessary multimodality data to maximize the chances of accurately declaring a target.

A typical sensing sequence starts by a wide area surveillance (WAS) platform, such as the Global Hawk unmanned aerial vehicle, covering several square kilometers until a target exceeds a detection threshold. The WAS will typically employ GMTI and SAR strip maps. Once a target has been detected, the on-board or off-board processing starts a track file to track the target carefully, using spot GMTI and spot SAR over a much smaller region than that initially covered when performing a WAS. It is important to recognize that a sensor system is not merely tracking a single target; several target tracks can be going on in parallel. Therefore, future networked sensor architectures rely on sharing the information to maximize the available resources.

To date, the most advanced capability demonstrated is based on passing target detections among several sensors using the Navy cooperative engagement capability (CEC) system. Multisensor tracks are formed from the detection inputs arriving at a central location. Although this capability has provided a significant advancement, not all the information available from multimodality sensors has been exploited. The limitation is with the communication and available distributed computing. Multimodality sensor data together with multiple look angles can substantially improve the probability of correct classification versus false alarm density. In addition to multiple modalities and multiple looks on the target, it is also desirable to send complex (amplitude and phase) radar GMTI data and SAR images to permit the use of high-definition vector imaging (HDVI) [3]. This technique permits much higher resolution on the target by suppressing noise around it, thereby enhancing the target image at the expense of using complex video data and much higher computational rates.

Another important tool to improve the probability of correct classification with minimal false alarm is high-range resolution (HRR) profiles. With this tool, the sensor bandwidth or, equivalently, the size of the resolution cell must be small resulting in a large data rate. However, it has been demonstrated that HRR can provide a significant improvement [4]. Therefore, next-generation sensors depend on available communication pipes with enough bandwidth to share the individual sensor information effectively across the network. Once the data are posted on the network, the computational resources must exist to maintain low latencies from the time data become available to the time a target geoposition and identification are derived. The next subsection discusses the long-term architecture to implement netting of multiple sensor data efficiently.

11.1.2 Long-Term Architecture

In the future, it will be desirable to minimize the infrastructure (foot print) forwardly deployed in the battlefield. It is most desirable to leverage high-speed satellite communication links to bring sensor data back to a combined air operations center established in the continental United States (CONUS).

The technology enablers for the long-term architecture shown in Figure 11.2 are high-speed, IP-based wireless, and fiber communication networks, together with distributed grid computing. The in-theater commander's ability to task his organic resources to perform reconnaissance and surveillance of the opposing forces, and then to relay that information back to CONUS, allows a significant reduction in the complexity, level, and cost of in-theater resources. Furthermore, this approach leverages the diverse analysis resources in CONUS, including highly trained personnel to support the rapid, accurate identification, and localization of targets necessary to enable the time-critical engagement of surface mobile threats.

Space, air, and surface sensors will be deployed quickly to the battlefield. As shown in Figure 11.3, the stage in the processing chain at which the sensor data are tapped off to be sent via the network will dictate the amount of data transferred. For example, in a few applications, one needs to send the data directly out of the analog-to-digital converters

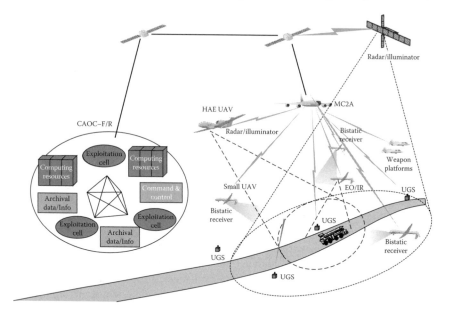

FIGURE 11.2   Postulated long-term architecture.

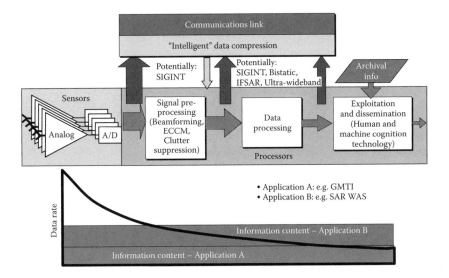

FIGURE 11.3 Sensor signal processing flow.

(A/D) to exploit the coherent data combining from multiple sensors. Most commonly, it is preferable to perform on-board signal preprocessing to minimize the amount of data transferred. However, one must still be able to preserve content in the transferred data that is required to exploit features in the data not available from processing a signal sensor end to end. For example, one might be interested in transmitting WAS data from SAR with high resolution to be followed by multiaspect SAR processing (shown in Figure 11.3 as application B). The data volume will be larger than the second example shown in Figure 11.3 as application A, in which most of the GMTI processing is done on-board. In any of these applications, it is paramount that "intelligent" data compression be done on-board before data transmission to send only the necessary parts of the data requiring additional processing off-board.

Each sensor will be capable of generating on-board processed data greater than 100 Mbits/s (million bits per second). Figure 11.4 shows the trade-off between communication

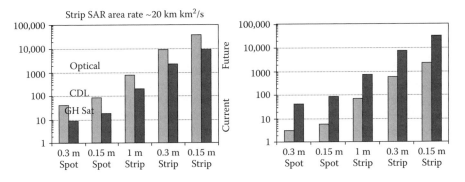

FIGURE 11.4 SAR data rate and computational throughput trade.

link data rates versus on-board computation throughputs for different postulated levels of image resolution (for spot or strip map SAR modes). For example, for an assumed 1-m strip map SAR, one can send complex video radar data to perform super-resolution processing off-board. This approach would require sending between 100 and 1000 Mbits/s. Another option is to perform the super-resolution processing on-board, requiring between 100 billion floating-point operations per second (GFLOPS) to 1 trillion floating-point operations per second (TFLOPS).

Specialized military equipment, such as the common data link, can achieve data rates reaching 274 Mbits/s. If higher communication capacity were available, one would much prefer to send the large data volume for further processing off-board to leverage information content available from multiple sensor data. As communication rates improve in the forthcoming years, it will not matter to the in-theater commander if the data are processed off-board with the benefit of allowing exploitation of multiple sensor data at much rawer levels than is possible to date.

11.2 Goals for Real-Time Distributed Network Computing for Sensor Data Fusion

Several advantages can be gained by utilizing real-time distributed network computing to enable greater sensor data fusion processing. Distributed network computing potentially reduces the cost of the signal processing systems and the sensor platform because each individual sensor platform no longer needs as much processing capability as a stove-piped stand-alone system (although each platform may need higher bandwidth communications capabilities). Also, fault tolerance of the processing systems is increased because the processing and network systems are shared between sensors, thereby increasing the pool of available signal processors for all of the sensors. Furthermore, the granularity of managed resources is smaller; individual processors and network resources are managed as independent entities rather than managing an entire parallel computer and network as independent entities. This affords more flexible configuration and management of the resources.

To enable collaborative network processing of sensor signals, three technological areas are required to evolve and achieve maturity:

- Guaranteed *communication, storage buffer,* and *computation resources* must keep up with the high-throughput streams of data coming from the sensors. If any stage of the processing falls behind due to a network problem or interruption in the processor, buffering the data will become a problem quickly as increasing volumes of data must be stored to accommodate the delayed processors. Section 11.3 addresses technological possibilities to mitigate these resource availability issues.
- *Middleware* in the network of processors must be developed to accommodate a heterogeneous mixture of computer and network resources. This middleware consists of a task control interface, which facilitates the communication between network resource management agents and entities, and an application programming interface for programming applications executed on the collaborative network processors. Section 11.4 will address these middleware interfaces.

- A *network resource manager* (NRM) system is necessary for orchestrating the execution of the application components on the computation and communication resources available in the collaborative network. Section 11.5 will discuss the components and functionality of the NRM.

11.3 The Convergence of Networking and Real-Time Computing

To date, networking of sensors has been demonstrated primarily using localized- and limited-capacity data links. As a result, the data available on the network from each sensor node typically represent the product of extensive prior processing of the radar data carried at the individual sensor. For example, the Navy CEC system, a relatively advanced current system, uses detection reports from independent sensors in the network to build composite tracks of targets. Access to raw (or possibly minimally preprocessed) multisensor data opens the opportunity for more effective exploitation of these data through an integrated sensor data processing. The future network-centric ISR architecture will likely employ worldwide wideband communication networks to interconnect sensors with distributed processing and fusion sites. The resulting distributed database will provide a common operational picture for deployed forces. The sensor data will return to a CONUS entry point and pass over a wideband fiber network to the various processing centers where the sensor data will be fused. The data link from the theater to CONUS is expected to be optical to achieve very high link capacity [5].

This section discusses technologies that will guarantee that wireless and terrestrial network resources, storage buffer resources, and computational resources are available for sensor signal processing.

11.3.1 Guaranteeing Network Resources

Sensor data will traverse wireless and terrestrial (e.g., optical, twisted-copper) networks in which bit errors, packet loss, and delay could adversely affect the quality and timeliness of the ultimate result. The goal then is to choose a network and processing architecture to ameliorate the deleterious effects of data loss and network delay in the data fusion process. Due to the costs associated with developing, deploying, and maintaining a fixed terrestrial infrastructure, as well as inventing wholly new modulation protocols and standards for wireless and terrestrial signaling, it is cost-effective and expedient for military technology to ride the "commercial wave" of technical investment and progress in communication technologies.

With a fixed network infrastructure consisting primarily of commercial components, combating data loss, and delay in terrestrial networks involve choosing the right protocols so that the network can enforce quality of service (QoS) demands; in wireless networks, this involves aggressive coding, modulation, and "lightweight" flow control for efficient bandwidth utilization. With sufficient complexity and bandwidth, it is possible with today's IP-based protocols to differentiate high-priority data to impart the mandated QoS for time-critical applications.

11.3.1.1 Terrestrial Networks

Reserving bandwidth on an IP-based network that is uniformly recognized across administrative domains involves employing protocols like RSVP-TE [6] or CR-LDP [7]. Although having sufficient communication bandwidth is an important aspect of processing sensor data in real time on a distributed network of resources, it does not guarantee real-time performance. For example, time-critical applications mapped onto networked resources should not have processing interrupted to service unmanaged traffic or be subject to a computational resource's resident operating system switching contexts to a lower priority task. For data that originate from sensors at very high streaming rates, a storage solution, as discussed in Section 11.3.2, is needed that is capable of recording sensor data in real time as well as robust in the face of network resource failures; this insures that a high-priority application can continue processing in the presence of malfunctioning or compromised networked equipment. However, adding a buffering storage solution only alleviates part of the problem; it does not mitigate the underlying problem of losing packets during network equipment failures or periods of network traffic that exceed network capacities.

For an IP-based network, one solution to this problem is to use remote agents deployed on primary compute resources or networked terminals located at switches that can dynamically filter unmanaged traffic. This is implemented by programming computer hardware specifically tasked with packet filtering (e.g., next generation gigabit Ethernet card) or dynamically reconfiguring the switch that directly connects to the compute resource in question by supplying an access control list (ACL) to block all packets except those associated with time-critical targeting. The formation of these exclusive networks using agents has been dubbed *dynamic private networks* (DPNs)—in effect, mechanisms for virtually overlaying a circuit switch onto a packet-switched network.

11.3.1.2 Wireless Networks

Unlike terrestrial networks, flow control and routing in mobile wireless sensor networks must contend with potentially long point-to-point propagation delays (e.g., satellite to ground) as well as a constantly changing topology. In a traditional terrestrial network employing link-state routing (e.g., OSPF), each node maintains a consistent view of a (primarily) fixed network topology so that a shortest path algorithm [8] can be used to find desirable routes from source to destination. This requires that nodes gather network connectivity information from other routers.

If OSPF were employed in a mobile wireless network, the overhead of exchanging network connectivity information about a transient topology could potentially consume the majority of the available bandwidth [9]. Routing protocols have been specifically designed to address the concerns of mobile networks [10]; these protocols fall into two general categories: proactive and reactive. Proactive routing protocols keep track of routes to all destinations, while reactive protocols acquire routes on demand. Unlike OSPF, proactive protocols do not need a consistent view of connectivity; that is, they trade optimal routes for feasible routes to reduce communication overhead. Reactive routes suffer a high initial overhead in establishing a route; however, the overall overhead of maintaining network connectivity is substantially reduced. The category of routing used is highly dependent upon how the sensors communicate with one another over the network.

Traditional flow control mechanisms over terrestrial networks that deliver reliable transport (e.g., TCP) may be inappropriate for wireless networks because, unlike wireless networks, terrestrial networks generally have a very low bit error rate (BER) on the order of 10^{-10}, so errors are primarily due to packet loss. Packet loss occurs in heavily congested networks when an ingress or egress queue of a switch or router begins to fill, requiring that some packets in the queue be discarded [11]. This condition is detected when acknowledgments from the destination node are not received by the source, prompting the source's flow control to throttle back the packet transmit rate [12].

In a wireless network in which BERs are four to five orders of magnitude higher than those of terrestrial networks, packet loss due to bit errors can be mistakenly associated with network congestion, and source flow control will mistakenly reduce the transmit rate of outgoing packets. Furthermore, when the source and destination are far apart, such as the communication between a satellite and ground terminal, where propagation delays can be on the order of 240 ms, delayed acknowledgments from the destination result in source flow control inefficiently using the available bandwidth. This is due to source flow control incrementally increasing the transmit rate as destination acknowledgments are received even though the entire frame of packets may have already been transmitted before the first packet reaches the receiver [13]. Therefore, to use bandwidth efficiently in a wireless network for reliable transport, flow control must be capable of differentiating BER from packet loss and account for long-haul packet transport by more efficiently using the available bandwidth. Some work in this area is reflected in RFC 2488 [14], as well as proposals for an explicit congestion warning, where, for example, the destination site would respond to packet errors with an acknowledgment that it received the source packets with a corruption notification.

At the physical layer, high data rates for a given BER have been realized by employing low-density parity check codes, such as turbo codes, in conjunction with bandwidth efficient modulation to achieve spectral efficiencies to within 0.7 dB of the Shannon limit [15]. Furthermore, extremely high spectral efficiencies have been demonstrated using multiple input, multiple output (MIMO) antenna systems whose theoretical channel capacity increases linearly with the number of transmit/receive antenna pairs [16]. Although turbo codes are advantageous as a forward error correction mechanism in wireless systems when trying to maximize throughput, MIMO systems achieve high spectral efficiencies only when operating in rich scattering environments [17]. In environments in which little scattering occurs, such as in some air-to-air communication links, MIMO systems offer very little improvement in spectral efficiency.

11.3.2 Guaranteeing Storage Buffer Resources

For a variety of reasons, it may be very desirable to record streaming sensor data directly to storage media while simultaneously sending the data on for immediate processing. For sensor signal processing applications, this enables multimodality data fusion of archived data with real-time (perishable) data from in-theatre sensors for improved target identification and visualization [18]. Storage media could also be used for rate conversion in cases in which the transmission rate exceeds the processing rate and for time-delay buffering for real-time robust fault tolerance (discussed in the next section). The storage media buffer reuse is deterministic and periodic so that management of the buffer is straightforward.

A number of possible solutions exist:

- *Directly attached storage* is a set of hard disks connected to a computer via SCSI or IDE/EIDE/ATA; however, this technology does not scale well to the volume of streaming sensor data.
- *Storage area networks* are hard disk storage cabinets attached to a computer with a fast data link like Fibre Channel. The computer attached to the storage cabinet enjoys very fast access to data, but because the data must travel through that computer, which presents a single point of failure, to get to other computers on the network, this option is not a desirable solution.
- *Network-attached storage* connects the hard disk storage cabinet directly to the network as a file server. However, this technology offers only midrange performance, a single point of failure, and relatively high cost.

A visionary architecture in which data storage centers operate in parallel at a wide-area network (WAN) and local area network (LAN) level is described in Cooley et al. [19]. In this architecture, developed by MIT Lincoln Laboratory, high-rate streaming sensor data are stored in parallel across a partitioned network of storage arrays, which affords a highly scalable, low-cost solution that is relatively insensitive to communications or storage equipment failure. This system employs a novel and computationally efficient encoding and decoding algorithm using low-density parity check codes [20] for erasure recovery. Initial system performance measures indicating the erasure coding method described in Cooley et al. [19] has a significantly higher throughput and greater reliability when compared to Reed–Solomon, Tornado [21], and Luby [20] codes. This system offers a promising low-cost solution that scales in capability with the performance gains of commodity equipment.

11.3.3 Guaranteeing Computational Resources

The exponential growth in computing technology has contributed to making viable the implementation of advanced sensor processing in cost-effective hardware with form factors commensurate with the needs of military users. For example, several generations of embedded signal processors are shown in Figure 11.5. In the early 1990s, embedded signal processors were built using custom hardware and software. In the late 1990s, a move occurred from custom hardware to COTS processor systems running vendor-specific software together with application-specific parallel software tuned to each specific application. Most recently, the military embedded community is beginning to demonstrate requisite performance employing parallel and portable software running on COTS hardware.

Continuing technology advances in computation and communication will permit future signal processors to be built from commodity hardware distributed across a high-speed network and employing distributed, parallel, and portable software. These

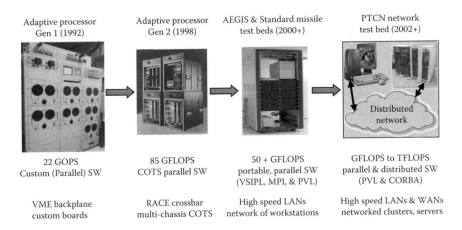

Adaptive processor Gen 1 (1992)	Adaptive processor Gen 2 (1998)	AEGIS & Standard missile test beds (2000+)	PTCN network test bed (2002+)

22 GOPS Custom (Parallel) SW	85 GFLOPS COTS parallel SW	50 + GFLOPS portable, parallel SW (VSIPL, MPI, & PVL)	GFLOPS to TFLOPS parallel & distributed SW (PVL & CORBA)
VME backplane custom boards	RACE crossbar multi-chassis COTS	High speed LANs network of workstations	High speed LANs & WANs networked clusters, servers

FIGURE 11.5 Embedded signal processor evolution.

computing architectures will deliver 10^9 to 10^{12} floating point operations per second (GFLOPs to TFLOPs) in computational throughput. The distributed nature of the software will apply to on-board sensor processing as well as off-board processing. Clearly, on-board embedded processor systems will need to meet the stringent platform requirements in size, weight, and power.

Wireless and terrestrial network resources are not the only areas in which delays, failures, and errors must be avoided to process sensor data in a timely fashion. The system design must also guarantee that the marshaled compute nodes will keep up with the required computational throughput of streaming data at every stage of the processing chain. This guarantee encompasses two important facets: (i) keeping the processors from being interrupted while they are processing tasks and (ii) implementing fail-over that is tolerant of fault.

11.3.3.1 Avoiding Processor Interruption

It is easy to take for granted that laptop and desktop computers will process commands as fast as the hardware and software are capable of doing so. A fact not generally known is that general computers are interrupted by system task processes and the processes of other applications (one's own and possibly from others working in the background on one's system). System task processes include keyboard and mouse input; communications on the Ethernet; system I/O; file system maintenance; log file entries; and so on. When the computer interrupts an application to attend to such tasks, the execution of the application is temporarily suspended until the interrupting task has finished the execution. However, because such interruptions often only consume a few milliseconds of processing time, they are virtually imperceptible to the user [22].

Nevertheless, the interruptions are detrimental to the execution of real-time applications. Any delay in processing these streams of data will instigate a need for buffering the data that will grow to insurmountable size as the delays escalate. A solution for these interrupt issues is to use a real-time operating system (RTOS) on the computation processors.

Simply put, RTOSs give priority to computational tasks. They usually do not offer as many operating system features (virtual memory, threaded processing, etc.) because of the interrupting processing nature of these features [22]. However, an RTOS can ensure that real-time critical tasks have guaranteed success in meeting streamed processing deadlines. An RTOS does not need to be run on typical embedded processors; it can also be deployed on Intel and AMD Pentium-class or Motorola G-series processor systems. This includes Beowulf clusters of standard desktop personal computers and commodity servers. This is an important benefit, providing a wide range of candidate heterogeneous computing resources.

A great deal of press has been generated in the past several years about RTOSs; however, the distinction between soft real-time and hard RTOSs is seldom discussed. Hard real-time systems guarantee the completion of tasks in a deterministic time period, while soft real-time systems give priority to critical tasks over other tasks but do not guarantee the completion of tasks in a deterministic time period [22]. Examples of hard RTOSs are VxWorks (Wind River Systems, Inc. [23]); RTLinux/Pro (FSMLabs, Inc. [24]); and pSOS (Wind River Systems, Inc. [23]), as well as dedicated massively parallel embedded operating systems like MC/OS (Mercury Computer Systems, Inc. [25]). Examples of soft RTOSs are Microsoft Pocket PC; Palm OS; certain real-time Linux releases [24,26]; and others.

11.3.3.2 Working through System Faults

When fault tolerance in massively parallel computers is addressed, usually the solution is parallel redundant systems for fail-over. If a power supply or fan fails, another power supply or fan that is redundant in the system takes over the workload of the failed device. If a hard disk drive fails on a redundant array of independent disks system, it can be hot swapped with a new drive and the contents of the drive rebuilt from the contents of the other drives along with checksum error correction code information. However, if an individual processor fails on a parallel computer, it is considered a failure of the entire parallel computer, and an identical backup computer is used as a fail-over. This backup system is then used as the primary computer, while the failed parallel computer is repaired to become the backup for the new primary eventually.

If, however, it were possible to isolate the failed processor and remap and rebind the processes on other processors in that computer—in real time—it would then be possible to have only a number of redundant processors in the system rather than entire redundant parallel computers. There are two strategies for determining the remapping as well as two strategies for handling the remapping and rebinding; each has its advantages and disadvantages.

To discuss these fail-over strategies, it is necessary to define the concepts of tasks and mappings. A signal processing application can be separated into a series of pipelined stages or tasks that are executed as part of the given application. A mapping is the task-parallel assignment of a task to a set of computer and network resources. In terms of determining the fail-over remapping, it is possible to choose a single remapping for each task or to choose a completely unique secondary path—a new mapping for each task that uses a set of processors mutually exclusive from the processors in the primary mapping path. If task backup mappings are chosen for each task, the fail-over will complete faster than a full processing chain fail-over; however, the rebinding fail-over for a failed task mapping is more difficult because the mappings from the task before and the task after the failed task mapping must be reconfigured to send data and receive data from the new mapping. Conversely, if a completely unique secondary path is chosen as a fail-over, then fail-over completion will have a longer latency than performing a single task fail-over. However, the fail-over mechanics are simpler because the completely unique secondary path could be fully initialized and ready to receive the stream of data in the event of a failure in the primary mapping path.

In terms of handling the remapping and rebinding of tasks, it is possible to choose the fail-over mappings when the application is initially launched or immediately after a fault occurs. In either case, greater latency is incurred at launch time or after the occurrence of a fault. For these advanced options, support for this fault tolerance comes mainly from the middleware support, which is discussed in the next section, and from the NRM discussed in Section 11.5.

11.4 Middleware

Middleware not only provides a standard interface for communications between network resources and sensors for plug-and-play operation, but also enables the rapid implementation of high-performance embedded signal processing.

11.4.1 Control and Command of System

Because many systems use a diverse set of hardware, operating systems, programming languages, and communication protocols for processing sensor data, the manpower and time-to-deployment associated with integration have a significant cost. A middleware component providing a uniform interface that abstracts the lower level system implementation details from the application interface is the common object request broker architecture (CORBA) [27]. CORBA is a specification and implementation that defines a standard interface between a client and server. CORBA leverages an interface definition language that can be compiled and linked with an object's implementation and its clients. Thus, the CORBA standard enables client and server communications that are independent of the host hardware platforms, programming language, operating systems, and so on. CORBA has specifications and implementations to interface with popular communication protocols such as TCP/IP. However, this architecture has an open specification, general interORB protocol (GIOP) that enables developers to define and plug in platform-specific communication protocols for unique hardware and software interfaces that meet application-specific performance criteria.

For real-time and parallel-embedded computing, it is necessary to interface with RTOSs, define end-to-end QoS parameters, and enact efficient data reorganization and queuing at communication interfaces. CORBA has recently included specifications for real-time performance and parallel processing, with the expectation that emerging implementations and specification addendums will produce efficient implementations. This will enable CORBA to move out of the command and control domain and be included as a middleware component involved in real-time and parallel processing of time-critical sensor data.

11.4.2 Parallel Processing

The ability to choose one of the many potential parallel configurations enables numerous applications to share the same set of resources with various performance requirements. What is needed is a method to decouple the mapping, that is, the parallel instantiation of an application on target hardware, from generic serial application development. Automating the mapping process is the only feasible way of exploring the large parameter space of parallel configurations in a timely and cost-effective manner.

MIT Lincoln Laboratory has developed a C++-based library known as the parallel vector library (PVL) [28]. This library contains objects with parameterized methods deeply rooted in linear algebraic expressions commonly found in sensor signal processing. The parameters are used to direct the object instance to process data as one constituent part of a parallel whole. The parameters that organize objects in parallel configurations are run-time parameters so that new parallel configurations can be instantiated without having to recompile a suite of software. The technology of PVL is currently being incorporated into the parallel vector, signal, and image processing library for C++ (parallel VSIPL++) standard library [29].

11.5 Network Resource Management

Given the stated goals for distributed network computing for sensor fusion as outlined in Section 11.3, the associated network communication, storage, and processing challenges in Section 11.3, and the desire for standard interfaces and libraries to enable application parallelism and plug-and-play integration in Section 11.4, an integrated solution is needed that bridges network communications, distributed storage, distributed processing, and middleware. Clearly, it is possible for a development team to implement a "point" solution, but this is inherently not scalable and very difficult to maintain. Therefore, an additional goal is to fully automate the process of configuring network communication, storage, and computational resources to process data for sensor fusion applications in real time, provide robust fault tolerance in the face of network resource failures, and impart this service in a highly dynamic network in the face of competing interests.

To address these needs, the NRM was developed. The novelty and potency of the NRM is its capability of taking a sensor signal processing application designed and tested on single target processing element (PE) and mapping it in a task- and a data-parallel fashion across a network of computational resources to achieve real-time performance [30]. Figure 11.6 is an object-oriented model of the components that constitute the NRM. A high-level overview of the NRM follows, and details will be provided in the following subsections. The task of building a model from which the NRM launches parallel applications is broken into three distinct phases:

1. Map generation involves breaking an application into various task- and data-parallel components.
2. Map timing collects performance metric information associated with the components (or tasks) running on host resources. Using the performance metrics, the

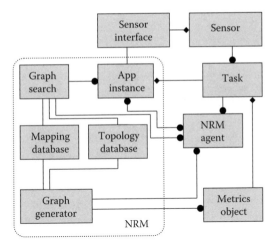

FIGURE 11.6 Object model for network resource manager (NRM).

NRM creates a weighted graph-theoretic view of various permutations of an application mapped in parallel across networked resources.

3. Map selection finds the path through the graph that best meets system and application performance requirements.

The graph generator and graph search objects will heavily leverage PVL (discussed earlier) objects in the instantiation of task- and data-parallel configurations of applications on host resources. It should be noted, however, that the NRM's capabilities are fully general and independent from those of PVL and could work with other applications that are not developed using PVL to instantiate task and data parallelism.

11.5.1 Graph Generator

As noted previously, PVL uses run-time parameters to generate new parallel configurations. This enables the NRM to launch applications in arbitrary parallel configurations using software developed for a single target PE without having to recompile the application software suite. The central challenge is to select a subset of the potentially astronomical number of permutations of parallel configurations as candidate parallel mappings. It is expected that the NRM will receive guidance in the form of performance and resource utilization bounds to help it avoid choosing undesirable configurations. It will also be given a series of constituent tasks that comprise an application, so that its primary objective is to choose candidate data-parallel configurations for each of the individual tasks. Using a graph-theoretic model, the application space may be broken up as shown in Figure 11.7.

Each column in the graph is populated with vertices; each vertex corresponds to a mapping of the task corresponding to the given column to a potentially unique set of computational resources in the system. Each vertex has edges entering and exiting: entering edges correspond to communications with preceding tasks and exiting edges

Task 1   Task 2   Task c−1   Task c
(stage 1)   (stage 2)   • • •   (stage c−1)   (stage c)

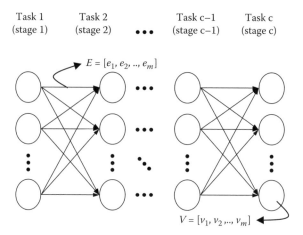

FIGURE 11.7   Sample graph with edge and vertex weights.

correspond to communications with succeeding tasks. Sensor signal processing applications may be represented as a stream signal processing flow, in which data move in one direction from task to task as they are processed. In this graph-theoretic model, task parallelism is represented along the horizontal axis of the graph, that is, pipelined, overlapping execution intervals, while data parallelism is represented by the mapping of each task in the application onto one or more parallel computational resources of each vertex. The graph-theoretic representation of data- and task-parallel applications and the corresponding flow of communication enable the graph generator of the NRM to capture the potentially astronomical number of combinations of application-to-resource mappings in a concise and efficient fashion.

Finally, the graph generator is also responsible for launching the executable for each task mapping (vertex) on target resources so that performance metrics can be collected as discussed in the next subsection.

11.5.2 Metrics Object

The metrics object (MO) is responsible for collecting performance metrics of tasks launched by the graph generator. The MO works closely with the graph generator to weight the graph. Each of the resources that hosts a task is time synchronized; metric agents (see NRM agents in Section 11.5.4) on each of the resources will provide the MO measurements for it to formulate the following performance parameters associated with graph weights: throughput; latency; RAM memory; and PE utilization. The MO will calculate another metric known as processor cost, which is a ratio of compute horsepower used in the mapping to the overall processing horsepower available in the network.

Link utilization percentages within each mapping are also measured, as well as inter-task utilization percentages. Map generation uses task column pairs to gather performance metrics in order to reduce the effort and time involved drastically. This is possible because the graph search algorithm will use a running tabulation of resource utilization percentages to ensure that simple linear superposition of path weights hold, given that these percentages remain under a given threshold. This is explained further in the next subsection. Once above the threshold, weight modifiers will be applied to subsequent stages during search. Finally, the MO will calculate a *network cost*, analogous to processor cost, which is a ratio of communications bandwidth used by a mapping pair with respect to the overall bandwidth available in the network.

11.5.3 Graph Search

The NRM must choose a path through the graph that determines the task mappings with which an application is launched on network resources. The choice of a path by the NRM is constrained by the time to result and the mandate to use a minimum set of networked resources. The data rate of the sensor data stream will drive required throughput for each task column in the graph; overall latency, which represents the total pipeline delay, is defined as the time period after which all data have been transmitted that a result is generated. To minimize any one application's impact on resource consumption, the path through the graph could be chosen to minimize the overall usage of computational or communication resources. This choice will depend on whether an application is launched in a network that is compute resource or communication bandwidth limited.

The graph search problem may be formalized as a discrete and constrained optimization problem: given a set of hard constraints, minimize (or maximize) a given objective function. As described in Section 11.5.2, the NRM may choose constraints and an objective function from the set of weights shown in Table 11.1.

Scalar weights are singular—that is, only one is associated with a given vertex or edge; vector weights may include many elements in an edge or vertex association. Because each vertex and edge may represent the combination of many PE and network communication elements associated with a mapping pair, processor, and network utilization may constitute weight vectors with many elements.

Although all weights tabulated previously may be chosen as constraints, memory, throughput, and network and PE utilization are not parameters that can be chosen as an objective function to optimize. This is because throughput is only a function of data rate; maximizing throughput has no impact on performance. Utilization also has no impact

TABLE 11.1 Graph Weights Associated With Individual Edges and Vertices, and Corresponding Sizes (types)

Weight	Type
Latency	Scalar
Throughput	Scalar
PE utilization	Vector
Processor cost	Scalar
Network utilization	Vector
Network cost	Scalar
Memory	Scalar

on performance and is only a measure of the validity of the solution. That is, subsequent stages in the graph may include resources from earlier stages, so keeping a running tabulation of utilization gives an indication of the onset of usage exceeding capacity and thereby degrading performance.

Network utilization and cost, PE utilization and cost, and memory are weights derived and constrained by the NRM, while data rate (throughput) and latency are application dependent and imposed by the sensor. The objective function that the NRM uses is chosen based on the desire to minimize an application's impact on resource usage or minimize the latency associated with an application's execution. For example, in a bandwidth-limited network, the graph search problem may be formulated as follows. While meeting application latency and throughput constraints, using less than 80% of the bandwidth available in the chosen network conduits and PEs and less than 100% of the available local PE-RAM memory, and using only a fraction of the overall processing bandwidth available network wide, select a parallel configuration for the application and the associated host resources using the smallest fraction of overall network bandwidth available. Even for moderately sized graphs (e.g., 1000 vertices by 10 stages), this is a complex combinatorial optimization problem; the general problem is NP complete. The authors have developed an iterative heuristic algorithm that has shown favorable performance for this class of problem in the quality of the solution and time to solution compared to other popular combinatorial optimization algorithms [31].

11.5.4 NRM Agents

The NRM agents are information and service links between the NRM and each of the resources. Agents must first register and be authenticated (e.g., using Kerberos [32]) before an NRM will invoke their services. This registration includes a characterization of the resource capabilities and services. When registered, the NRM will use these remotely deployed agents on computational resources to download and launch parameterized executables and modify the ACL of switches and routers under its control in the formation of DPNs. Agents also provide a mechanism for centralized software maintenance and configuration by acting as transaction managers in the download and installation of applications, databases, middleware, and so on. As stated earlier, the agents also provide a measurement object that is instantiated by applications to provide the NRM's MO with performance metrics during graph generation. Finally, agents give the NRM a view of the network state, periodically sending diagnostic messages indicating its operational status.

11.5.5 Sensor Interface

Sensors can be thought of as resources much like computational and communication resources, which are served by the NRM agents; thus, the sensor interface can be thought of as another type of NRM agent. Because many different sensor platforms could be served by an NRM-managed resource network, the sensor interface provides a common, abstract mechanism for communication between the NRM and the sensor platforms.

Sensors will request services through the sensor interface from the NRM using a well-defined middleware interface such as CORBA. This request for services involves requesting the proper application for the data stream that the sensor will be delivering to the network of resources as well as a request for the required metric constraints, such as throughput and latency (discussed in Section 11.5.2), needed to process the sensor data stream effectively. The determination of required constraints could involve negotiations between the sensor and the NRM through the sensor interface. The NRM uses the sensor interface to direct the sensor platform to start sending a data stream once the NRM has marshaled the resources that the sensor will need to satisfy the request. Finally, the sensor interface also facilitates communications between the sensor platform and the NRM regarding flow control, application shutdown, and so on.

11.5.6 Mapping Database

This mapping database is populated with data structures generated by the graph generator and MO; it represents the weighted graph-theoretic characterization of the various parallel permutations of an application that is mapped to networked resources. Graph search uses the mapping database to reconstitute a weighted graph for each application for which it is asked to find resources and the degree and form of parallelism needed to meet real-time constraints.

11.5.7 Topology Database

The topology database stores the current state of each of the resources; the graph generator and graph search use this database. Graph generator uses the topology database to determine which resources are available and most appropriate for candidate task-application mappings. Graph search uses this database to verify that resources are functional before a set of resources is chosen to host an application, as well as for generating and modifying weights associated with resource utilization. The topology database is generated during the discovery phase when the NRM first comes online (e.g., see Breitbart et al. [33] and Astic and Foster [34]). Alternatively, an administrator could choose to generate a topology database for the NRM that enumerates connectivity and capability among all computation and storage resources under its control. Agent reports (or lack thereof) will affect state changes in this database indicating whether the resource is online or offline.

11.5.8 NRM Federation

In a large network with a sizeable number of resources, using a single NRM may not be the most effective solution. In such a scenario, multiple NRMs are organized in a bilevel hierarchy; WAN NRMs interface with sensors and administer backbone communication resources, underneath which LAN NRMs administer and allocate compute resources for regional compute centers (RCCs). The primary responsibility of a WAN NRM is to choose a location on the network at which distributed computing is conducted for each application and to allocate WAN bandwidth for data flow between sensors and LAN resources. The objective of the WAN NRM is to load balance WAN traffic and computational load, taking into account the relative overall processing capability of each RCC. Each LAN NRM advertises its current processing capability using standardized metrics.

Each NRM is a federated collection, using a voting mechanism to elect an executor independently at the LAN and WAN levels. Each federation monitors the health of its executor by inspecting periodic diagnostic reports that the executor broadcasts. In response to an executor's diagnostic report (or lack thereof), the federation may choose to relieve the current executor of its responsibility and elect a new one. This prevents any one NRM failure from rendering resources unusable or disabling a sensor from contracting for network services.

Earlier paragraphs have detailed the LAN NRMs graph-theoretic representation of network resources, as well as its construction, weighting, and search criteria. The WAN NRM graph-theoretic representation and weighting are somewhat different from that of an LAN NRM; however, its construction and search criteria are formulated in an identical manner. The vertices in a WAN graph represent RCCs and each column corresponds to an application, while the concatenation of applications across the columns in a WAN NRM graph spans a mission. This is in contrast to an LAN NRM, in which the concatenation of tasks in its graph spans an application.

11.5.9 NRM Fault Tolerance

The absence of a heartbeat or the delivery of an error report by an agent alerts the NRM to a system fault. The NRM's fault tolerance policy is application dependent and is derived from a mandate by the developer and/or client. The policy is a trade-off between resource usage and seamless fail-over and includes redundant processing, surgical replacement, or restart of the application. Redundant processing is the most robust fail-over mechanism; the NRM simply assigns duplicate sets of resources to process the same data. If one set of resources fails, results are obtained from one of the duplicate sets. Redundant processing has the highest resource cost of all fault tolerant policies.

Conversely, the NRM may choose to replace the failed component dynamically so that processing is able to continue. In this case, the NRM may have allocated distributed network storage to act as a time-delay buffer in the event of resource failure. This would enable the application, if so instrumented, to pick up processing at the point at which the failure occurred. Finally, the NRM could simply choose to halt execution of the application and start over with a new set of processing resources, although a certain amount of data and the corresponding results may be lost irrevocably.

11.6 Experimental Results

A proof-of-concept experiment has been conducted at the MIT Lincoln Laboratory in which the NRM allocates distributed networked resources for a sensor data

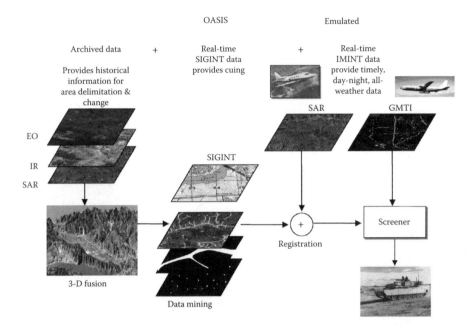

FIGURE 11.8 OASIS ATR and visualization.

fusion application in various scenarios [35]. The sensor fusion application is OASIS (operator assisted integrated systems), which is an automatic target recognition and visualization suite (see Figure 11.8). OASIS processes real-time SAR data and archived data generated by sensors with different modalities like EO and IR [36]. A block diagram of the experimental test bed is shown in Figure 11.9. The experimentation resource network consisted of three SGI O2 workstations, an eight-processor SGI Origin, an eight-node, dual Pentium3 class Beowulf cluster, and a PC workstation, which hosted the NRM.

For this experiment, two SGI O2s were used as sensor surrogates to transmit unprocessed complex SAR imagery generated with range and cross-range resolutions of 1 and 1/4 m, respectively. The sensor surrogates fed data into the OASIS processing chain. To keep the complexity of the system manageable, only the most computationally intensive stage was made remappable. This stage, the HDVI processing [3] (stage 3 in Figure 11.10), had six options for the NRM ranging from a single SGI processor to six Pentium3 class cluster processors. The HDVI processing was conducted on targets detected on the two images at both resolutions, and image formation was conducted on processors in the LAN. The performance metrics for the OASIS applications were determined with a combination of actual performance measurements and modeled performance analyses. Table 11.2 is a tabulated synopsis of the expected performance of the NRM and Table 11.3 shows the actual performance of the NRM. The expected and actual performance values compared very well.

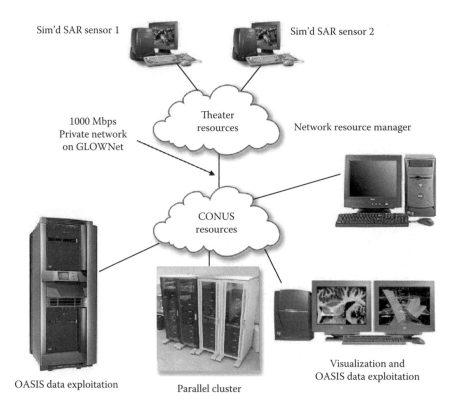

Sim'd SAR sensor 1

Sim'd SAR sensor 2

1000 Mbps
Private network
on GLOWNet

Theater
resources

Network resource manager

CONUS
resources

OASIS data exploitation

Parallel cluster

Visualization and
OASIS data exploitation

FIGURE 11.9   Experimentation resource network.

Because this network was PE resource limited, the objective of the NRM was to use the smallest fraction of PE bandwidth available across the network while meeting network conduit, PE utilization, latency, throughput, and network-wide bandwidth usage constraints. It is clear from the results that the NRM was able to tailor the communication and computation solution it delivered based on the particular application needs and the constraints imposed. The successful completion of this experiment has initiated further research and development to give the NRM greater functionality, automation, and flexibility.

Acknowledgments

The authors thank the members of the Precision Targeting via Collaborative Networking team at the MIT Lincoln Laboratory for formulating many of the concepts discussed in this chapter. The authors also thank Dr. Mari Maeda, formerly of DARPA/ITO, and Dr. Gary Koob of DARPA/IPTO for their encouragement and support of this project. This work is sponsored by the United States Air Force under Air Force contract F19628-00-C-002. Opinions, interpretations, conclusions, and recommendations are those of the authors and are not necessarily endorsed by the U.S. government.

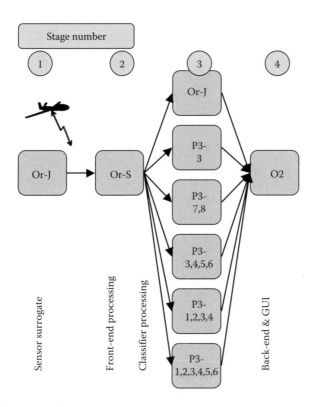

FIGURE 11.10　Graph of OASIS application onto the experimental resources.

TABLE 11.2　Synopsis of NRM Expected Performance

Experimental Configuration	Max Comm BW Requirement (MB/s)	Max Throughput Requirement (GFLOPS)	Processors Employed	Result Turn-Around Time
1 m data	26	0.7	1	1.6
1 m data with HDVI	26	2.2	2	2.6
1/4 m data	410	2.5	2	2.8
1/4 m data with HDVI	410	10	10	7

TABLE 11.3　Synopsis of NRM Performance

Experimental Configuration	Comm BW Measured (MB/s)	Throughput Measured (GFLOPS)	Processors Employed	Result Turn-Around Time
1 m data	26	0.7	1	1.4
1 m data with HDVI	26	2.2	2	2.5
1/4 m data	410	2.5	2	2.7
1/4 m data with HDVI	410	10	8	7.8

References

1. Usoff, J., Beavers, W., and Cox, J., Wideband networked sensors processing, in *Proceedings of High Performance Embedded Computing Workshop*, November 2001.
2. Cuomo, K. M., Pion, J. E., and Mayhan, J. T., Ultrawide-band coherent processing. *IEEE Trans Antenna Propag*, 1999; 47: 1094.
3. Benitz, G. R., High-definition vector imaging. *MIT Lincoln Lab J, Special Issue Super-Resolution*, 1997; 10(2): 147.
4. Nguyen, D. H., et al., Super-resolution HRR ATR Performance with HDVI. *IEEE Trans Aerospace Electron Syst*, 2001; 37(4): 1267.
5. Chan, V. W. S., Optical space communications. *IEEE J Select Topics Quantum Electron*, 2000; 6(6): 959.
6. Awduche, D., et al., RSVP-TE: extensions to RSVP for LSP tunnels, RFC 3209, http://www.faqs.org/rfcs/rfc3209.html, accessed December 2001.
7. Ash, J., et al., Applicability statement for CR-LDP, RFC 3213, http://www.faqs.org/rfcs/rfc3213.html, accessed January 2002.
8. Cormen, T. H., Leiserson, C. E., and Rivest, R. L., *Introduction to Algorithms*. New York: McGraw-Hill, 1993.
9. Strater, J., and Wollman, B., OSPF modeling and test results and recommendations. Mitre Technical Report 96W0000017, Mitre Corporation, 1996.
10. Perkins, C., *Ad Hoc Networking*. Boston: Addison-Wesley, 2001.
11. Floyd, S., and Jacobson, V., Random early detection gateways for congestion avoidance. *IEEE/ACM Trans Netw*, 1993; 1(4): 397.
12. Stevens, W., TCP slow start, congestion avoidance, fast retransmit and fast recovery algorithms, RFC 2001, http://www.faqs.org/rfcs/rfc2001.html, accessed January 1997.
13. Stadler, J. S., Performance enhancements for TCP/IP on a satellite channel, in *Proceedings of IEEE Military Communication Conference 1998 (MILCOM98)*, vol. 1, p. 270, Boston, MA, USA, October 1998.
14. Allman, M., Glover, D., and Sanchez, L., Enhancing TCP over satellite channels using standard mechanisms. RFC 2488, http://www.faqs.org/rfcs/rfc2488.html, accessed January 1999.
15. Berrou, C., Glavieux, A., and Thitimajshima, P., Near Shannon limit error-correcting coding and decoding: turbo codes. 1, in *Conference Rec. IEEE International Conference on Communication 1993 (ICC 93)*, vol. 2, p. 1064, Geneva, Switzerland, May 1993.
16. Foschini, G. J., Layered space-time architecture for wireless communication in a fading environment when using multiple antennas. *Bell Labs Tech J*, 1996; 1(2): 41.
17. Raleigh, G. G., and Cioffi, J. M., Spatio-temporal coding for wireless communications, in *Proceedings of IEEE Global Telecommunication Conference 1996 (GLOBECOM 96)*, vol. 3, p. 1405, November 1996.
18. Sisterson, L. K., et al., An architecture for semi-automated radar image exploitation. *Lincoln Lab. J.*, 1998; 11(2): 175–204.

19. Cooley, J. A., et al., Software-based erasure codes for scalable distributed storage, in *Proceedings of 20th IEEE Symposium Mass Storage Systems,* pp. 157–164, April 2003.

20. Luby, M. G., et al., Practical loss-resilient codes, in *Proceedings of 29th ACM Symposium on Theory Computing,* pp. 150–159, 1997.

21. Byers, J. W., Luby, M. G., and Mitzenmacher, M., Accessing multiple mirror sites in parallel: using tornado codes to speed up downloads, in *Proceedings of IEEE INFOCOM 1999,* pp. 275–283, March 1999.

22. Silberschatz, A., and Galvin, P., *Operating System Concepts,* 5th edn., Reading, MA: Addison-Wesley, 1998.

23. Wind River Systems, Inc. http://www.windriver.com/, accessed July 2003.

24. FSMLabs (Finite State Machine Labs), Inc. http://www.fsmlabs.com/, accessed July 2003.

25. Mercury Computer Systems, Inc. http://www.mc.com/, accessed July 2003.

26. Abbott, D., *Linux for Embedded and Real-Time Applications.* Amsterdam: Newnes, 2003.

27. Object Management Group. http://www.omg.org/, accessed July 2003.

28. Hoffmann, H., Kepner, J., and Bond, R., S3P: Automatic, optimized mapping of signal processing applications to parallel architectures, in *Proceedings of High Performance Embedded Computing Workshop 2001,* September 2001.

29. The vector, signal, and image processing library. http://www.vsipl.org/, accessed July 2002.

30. Reuther, A. I., and Goodman, J. I., Resource management for digital signal processing via distributed parallel computing, in Proceedings High *Performance Embedded Computing Workshop 2002,* September 2002.

31. Goodman, J. I., et al., Discrete optimization using decision-directed learning for distributed networked computing, in *Proceedings of IEEE Asilomar Conference on Signal, System Computers,* pp. 1189–1196, November 2002.

32. Neuman, B. C., and Ts'o, T., Kerberos: an authentication service for computer networks, *IEEE Commun.,* 1994; 32(9): 33.

33. Breitbart, Y., et al., Topology discover in heterogeneous IP networks, in *Proceedings of IEEE INFOCOM 2000,* pp. 265–274, March 2000.

34. Astic, I., and Foster, O., A hierarchical topology discovery service for IPv6 networks, in *Proceedings of 2002 Network Operations Manage. Symposium,* pp. 497–510, April 2002.

35. Reuther, A. I., and Goodman, J. I., dynamic resource management for a sensor-fusion application via distributed parallel grid computing, in *Proceedings of High Performance Embedded Computing Workshop 2003,* 2003.

36. Avent, R. K., A multi-sensor architecture for detecting high-value mobile targets, in *Proceedings of 2002 SIAM Conference on Imaging Science (IS02),* March 2002.

II

Architecture

12

Interoperability in Pervasive Environments

Imen Ben Lahmar
Institut Telecom SudParis

Hamid Mukhtar
*National University of
Sciences and Technology*

Djamel Belaïd
Institut Telecom SudParis

12.1 Introduction

Pervasive environments, according to Mark Weiser, are characterized by high dynamicity of the environment due to the mobility of the devices and the users and by high heterogeneity of integrated technologies in terms of networks, devices, and hardware/software infrastructures. To cope with such dynamics and diversity, pervasive computing systems should have the capacity to be deployed and executed in an *ad hoc* manner, integrating the available hardware and software resources. Such systems are characterized by their lack of required infrastructure and ease of formation. Moreover, the devices of a pervasive environment should be able to interoperate with one another in order to fulfill the service discovery and integration. Thus, supporting interoperability between heterogeneous system components becomes a key issue. Consider the following scenarios:

- While walking in a shopping mall, you suddenly feel hungry. You take out your mobile device to find nearby restaurants on the current floor of the mall, choose your favorite restaurant and order your food while you walk toward it.

- During a conference you take photos with some of your colleagues using your mobile phone. To share the photos with them, your mobile phone discovers the devices of your colleagues and sends the selected photos to each one of them on a single click.
- When you are away from home, your mobile device automatically connects to the social networks where you and your family members are connected and keeps you updated about their activities on the social network. The application also sends information about your traveling and location to your family members each time you change your place.

In all of the above-mentioned scenarios, the mobile device communicates with the services provided by the third parties or other devices in the proximity. The communication between different services and devices is possible only if they are interoperable. As it is evident, different devices should not only be able to communicate using lower level technologies forming *ad hoc*, peer-to-peer networks, but they should also be able to discover, advertise, or use services at a higher level.

In this chapter we survey interoperability issues in pervasive environments by investigating how different state-of-the-art solutions have been applied; these serve to illustrate to the reader how to perform real-world software and device interoperability. But before that, we provide a brief overview of the interoperability problems.

12.2 Interoperability

The IEEE Glossary defines interoperability as "the ability of two or more systems or components to exchange information and to use the information that has been exchanged" [1]. Thus, interoperability refers to the ability to access any available service provided by any device at anytime and anywhere. In pervasive environments, we need to integrate diversity of components programmed using different languages into an infrastructure that can successfully interact and cooperate [2]. However, this is not possible if the devices found in the pervasive environment are not able to interoperate among them.

Interoperability issues may arise at different levels: device/network level, protocol level, and service level. A device is said to be an interoperable resource if it is able to discover and interact with another one whose software, hardware, and network capacities are different. Issues related to software include differences of operating systems; issues related to hardware include difference in hardware architecture of two devices; and issues related to differences at network layer are the use of different network-layer protocols, that is, Bluetooth versus Wi-Fi. Service interoperability, on the other hand, corresponds to the capacity of services to interact and to compose with other services despite their heterogeneous interaction protocols. Service interoperability can be achieved only if the protocols used by the services are also interoperable (protocol interoperability). It is important to have interoperability at all these levels for successful communication between services found on different devices. If two services are interoperable but their respective devices are not, then they will not be able to exchange information with each other.

12.2.1 Service-Oriented Architecture: Toward Service Interoperability

One appropriate paradigm for building applications for pervasive computing systems is service-oriented architecture (SOA) [3]. Therein, applications are defined as a composition of re-usable software components, which implement the business logic of the application domain collectively by providing and requiring services to/from one another. The benefit of this approach lies in the looser coupling of the components making up an application. Services are provided by components and are platform independent, implying that a client from any communication device, using any computational platform, operating system, and any programming language, can use the service. This allows the separation of the business functionality (services) from its implementation (components). While different service providers and service clients may use different technologies for implementing and accessing the business functionality, the representation of the functionalities on a higher level (services) is same, resulting in service-level interoperability. The underlying assumption is that the various service providers and clients may communicate using some common set of well-known protocols, thus achieving protocol-interoperability at the same time.

SOA assumes that services can dynamically be invoked and composed. Thus, SOA is designed to allow developers to overcome many distributed computing challenges, including application interoperability and integration, and so on. However, the SOA paradigm alone cannot meet the interoperability requirements of pervasive computing systems. One of the most relevant drawbacks is that the interaction between services is based on their syntactic description. Thus, it does not take into account the semantic matching of services; hence, this may prevent the interoperability between services.

Zender et al. [4] cite some challenges related to the diversity of SOA technologies that represent a major issue for the interoperability between SOA services:

- First, SOA technologies differ by means of active or passive service discovery, that is, a request to services is made actively (as in Web services and Jini) or just wait for service announcements on a public channel (as in UPnP).
- Secondly, some SOA technologies depend on specific platforms or programming languages. For instance, Jini requires a Java Runtime Environment and utilizes Java for service execution. This means an application based on a technology other than Java is not able to interact with Jini clients or service providers.
- Thirdly, SOA technologies vary in the use of brokers that contain the published services. Brokers can be implemented as dedicated and sometimes very complex software components (e.g., UDDI brokers for Web services), simple files (like XML-based WS-Inspection documents), or even as special services (e.g., Reggie for Jini).

Thus, SOA is not sufficient to resolve the interoperability requirements. Hence, we require a more theoretical and formal level to consider the interoperability issues.

12.2.2 Interoperability Challenges

The interoperability of the pervasive computing systems is challenged by the diversity of the *service discovery protocols* (SDPs) (like Jini, service location protocol (SLP),

Universal Plug-n-Play (UPnP), etc.) since devices made by different manufacturers can implement standardized SDPs differently. Moreover, the published services can support heterogeneous *interaction protocols* (like SOAP, RMI, etc.) Thus, applications are required to be aware of the different used protocols in order to integrate the discovered services. Furthermore, services may be described in different syntactic ways, implying that they have different signatures. Toward this heterogeneity, we require a mechanism that allows services to interoperate despite their syntactic heterogeneity. For example, consider that two services named Printer Service and Printing Service are similar in their semantics—both of them provide the capability to print documents—they may be treated as different services if they are to be judged on syntactic similarities. Therefore, we need an approach for eliminating syntactic heterogeneity. Beyond the ability of two or more services to exchange information, there is a need for semantic interoperability to interpret the exchanged information. For example, a discovered service will correspond syntactically to the required one; however, it may not match the required concept. Thus, there is a semantic mismatching between the required and provided services that will prevent services to interoperate together.

12.3 Frameworks and Middleware for Interoperability in Pervasive Environments

In the literature, we find a number of approaches for achieving interoperability at various levels: services, protocols, devices, and network. Various research works at different places around the world have led to the development of a number of middleware and frameworks for interoperability. Next, we are going to represent them under the categorization of service interoperability, service protocols interoperability, and devices or network interoperability.

12.3.1 Service Interoperability

The most common approach to deal with syntactic heterogeneity is to use semantic-based techniques. Semantic interoperability represents a crucial need to facilitate system components integration. Semantics of an entity encapsulate the meaning of this entity. This meaning is captured in a vocabulary of terms and their interrelationships known as ontology. Two different terms matching to the same concept will be treated to be semantically equivalent. In this way, semantics of entities become machine interpretable, enabling machine reasoning on them.

In the following, we present some approaches that tackle the semantic interoperability in pervasive environment by using explicit semantic descriptions of services. Some of these approaches require service interoperability for achieving service composition. *Service composition* refers to combining two or more of the existing services to form a new service. Composing services together is a challenge for SOA applications in pervasive environments [5] and interoperability is among one of these challenges, such as cross-domain semantic search, browsing, and recommending.

12.3.1.1 Amigo

Amigo is a service-oriented system architecture designed to develop a networked home system enabling the integration of devices and their hosted services in a pervasive computing environment [6]. Figure 12.1 presents an abstract reference service architecture for the Amigo system. The proposed architecture is divided into three layers: application, middleware, and platform. At the application layer, services are functionally described based on a common syntactic service description language in order to enable service discovery and invocation independently of service implementation details. The middleware layer includes essential mechanisms for service discovery and service communication, while the platform layer offers base system and network support.

To overcome the service interoperability issues, the Amigo approach builds upon the semantic modeling of information and functionality. Amigo supports semantic modeling of both functional and non-functional aspects of the software system. For this purpose, the Amigo architecture uses the concept of *component* and *connector* to represent services and their interactions, respectively. As such, the description of components and connections are based on semantic concepts. Such concepts come from established ontologies. The interoperability mechanisms among various components and connectors comprise *conformance relations* and *interoperability methods*. Conformance relations aim at checking conformance (matching) between services for assessing their capacity to interoperate and enable reasoning on service composability. Interoperability methods aim at enabling integration of partially conforming services. Conformance relations together with interoperability methods establish semantic-based service interoperability.

12.3.1.2 COCOA

COCOA (COnversation-based service COmposition middlewAre) [7] is a semantic middleware, that deals with several aspects of service composition such as services discovery, registration, composition, and QoS evaluation. The term *user task* is used to describe the service composition. The middleware is capable of composing services out of heterogeneous services described in different description languages and supports interoperability between them at syntactic and semantic levels. This is carried out by specifying service

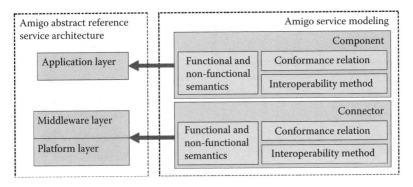

FIGURE 12.1 Amigo service modeling.

conversations as finite state automata, which enables the automated reasoning about service behavior independently from the underlying conversation specification language.

To manage interoperability among different service description mechanisms, COCOA introduces the architecture of a semantic service registry for pervasive computing. This registry allows heterogeneous service capabilities to be registered and retrieved by translating their corresponding descriptions to a predefined service model through the *Description Translator*. This service model is defined in OWL-S.

Figure 12.2 shows the service composition process in COCOA. For service composition, the first step COCOA performs is a semantic matching of interfaces, called *Service Matching*, which leads to the selection of the set of services that may be useful during the composition. Then, COCOA performs a conversation matching starting from the set of previously selected services, thus obtaining a conversation composition that behaves as the task's conversation. The matching is based on a mapping of OWL-S conversations to finite state automata. This mapping facilitates the conversation composition process, as it transforms this problem to an automaton equivalence issue. Once the list of sub-automata that behaves like the task automaton is produced, a last step consists in checking—through the *Service Conformance* and *Service Coordination*—whether the atomic conversation constraints have been respected in each sub-automaton. After rejecting those sub-automata that do not verify the atomic conversation constraints, COCOA selects arbitrarily one of the remainders, as they all behave as the user task. Using the subautomaton that has been

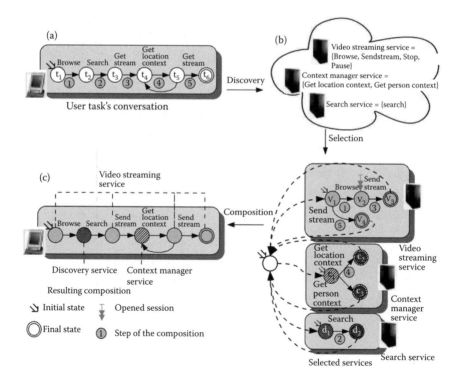

FIGURE 12.2   Service composition in COCOA.

selected, an executable description of the user task that includes references to existing environment's services is generated, and sent to the *Service Discovery & Invocation* that executes this description by invoking the appropriate service operations.

12.3.1.3 PerSeSyn Platform

The PerSeSyn (Pervasive Semantic Syntactic) service discovery platform [8] is defined to be a part of a middleware for ambient computing environments that processes service descriptions from both semantic and syntactic service description languages.

It is composed of the PerSeSyn service description model (PSDM), and its instantiation, the PerSeSyn Service Description Language (PSDL). PSDM is a conceptual model for enabling semantic mapping between heterogeneous service description languages. PSDL, which is an instantiation of PSDM, is a combination of existing standards for service specification, namely SAWSDL and WS-BPEL. PSDL is then employed as the common representation for service descriptions and requests. Based on PSDM and PSDL, they define a set of conformance relations for matching heterogeneous service descriptions to support the semantic interoperability of services.

The matching between services is performed at various degrees of conformance between services and their clients; services are ranked with respect to their suitability for a specific client request. Out of various candidate services for a given client request, the most promising (highest ranked) service is selected. The different cases of matching of heterogeneous service descriptions are outlined in the table in Figure 12.3. In this table a service request and a service advertisement can be described as: (i) a syntactic capability name; (ii) a list of syntactic capabilities; (iii) a list of semantic capabilities. Additionally, service advertisements can be further described as (iv) a syntactic capability with an associated syntactic conversations and (v) a semantic capability with an associated semantic conversation. To deal with these different cases, a number of matching algorithms have been outlined and can be found in [8].

12.3.1.4 My Service Integration Middleware (MySIM)

Ibrahim et al. [9] tackles the interoperability for service composition in pervasive environments. They propose a generic service composition middleware model to describe the service composition based on semantic interoperability of services.

		Request		
		Syntactic capability name	List of syntactic capabilities	List of semantic capabilities
Service	Syntactic capability name	SynNameMatch	SynNameMatch	SynNameMatch
	List of syntactic capabilities	SynNameMatch	SynSigMatch	SynSigMatch
	List of semantic capabilities	SynNameMatch	SynSigMatch	SemSigMatch
	Syntactic conversation	SynNameMatch + ExeConv	SynSigMatch + Execonv	Synsig match + ExeConv
	Semantic conversation	SynNameMatch + ExeConv	SynSigMatch + ExeConv	SemSigMatch + ExeConv

FIGURE 12.3 Interoperable matching of services capabilities in PerSeSyn.

The middleware is defined as a framework containing tools for composing services. However, their proposed model is an abstract layer that allows describing the service composition middleware independently of a service technology, language, platform, and so on.

MySIM is split into four components: the translator, the generator, the evaluator, and the builder, as shown in Figure 12.4. The request descriptions are translated to a system comprehensible language in order to be used by the middleware. Once translated, the request specification is sent to the generator. The generator will try to provide the needed functionalities by composing the available service technologies, and hence composing their functionalities. It tries to generate one or several composition plans with the same or different technology services available in the environment.

Composing service is technically performed by chaining interfaces using a syntactically or semantically method matching. The interface chaining is usually represented as a graph or described with a specific language. The evaluator chooses the most suitable composition plan for a given context. This selection is done from all the plans provided by the generator. In pervasive environments, this evaluation depends strongly on many criteria like the application context, the service technology model, the quality of the network, the nonfunctional service QoS properties, and so on.

The builder executes the selected composition plan and produces an implementation corresponding to the required composite service. It can apply a range of techniques to realize the effective service composition.

12.3.2 Services Protocols Interoperability

Service Discovery Protocols (SDPs) enable finding and using networked services without any previous knowledge of their specific location. They are essential to enable devices and services to properly discover and communicate with each other. Several SDPs such

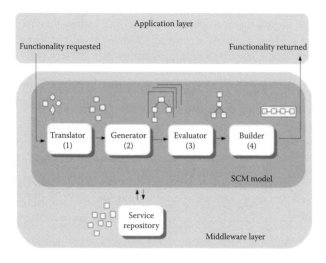

FIGURE 12.4 Architectural model of MySIM.

as Jini, SLP, UPnP, Bonjour, and Salutation are available and widely used in different systems. Some protocols like Jini are restricted to Java-based technologies only. Moreover, the published services can support heterogeneous interaction protocols (like SOAP, RMI, etc.).

Due to these differences, the existing protocols present many heterogeneity issues ranging from protocol design and implementation to its use. This means that components and services cannot discover each other if they do not use a common protocol and they cannot interact between them if they support different interaction protocols. Therefore, there is a need to tackle the service protocols (discovery and interaction) interoperability to enable communication among services. Some of the approaches that address service protocols interoperability problem are discussed next.

12.3.2.1 INDISS System

The INDISS (INteroperable DIscovery System for networked Services) [10] system is based on event-based parsing techniques to provide full-service discovery interoperability to any existing middleware. The system adapts itself to both its environment and its host to offer interoperability anytime and anywhere. Hosting INDISS enables the networked home system to discover and interpret all the services available in the home environment, independent of underlying middleware technologies. As a result, service discovery interoperability is provided to applications without altering them: applications are not aware of the existence of INDISS.

The basic principle behind INDISS is its ability to listen to various SDPs and then forward it to the appropriate SDP unit. Inside an SDP unit, a parser component generates semantic events from input protocol messages, and a composer component generates output protocol messages from semantic events. The principle of this approach is to translate SDPs messages into sets of events exchanged between the system's components. The parser extracts the events from the message and passes it to the event bus. Any service discovery system that is attached to the event bus gets these events and composes events relevant to its protocol format. The events are then sent over the network.

Parsers and composers are dedicated to specific SDP protocols. To support more than one SDP, several parsers and composers must be embedded into the system. Hence, interoperability between two SDPs is realized by combining a parser and a composer from each SDP in either direction.

The detection of an SDP is ensured by a Monitor component of the system that is able to subscribe to several SDP multicast groups, and to listen to all their respective ports. As an example, in Figure 12.5, the Monitor receives an incoming UPnP message from the service, and forwards it to the parser of the UPnP unit. The UPnP parser then translates the message into a set of events. The SLP composer receives the relevant (subset of) events according to its event filter, and composes the adequate SLP messages. This message is then sent onto the SLP multicast group and received by the client.

12.3.2.2 MSDA

While INDISS is a transparent approach—since the interoperability layer is located close to the network and directly translates SDPs messages to/from the various SDPs—MSDA (Multi SDP) is an explicit one based on protocol integration. The MSDA middleware

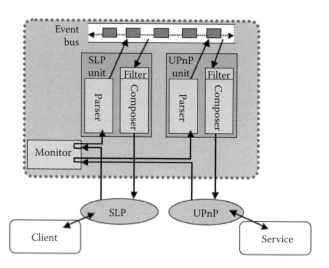

FIGURE 12.5 INDISS architecture.

platform aims to manage the dynamic composition of the networks in the environment [11]. It integrates the existing middleware protocols, and provides a generic service to clients for performing service discovery and access in the environment. MSDA is an additional layer on top of the existing SDPs. MSDA is instantiated independently in each network of the environment, and each instance registers as a service (the MSDA service) with the active SDPs in the network. The MSDA service provides a pull-based service discovery interface (i.e., clients issue discovery requests and are returned the matching services).

As shown in Figure 12.6, the main components of MSDA are: (i) the MSDA Manager that manages discovery and access requests within the network for local and remote clients; (ii) Plug-ins that interact with specific SDPs to collect service information, register the MSDA service to be used by local clients, and perform service access on behalf of remote clients; (iii) Transformers that extend service descriptions with context information; and (iv) MSDA Bridges that assist MSDA Managers in expanding the service discovery and service access to other networks in the whole pervasive environment.

In MSDA, services are described using the MSDA Description format, which is a generic and modular service description format. In addition to the service information collected from the SDP description, the MSDA description also contains information to assist the remote access to the service (e.g., network path leading to the destination network, access protocols supported) and information to control the dissemination of the description (e.g., minimum bandwidth requirements). A given SDP Plug-in first generates an initial MSDA description based on the SDP description it receives, and forwards it to its Transformers that will extend the MSDA description with service-specific context information. The MSDA Description is then forwarded to the MSDA Manager that adds network-specific context information. Each MSDA Bridge that forwards a description also extends it with context and propagation information. Discovery requests in MSDA are created by MSDA Managers on behalf of client applications and are similar to service descriptions.

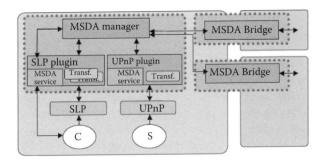

FIGURE 12.6 MSDA architecture.

12.3.2.3 Pervasive Middleware Architecture

Uribarren et al. [14] presents a pervasive middleware architecture that aims at integrating devices and services. It is based on extracting the functional description of the available services, representing them using a unified service specification, and exposing them using standard service technologies (UPnP, WS, Jini, etc.).

Figure 12.7 depicts an overview of the middleware model. The Drivers layer is responsible for giving access to a device, extracting the functional description of the available services, and generating a full description of each service using a unified syntax. Thus, a device is extended by drivers of different technologies available in the environment.

A UPnP-based application will search for UPnP devices in the network to be able to use the services that they provide. In the same way, a Web service-based application will consume the available Web services. Both applications can access the same service, but using different discovery and interaction protocols. This is achieved by the instantiation of service instances in the corresponding service space.

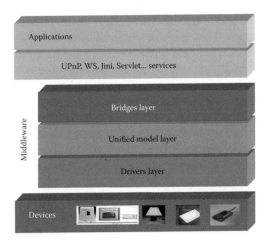

FIGURE 12.7 Overview of pervasive middleware.

The Bridges layer is composed of a set of components that act as gateways to standard-ized service spaces. A bridge, using the unified description of a service, is responsible for procuring an interface for the corresponding service space. For example, a bridge to UPnP should instantiate an UPnP device that offers a service based on the unified description. A bridge to Web services should also instantiate a Web service based on the same unified description. Thus, new device technologies can be incorporated, just add-ing new drivers. In a similar way, new service technologies could be integrated by means of new bridges. Hence, the service technologies would be extended by a bridge to map the service described by the unified model to a specific service using the service technologies.

12.3.3 Devices and Networks Interoperability

Given the various inter-device communication technologies (Bluetooth, Wi-Fi, GPRS, etc.), network heterogeneity becomes an important issue to consider for achieving interoperability. Two device fulfilling service and protocol interoperability will not be able to communicate with each other if there is an interoperability issue at the network level. In this section, we discuss some approaches for interoperability in pervasive envi-ronments at a device or network level as defined previously. Interoperability at device or network layer caters for lower level heterogeneities hardware, software, and network. Two approaches are presented here.

12.3.3.1 Middleware for Network Heterogeneity

Mukhtar et al. [15] present a graph-theoretic approach for dealing with the problem of network heterogeneity. A composition of services, which is described at a higher level, is converted to a graph representation such that the various services and their interactions are represented by nodes and their edges, respectively, in the graph. In parallel, the underlying network is also modeled as a graph such that devices and their interconnec-tions are represented by nodes and their edges, respectively, in the graph. Two devices in the latter graph will interconnect if and only if they have a common protocol for com-munication. For example, two devices both supporting Wi-Fi will be represented by two nodes connected by an edge in the graph. On the other hand, there will be no edge between two nodes representing a Wi-Fi- and a Bluetooth-enabled device.

For a client requesting a service, the required service may be available on any device including those that support a different protocol than the client. The problem is to ensure that the client should be able to use a service of a device that speaks the same protocol. For this purpose, the service graph is mapped onto the network graph. All the services can be mapped on different devices if and only if there is a one-to-one match between the two graphs. An example application showing various services in a heterogeneous environment is shown in Figure 12.8.

When selecting a device, the middleware also evaluates device capabilities in terms of hardware and software. For example, if the client requires the usage of audio unit, then the device providing the service should be capable of providing audio. This fact is con-sidered during matching of service and network graphs.

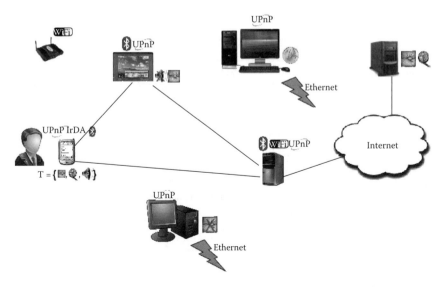

FIGURE 12.8   Overcoming network heterogeneity using a graph-based approach.

12.3.3.2 ubiSOAP

The ubiSOAP communication middleware is specifically designed for resource-limited portable ubiquitous devices which can be interconnected through multiple wireless links (i.e., Bluetooth, Wi-Fi, GPRS, etc.) [16]. It is based on Web service standards for implementing ubiquitous services by extending the standard SOAP protocol with group messaging connectivity.

As shown in Figure 12.9, the lower layer is the ubiSOAP connectivity layer which selects the network based on user policies, as users may require the utilization of a certain type of network for personal reasons. This layer also identifies and addresses the applications in the networking environment. The upper layer is the ubiSOAP communication layer which extends the use of the standard SOAP protocol for messaging between the participating services by introducing SOAP multi-network multi-radio point-to-point transport and group (multicast) transport protocols.

12.3.3.3 Adaptation Patterns

Ben Lahmar et al. [17] have proposed to use adaptation patterns to overcome the mismatching that are related to software, hardware, and network characteristics of devices and that are detected at in it time or during the execution of the application.These captured mismatches imply that the discovered services/devices cannot interoperate between them. Thus, a certain adaptor is needed to overcome the particular mismatch. Several adaptation patterns have been defined for different situations. For example, a Proxy adaptation pattern is proposed to overcome the heterogeneity of network interfaces of devices. Each adaptor is described using an adapter template which is then instantiated according to the use case. The adaptor template consists of an adaptive logic component whose implementation is generated dynamically and an extra-functional

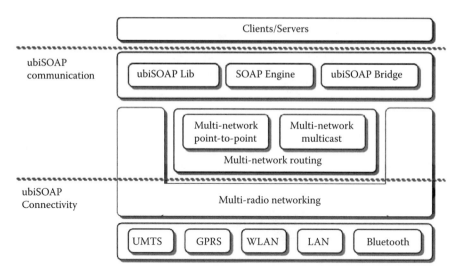

FIGURE 12.9   ubiSOAP architecture.

component that provides transformation services allowing, for example, encryption, compression, and so on.

Figure 12.10 gives a component-based description of the proxy pattern following the adapter template. As it can be seen, the proxy pattern represents a specific case of the adapter template. It contains only a proxy component representing the adaptive logic component that forwards the call of the service to the remote component. The adaptive logic component of the proxy pattern is created in intermediate C device that supports the both connection interfaces of the A and B devices.

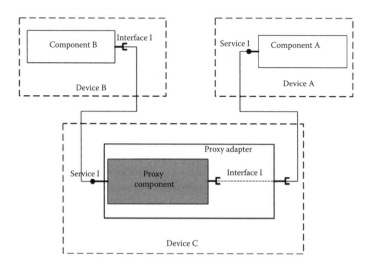

FIGURE 12.10   Proxy adaptation pattern.

12.4 Discussion

The interoperability between services or devices can be classified into three categories. The first category is related to service interoperability, that is, to allow a service requestor to be able to use a service offered by a service provider. A number of approaches were discussed that benefit from the use of semantic technology for service matching and interoperability.

The second category is related to interoperability between service discovery and interaction protocols. Devices made by different manufacturers can implement different protocols for device communications; however, it is not possible that each device will be required to embed all the available protocols in order to discover and to interact with new services. Few approaches for eliminating protocol disparity were presented in this chapter.

The third category concerns the heterogeneity arising at devices level due to network-level protocols such as Wi-Fi, Bluetooth, Infrared, and so on. Three different approaches to consider network heterogeneity have been discussed.

12.5 Conclusion

This chapter outlined various approaches toward interoperability in *ad hoc* pervasive computing environments. These approaches build on research background coming from the domains of service-oriented architectures, semantic web services, and software architectures. It is clear from the above discussion that interoperability in pervasive computing systems is still a major and open issue. It is hoped that as pervasive environments are becoming a reality, vendors, and manufacturers will need to address the issue of interoperability for better utilization of their hardware and software products.

References

1. Morell, L. J., A theory of fault-based testing. *IEEE Trans Software Eng* 1990; 16(8): 844–857.
2. Henricksen, K., Indulska, J., and Rakotonirainy, A., Infrastructure for pervasive computing: challenges, in *Workshop on Pervasive Computing Informatik*, Vienna, Austria, 2001.
3. Papazoglou, M., and van den Heuvel, W-J., Service oriented architectures: approaches, technologies and research issues. *VLDB J* 2007; 16(3): 389–415.
4. Zender, R., Lucke, U., and Tavangarian, D., SOA interoperability for large-scale pervasive environments, in *International Conference on Advanced Information Networking and Applications Workshops*, 2010, pp. 545–550.
5. Erl, T., *Service-Oriented Architecture: Concepts, Technology, and Design*. Upper Saddle River, NJ, USA: Prentice Hall PTR, 2005.
6. Georgantas, N., Mokhtar, S. B., Bromberg, D., Issarny, V., Kalaoja, J., Kantorovitch, J., Gérodolle, A., and Mevissen, R., The Amigo service architecture for the open networked home environment, in WICSA'05: *Proceedings of the 5th Working IEEE/IFIP Conference on Software Architecture*, Washington, DC, USA, 2005. IEEE Computer Society, pp. 295–296.

7. Mokhtar, S. B., Georgantas, N., and Issarny, V., COCOA: conversation based service composition for pervasive computing environments, in *2006 ACS/IEEE International Conference on Pervasive Services*, 2006, pp. 29–38.
8. Mokhtar, S.B., Raverdy, P-G., Urbieta, A., and Cardoso, R. S., Interoperable semantic & syntactic service matching for ambient computing environments. *International Journal of Ambient Computing and Intelligence (IJACI)* 2009; 2: 13–32.
9. Ibrahim, N., Le Mouël, F., and Frénot, S., Mysim: a spontaneous service integration middleware for pervasive environments, in ICPS'09: *Proceedings of the 2009 International Conference on Pervasive Services*, New York, NY, USA, 2009, ACM Press, pp. 1–10.
10. Bromberg, Y-D., and Issarny, V., Indiss: interoperable discovery system for networked services, in *Middleware'05: Proceedings of the International Conference on Middleware*, Grenoble, France, 2005, pp. 164–183.
11. de La Chapelle, A., Issarny, V., Raverdy, P-G., and Chibout, R., The MSDA multi-protocol approach to service discovery and access in pervasive environments, in 6th International Middleware Conference, Grenoble, France, 2005.
12. Amigo Consortium. Amigo middleware core enhanced: Prototype implementation & documentation. Deliverable 3.3, European Amigo Project, October 2006.
13. Thomson, G., Sacchetti, D., Bromberg, Y-D., Parra, J., Georgantas, N., and Issarny, V., Amigo interoperability framework: dynamically integrating heterogeneous devices and services. *Constructing Ambient Intelligence* 2008; 11: 421–425.
14. Uribarren, A., Parra, J., Uribe, J. P., Makibar, K., Olalde, I., and Herrasti, N., Service oriented pervasive applications based on interoperable middleware, in The 1st International Workshop on Requirements and Solutions for Pervasive Software Infrastructures, Dublin, Ireland, 2006.
15. Mukhtar, H., Belaïd, D., and Bernard, G., A graph-based approach for *ad hoc* task composition considering user preferences and device capabilities, in *Workshop on Service Discovery and Composition in Ubiquitous and Pervasive Environments*, New Orleans, LA, USA, December 2008.
16. Caporuscio, M., Raverdy, P-G., and Issarny, V., ubiSOAP: a service oriented middleware for ubiquitous networking. *IEEE Transactions on Services Computing* 2010; 2010.
17. Ben Lahmar, I., Belaïd, D., and Mukhtar, H., Adapting abstract component applications using adaptation patterns, in *Proceedings of the Second International Conference on Adaptive asnd Self-adaptive Systems and Applications*. Lisbon, Portugal, Adaptive 2010, pp. 170–175.

13

P2P-Based VOD Architecture: A Common Platform for Provisioning of Pervasive Computing Services

Sami Saleh
Al-Wakeel
King Saud University

13.1 Introduction

Next generation network (NGN) provides multimedia services over broadband IP-based networks, which supports high-definition TV (HDTV), and DVD quality video content. Video-on-demand (VOD) is expected to be an increasingly popular media streaming service over these NGN. It will allow users to access various services and applications such as video information retrieval services, entertainment movies, interactive games, collaboration and conferencing systems, and distance learning, while maintaining its required levels of security, interactivity, and reliability [1]. To fulfill the requirements of an IP-based VOD (IP/VOD) service, IP/VOD systems require continuous data transfer over relatively long periods of time, media synchronization, very large storage, and special indexing and retrieval techniques adapted to multimedia data types. Therefore, IP-based video services are built through merging three industries: computing, communication, and broadcasting.

IP/Video content delivery also consumes large amounts of network bandwidth due to its scalability, that is, to support a large number of clients, thus imposing a heavy burden on the network and the system resources. A HD stream, for instance, may require 10 Mbps or more of bandwidth under MPEG-2 encoding. Therefore, any network link that handles many subscribers, each capable of demanding one or more IP/VOD streams, must have enough bandwidth to meet the users' demands. In addition, system client must comply with the necessary buffer size and video request rate for the IP/VOD delivery policy [2]. Thus, a cost-effective design for an IP/VOD system needs to evaluate a collection of various VOD system components [3]. In terms of the IP/VOD transmission network, the system design must guarantee the required bandwidth for video traffic, and this bandwidth must support the IP/VOD service quality needs and meet its IP packet loss polices.

Due to these technical and operational challenges, economic and design constraints, and due to the need for significant initial investments for full-service provision, more exploration of the large-scale deployment of IP-based IP/VOD is still required.

Recently, a peer-to-peer (P2P) pervasive network has been proposed to meet these challenges and is considered an appropriate candidate for designing a scalable VOD service distribution architecture. Besides, it allows an optimal use of the network resources by moving the computing and bandwidth requirements toward the side of the network clients.

This chapter is organized as follows. In next section, we describe briefly the main characteristics and the basic components and various architectures of the IP/VOD system. In Section 13.3, we present the P2P pervasive architecture and access mechanisms that have been proposed for VOD networks. In Section 13.4, we present the mathematical model developed to estimate the VOD network bandwidth requirements. In Section 13.5, the advantages of VOD P2P pervasive networks are illustrated by presenting a case study that introduces a comparison of a centralized system to the (P2P) VOD architecture bandwidth requirements as a function of the key parameters of the P2P pervasive network design. In the last section, we summarize the key issues and our main conclusions.

13.2 IP/VOD Network Components, Architecture, and Mechanisms

A typical IP/VOD broadband network consists of a number of remote client clusters that communicate their video requests via the network inbound links (client-to-server) and their video broadcasts via the network outbound links (server-to-clients). The components of the IP/VOD network comprises the service control point(s), intelligent peripherals (such as multimedia storage servers, set-top boxes, and cluster switches), and primary multimedia routers; these determine the system performance and communication costs. IP-based video delivery encodes all video, whether broadcast or VOD, into IP data packets and transmits them to subscribers over IP networks. IP/VOD system architecture designs range from the simplest centralized system to complex distributed systems. The architecture design for IP/VOD systems is based on the incorporation of such continuous media into a large array of extremely high-capacity storage devices, such as optical or magnetic disks, which are randomly accessible, with a short seek time, and are permanently on-line [4,5]. Video object delivery from the server to the client in general may be composed of multiple media streams, such as audio and video, whose retrieval must proceed so as to not only maintain continuity of playback of each of the constituent media streams, but also preserves the temporal relationships among them [6]. Various network architectures exist for producing IP/VOD designs that minimize network costs and fulfill the service quality constraints. In what follow, we focus on two main widely used architectures: centralized network and distributed local proxies.

13.2.1 Centralized Network Architecture

In centralized architecture, all remote clusters communicate with the network's centralized primary servers through a broadband channel, which represent the backbone of the network; there are no local servers. All client requests are received by the primary central server's router, which acts as a gateway where data decoding, de-multiplexing, regeneration, multiplexing, encoding, and carrier switching take place. The video primary servers then retransmit the video data information to the destination clients via the outbound links. The basic characteristic of the centralized VOD system, as depicted in Figure 13.1,

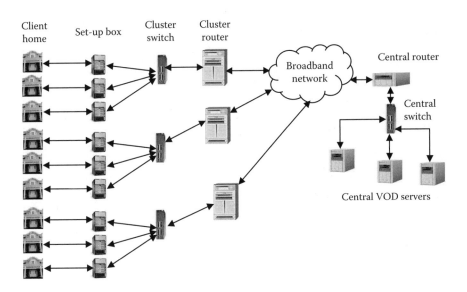

FIGURE 13.1 Centralized VOD network.

is that the multimedia information is always transported on demand from the central multimedia server to the subscribers through the network.

If the server fails or becomes incapable of supporting existing connections, these connections will be blocked [5]. This solution suffers from very significant scalability problems, especially when scaled up for millions of potential users. Providing access to a large library of pre-encoded content using this approach requires enormous servers with enormous network connections [7].

13.2.2 Distributed Local Proxy IP/VOD Architecture

In distributed local proxy architecture, local proxy servers are installed at strategic locations in the network (closer to the clients). Remote clusters can communicate with the network's centralized primary servers as well as with its local proxy servers. Each local cluster server can support a number of customers connected to it through a cluster switch. The customers are connected to the central server's location through the cluster router, which acts as an interface between the client cluster and the broadband network.

The main idea of distributed IP/VOD local proxies (as shown in Figure 13.2) is to distribute the centralized multimedia server functions within the network using the concept of local proxy storage. If the user cannot be served by the local proxy multimedia server for any reason, such as the blockage of the local multimedia server, or the multimedia information is not available in the local proxy server, then the request of the user will be transported to the centralized multimedia server. By locating the proxy server close to

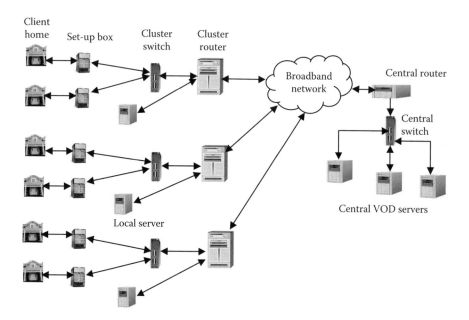

FIGURE 13.2   Distributed local proxy VOD network.

the user, it is expected that there will be significant reductions in the load on the system as a whole [1,5]. Another advantage of the distributed local proxy VOD system is that it can be expanded in a horizontal manner for system scalability and evolution. It can start from an initial two-level system (with a centralized multimedia server and one local video server) to a system with as many local servers as needed. Compared with the centralized multimedia server system, the distributed system may utilize a lower than average network bandwidth and have higher system reliability, but at the expense of needing a significant amount of local storage systems.

13.3 P2P VOD Pervasive Network

Traditionally, VOD services have been built based on centralized or distributed architectures. However, these architectures cannot provide the quality of service needed to serve a large population of users due to their limited outbound channel capacity from server to clients. Besides, both solutions suffer from very significant scalability problems, especially when scaled up for millions of potential users. Providing access to a large library of pre-encoded content using this approach requires enormous servers with enormous network connections [7]. For both system architectures, it was found that the incremental increase in the interactive traffic due to movie surfing compensates for any gain in the bandwidth made by using multicasting for movie delivery.

Recently, a P2P pervasive network has been proposed to meet the challenge of providing live and interactive video broadcast to a large number of clients over a wide area [8]. The P2P VOD architecture distributes the video files to the user's set-top boxes in a pervasive fashion at the network edge. Thus, a P2P-based pervasive network is an appropriate candidate for designing a scalable VOD service distribution architecture, as the computing and bandwidth requirements are pushed toward the side of the network clients. Besides, it allows the optimal use of the network resources by building multi-source streams from neighboring contributing clients to a requesting client. This in turn results in the minimization of VOD request rejection rates for a very large content library [9]. At the same time, the decreasing of processor cost and size at the VOD client location will enhance trends toward the vision of fully pervasive multimedia social computing and represents a big step along this path. From economic point of view, P2P pervasive network for VOD is a very cost-effective system, as it allows the network bandwidth to be provisioned in a managed manner, it allows an optimal use of the network bandwidth resources [9,10].

13.3.1 P2P Networking-Related Research

To respond to the main concern of VOD pervasive network providers, a large number of research works have concentrated on P2P VOD service distribution, and on VOD live P2P streaming to very large numbers of clients [11–18]. Many of these proposals were aimed toward solving the VOD system scalability problem [19]. The goals of other P2P VOD pervasive system research studies were to maximize the aggregate throughput among all the VOD peers, and to eliminate redundant VOD packet delivery using various coding techniques [8,20,21]. Other research literatures have been directed toward studying bandwidth and resources provisioning in VOD P2P network and cover the optimization of network bandwidth resources [22]. Reference [23,24] provides a detailed coverage of the diverse topics related to pervasive computing (also called ambient intelligence) and intelligent multimedia technologies.

In the remaining of this chapter, we study the bandwidth requirements problem in P2P VOD pervasive network architecture and present:

- A summary of P2P pervasive system architecture and the key parameters that can be used to analyze the performance of VOD pervasive system deployment.
- An analytical model for planning of P2P VOD pervasive service traffic.
- A numerical case study for estimation of the pervasive network bandwidth, to illustrate the advantages of VOD P2P pervasive networking.

13.3.2 P2P Pervasive Network Architecture

A P2P network architecture consists of the components shown in Figure 13.3. In this architecture, a large number of clients (or peers) interested in some video content cooperate with each other by exchange their stored video contents. The video content initially exists on a centralized head-end server. The network bandwidth from the centralized servers to the system clients are usually limited, and the inbound and outbound capacities of the clients are also typically asymmetric (i.e., the inbound rate is smaller than the download outbound rate). However, clients can enhance the system bandwidth by donating their own inbound and outbound bandwidths to the system. Thus, P2P architecture utilizes the numerous clients inbound and outbound links bandwidth and storage capacities available at clients set up boxes (STBs) to build an advanced VOD architecture based on multisource streaming from partners contributing STBs to a requesting STB [25,26].

The video distribution scheme acts as follows: the system central head-end server acts initially as a origin peer to search and download the requested content [25,26]. The file video content is divided by the head-end server, into a number of segments, and are further divided into blocks which streamed by a different contributing STB to the requesting STB. This will reduce the limited STB uplink capacities in asymmetrical broadband networks. An initial video sequence blocks of each title in the centralized library is downloaded to all STBs, which allows any end user to instantly start playing the initial part of any content while the rest of the sequence is received through multisource streaming from other STBs. If a client wants to access a given live stream, it will first question a video hub office (VHO) controller whose address is known to all clients. This controller provides the client with the subset of active clients (typically less than 30 in our study) who are members of the multicast mesh network associated with that requested VOD stream. The client then exchange content and control messages with each of these clients through a video switching office in its cluster area, and joins the mesh network established between these clients. Upon connection, each client nodes downloads all blocks it needs and receives complementary sub-streams from its multicast mesh network partners (peers) and has enough storage to keep all the blocks they have downloaded. The P2P mesh network membership changes as a

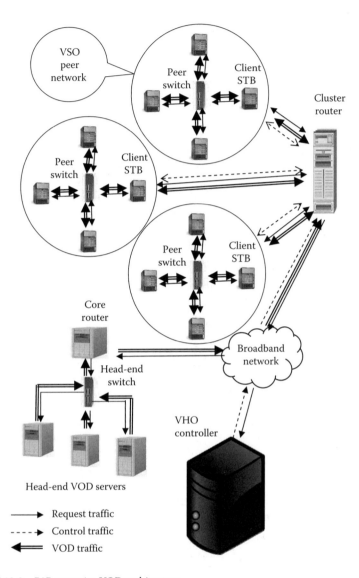

FIGURE 13.3 P2P pervasive VOD architecture.

result of client arrivals and departures, and because clients periodically try to find new partners to increase their download rates. As a result, P2P architecture allows the storage and streaming resources scale with the number of clients and provides a scalable video distribution solution able to cost-effectively support very large content libraries [12,25,26]. In addition, it is expected that P2P will also reduce the channel— or alternatively video stream—change time in change time in IP based TV (IPTV), which is a major obstacle in IPTV services' wide adoption [27].

13.4 P2P Pervasive Network Planning Study and Models

In this section, we describe various key parameters used for P2P pervasive network modeling study, based on a functional view of a typical P2P pervasive network architecture described before.

1. *Movie Request Traffic Modeling* There are two types of movie requests in the VOD network; the first one is the request for initializing or starting the video movie (labeled normal request in the study) which lasts for an average period of (t_n). The other type is the request for interactive service (e.g., stop/pause, jump forward, fast reverse, etc.) to be performed on the viewed movie (labeled interactive request in the study), and lasts for an average period of (t_1) minutes. Since each of these requests is independent from each other, and the arrival requests come from large numbers of client set-up terminals, the arrival process of normal requests, as well as of interactive requests, to each video storage server can be modeled as a Poisson process with average rates: λ_n and λ_1, respectively. With this assumption, the distribution of the sum of k of independent identically distributed random variables, representing the request inter-arrival times, is then the Erlang distribution.

2. *Bandwidth Provisioning Methodology* Bandwidth provisioning is used to determine the VOD system bandwidth required for a "no blocking" service. The model is based on the Erlang-B formula with different values of blocking probability. Our study aim is to estimate the number of server ports supported by the down channel from the video storage source to client setup box. In pervasive P2P network, the video source can be the origin central video server or another client setup box. Using the estimated number of source ports, determined from Erlang-B formula for the movie traffic, the VOD system bandwidth required is then simply determined by multiplying the movie rate (according to the movie being HD or standard definition (SD) VOD movies) by the number of source ports. The total VOD system bandwidth demand equals the sum of the server ports in use times the movie bandwidth per port stream.

3. *Movie Class Modeling* The distribution of VOD movie requests generally follows a Zipf-like distribution, where the VOD movies have two classes: popular and unpopular. The relative probability of a request for i (the most popular movie) is proportional to $1/i^a$, with $0 < a < 1$, and typically taking on some value less than unity [28]. The assumption here is that all blocks of a popular movie belonging to the popular class are stored in the mesh network of the client, and if needed by a given client, they can be downloaded from its mesh network partners.

   For Zipf-like distributions, the cumulative probability that one of the k popular movies class is accessed (i.e., the probability of a movie request from the client mesh network) is given asymptotically by:

$$\psi(k) = \Sigma\delta/\iota^\alpha = \delta\, k^{\,1-\alpha}/(1 - \alpha) \tag{13.1}$$

and

$$\delta = (1 - a)/V^{1-\alpha}$$

where V is the total number of movies in the system [19].

Next, we estimate the probability (P_o) of a request for a movie that belongs to the unpopular class (i.e., does not exist in the mesh network setup boxes), and therefore should be obtained from the central head-end video server. For a VOD system with V total movies and **k** popular ones at the mesh network, the probability of a request for an unpopular movie stored in the head-end server is

$$P_o = 1 - (k/V)^{1-\alpha} \qquad (13.2)$$

4. *Network Traffic Delivery Modeling* The analytical methods for provisioning network links in this study assume steady-state busy hour traffic for movie retrieval normal requests. In the steady state, multicasting is used by the network to reduce IP/VOD traffic volumes. The network needs to deliver only one video stream (one video server port) for a group of viewers (multicast group) watching the same video or broadcast program segment. The steady-state demand is therefore the total bandwidth of all video streams (or server ports) in use.

However, steady-state normal request demand is usually disrupted by service interactive request and video channel surfing. One way the network can make VOD interactive changes fast is to send unicast stream (one per viewer) streams at higher than usual rates [29]. While interactive requests may be short lived (say, a minute), each request superimposes a significant additional demand on top of the steady-state demand. Thus, interactive request traffic demands that capacity planning and engineering must include interactive transient effects.

13.4.1 System Bandwidth Requirements for P2P Pervasive VOD Architecture

In this section, we analyze the key parameters that have influence on the P2P pervasive network bandwidth requirement. We carry a traffic analysis to determine the number of servers for a P2P/VOD architecture. The assumption is that the client will download the unpopular movies from high-capacity centralized head-end servers while the popular materials are downloaded from its multicast mesh network partners (peers) setup boxes. We proceed as follows.

Let x be the number of VOD system cluster areas, Z the multicast factor (i.e., number of viewers who request the same multimedia movie within a short period of time, thus it can be served from the same server port), h the number of houses in VOD system cluster service area, P_o is probability of unpopular request. And is given by (13.2) as $P_o = 1 - (k/V)^{1-\alpha}$, M is the number of mesh networks per cluster area, λ_n the average number of normal request attempts per movie per period per household, λ_1 the average number of interactive request attempts per movie per period per household, t_n the holding time of a normal request for a movie in minutes, t_1 the holding time of an interactive request in minutes, T the peak busy period in minutes, p the penetration of service in a VOD system cluster area, D the diversity factor between mesh networks in requesting unpopular movie. The maximum value is M and the average value is $D = M/2$.

We calculate the P2P parameters as follows:

n = Number of active clients per mesh network = $(h/M) * p$.

M_m = Traffic supported by mesh network peers in Earlang, calculated as follows:

$$M_m = \left[\frac{n \times \lambda_n \times t_n}{Z \times T} + \frac{n \times \lambda_I \times t_I}{T}\right] \times (1 - P_o) \tag{13.3}$$

M_{Cm} = Mesh Network Traffic supported by the broadband network from centralized head-end servers, calculated as follows:

$$M_{cm} = \left[\frac{n \times \lambda_n \times t_n}{Z \times T} + \frac{n \times \lambda_I \times t_I}{T}\right] \times P_c \tag{13.4}$$

Now, the number of server ports (e.g., video streams) needed by the VOD mesh network (Sm) to support traffic to a requesting client by a mesh P2P can then be found using the Erlang-B formula and the mesh network peers traffic (M_m) with a given blocking probability P_B, where

$$P_B = \frac{(M_m)^{S_m} / S_m!}{\sum_{L=0}^{S_m} (M_m)^L / L!} \tag{13.5}$$

Similarly, using M_{cm} and the Erlang-B formula, we can find (S_{cm}) the number of server ports needed by the VOD mesh network to support traffic downloaded from centralized head-end servers through the broadband network. Thus,

S_T = total number of server ports per mesh network = $S_m + S_{cm}$

The bandwidth per mesh network is given by

$$\mathbf{W}_{onm} = S_T * r \tag{13.6}$$

where r is the movie stream rate (e.g., 3 Mbps for SD movie).
    To find the bandwidth per customer link, we proceed as follows:

N = number of connection links per mesh Network = $n(n-1)/2$,
        assuming fully mesh connected network;
    S_{mL} = number of server ports per customer connection link = S_T/N

and

$$W_{mL} = \text{rate per customer connection link} = S_{mL} * r \tag{13.7}$$

To find the bandwidth per cluster broadband link, we proceed as follows:

$$S_{sc} = \text{number of server ports per cluster area broadband link} = S_{cm} * M$$

The cluster link bandwidth downloaded from central servers W_{csm} can then be found by multiplying the corresponding number of ports with the movie bandwidth (r). Therefore

$$\text{cluster broadband link bandwidth} = W_{csm} = (S_{cm}*M)*r/D \tag{13.8}$$

Using $D = M/2$ as average value for diversity, then $W_{csm} = 2*S_{cm}*r$

Finally, the available bandwidth rate per household is given by

$$W_{mh} = r*(S_{cm} + S_m)/n = W_{mh} + W_{cmh} \tag{13.9}$$

Overall, the system bandwidth from the central servers is

$$TW_{mh} = x * W_{csm} \tag{13.10}$$

13.4.2 Traffic Provisioning Model for a Centralized Architecture System

To compare the P2P bandwidth requirement with the centralized VOD system, we need to estimate (S_c), the number of server ports supported by the VOD centralized network. S_c can then be found using the Erlang-B formula, from the total calculated centralized network traffic (M_c) with a given blocking probability P_B. The provisioning of the centralized VOD system bandwidth is determined as follows [30]: Let

M_c = Total network traffic in Erlang for a centralized system.

$$= \frac{x \times h \times p \times \lambda_n \times t_n}{Z \times T} + \frac{x \times h \times p \times \lambda_1 \times t_1}{T} \tag{13.11}$$

where p is penetration of service in a VOD system cluster area,

$$P_B = \frac{(M_c)^{S_c}/S_c!}{\sum_{N=0}^{S_c} (M_c)^N/N!} \tag{13.12}$$

and the required total centralized system bandwidth is

$$W_c = S_c * r \tag{13.13}$$

Bandwidth per cluster area can then be calculated by total system bandwidth W_c divided by number of clusters x

$$w_{clc} = \text{bandwidth per cluster area} = W_{clc} = S_c * r/x = W_c/x \tag{13.14}$$

13.4.2.1 Centralized VOD Customer Link Requirements

In this section, we analyze the traffic bandwidth per a centralized system customer link. Assuming fiber-to-the premises (FTTP) topology, as shown in Figure 13.4, each router in the path to customer can serve multiple routers of the type below it [20]. The VOD network core router and cluster router deliver content to the edge of the network. The cluster router then forwards the video streams to an optical line terminal (OLT), which in turn forwards the content to an optical network unit (ONU) located close to house hold. Since each cluster area has (h) houses, then each OLT will serve a maximum of h houses. The link from the cluster router to the OLT must therefore deliver content to h housing premises. While not all of them are customer to VOD network, it is reasonably expected to find ($p*h$) of VOD customers on an OLT, where h and p are as defined before.

Based on this architecture, we can estimate the required number of server ports per cluster link as:

$$S_{cc} = \text{number of server ports per cluster link (from VOD core}$$
$$\text{router to cluster router) in a centralized VOD system}$$
$$= S_c/x$$

and

$$S_{cL} = \text{number of server ports per ONU connection}$$
$$\text{link (from OLT to ONU)}$$
$$= S_c/M*x$$

where M is the number of ONUs in the cluster area.

The bandwidth per ONU is:

$$W_{onc} = \frac{r*S_c}{(M*x)} \tag{13.15}$$

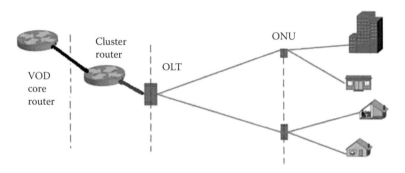

FIGURE 13.4   Centralized VOD FTTP architecture.

TABLE 13.1 Key Modeling Parameters Values

λ_n	1.5	λ_1	4.00	T	420.00
t_n	120.00	t_1	6.00	M	30.00
r	3.00	Z	30.00	H	600.00
P_B	0.01	X	250.00	P	0.40
n	8.00	N	28.00	P_o	.1

and the allocated server ports per household (S_{ch}) is given by

$$S_{ch} = \frac{S_c}{(x*h*P)}$$

and the allocated bandwidth per household (W_{ch}) is given by

$$W_{ch} = \frac{r*S_c}{(x*h*P)}$$

(13.16)

13.5 Results and Performance Analysis

In this section, we present the analysis results based on the developed traffic models. Our aim is to determine the VOD channel bandwidths per house hold required for a "no blocking" service. Our analysis assumes SD IP/VOD (video and audio) movie resolutions and the key parameter values shown in Table 13.1.

The effects of the system parameters (such as the multicast factor, movie holding time, and the average number of requests arriving to the system during the peak period) on the required system VOD channel bandwidth are shown in Table 13.2 for both of the centralized system and the P2P network.

TABLE 13.2 Total Bandwidth Required for Centralized IP/VOD System and P2P Network (Mb/s)

Multi Cast Factor, Z	Centralized VOD Architecture System			P2P VOD Architecture		
	Bandwidth per Cluster Link, W_{clc}	Bandwidth per Household, W_{ch}	Total Central Network Bandwidth, W_c	Bandwidth per Cluster Link, W_{csm}	Bandwidth per Household, W_{mh}	Total Network Bandwidth, TW_{mh}
1	347.20	1.45	86,799.32	8.95	4.27	4217.21
30	51.61	0.215	12,901.42	8.95	1.743	2237.49
35	50.15	0.209	12,536.56	8.88	1.724	2220.09
40	49.05	0.204	12,262.90	8.83	1.710	2206.86
45	48.20	0.201	12,050.03	8.79	1.699	2196.46
50	47.52	0.198	11,879.73	8.75	1.690	2188.06
55	46.96	0.196	11,740.39	8.72	1.683	2181.15

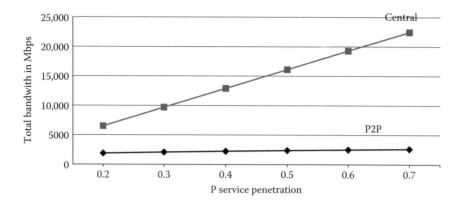

FIGURE 13.5 Total bandwidth versus service penetration for P2P and VOD centralized system

It is clear that the VOD system with P2P streams for this sample can reduce the required total central servers bandwidth to 5% (4217.21/86,799.32 = 4.88%) of the bandwidth that a unicast centralized system (at $z = 1$) would use. Even with multicast streams ($z = 30$) for this sample, P2P can result good saving of the required central bandwidth further to 83% (2237.49/12,901.42= 17.3%). In Figures 13.5 and 13.6, we plot the relationship between service penetration rate, the total network bandwidth rate, and the bandwidth per cluster link for P2P and central system. As shown, P2P allows great saving in both of the total servers' bandwidth and the cluster link bandwidth needed to serve VOD system customers' requests.

In Figure 13.7, we plot the relationship between the cluster channel bandwidth and the interactive traffic request rate. As shown, the increment of the interactive traffic (i.e., increment of λ_I due to movie surfing), has a stronger impact on central system cluster bandwidth, compared to P2P system.

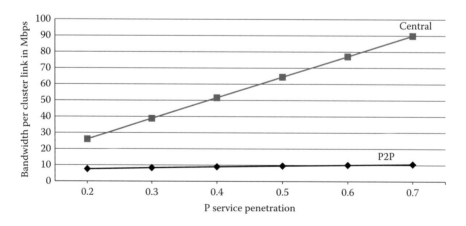

FIGURE 13.6 Cluster link bandwidth versus service penetration for P2P and centralized system.

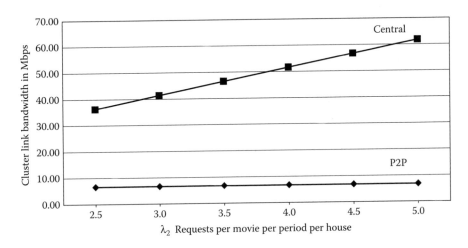

FIGURE 13.7   Cluster link bandwidth versus interactive traffic for P2P and centralized system.

In Figure 13.8, we plot the relationship between the available bandwidth per ONU/ mesh community and the service penetration rate for central and P2P systems, respectively. For this, the central system has a lower bandwidth compared to P2P system due to the P2P architecture utilization of the numerous clients links bandwidth and storage capacities available at clients set up boxes.

In Figure 13.9, the relationship between the available bandwidth per household and the service penetration rate for central and P2P systems, respectively. For this, the central system has a lower bandwidth compared to P2P.

The bandwidth per house hold of the system is also dependent on the interactive traffic rate for a given service penetration and multicasting as shown in Figure 13.10.

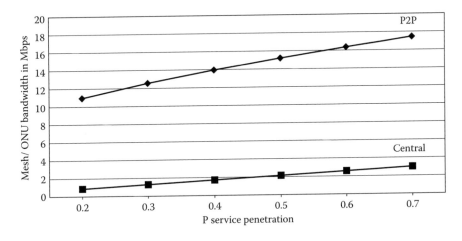

FIGURE 13.8   Bandwidth per ONU/mesh network versus service penetration rate.

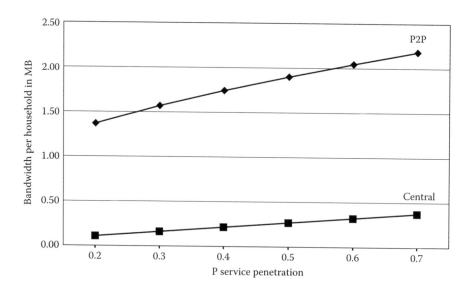

FIGURE 13.9   Bandwidth per house hold versus penetration service rate.

13.5.1 Comparison of Architectures Results and Analysis

The comparison of the centralized system bandwidth requirements to the P2P pervasive system as a function of the multicast factor can be deduced from Table 13.2. From the numerical results shown in the table, we can see that a VOD centralized system, with more interactive unicast streams, requires a much larger overall system bandwidth

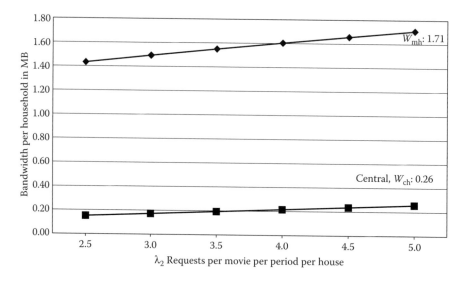

FIGURE 13.10   Bandwidth per house hold versus movie interactive rate.

channel when compared to the P2P pervasive VOD system. Even with a large number of multicast streams, P2P pervasive networking can result good saving of more than 80% of the required central bandwidth. The P2P architecture, however, is less efficient when looked at from the required bandwidth per household. However, this is not reflected on a huge bandwidth requirement on cluster links nor on the overall system bandwidth due to content exchange between mesh network clients. We therefore, conclude that to produce designs that aim toward minimizing network costs and that also respect quality constraints, a P2P pervasive system is the better choice.

13.6 Conclusion

This chapter presents P2P pervasive VOD system architecture. The key parameters that can be used to analyze the performance of VOD pervasive system deployment are discussed in detail. The chapter also presents a planning and bandwidth provisioning methodology for a P2P pervasive network based on a functional view of a typical P2P pervasive network architecture with a steady-state peak hour traffic. Traffic models based on Erlang analysis were developed for the proposed P2P pervasive system, and for the centralized servers systems. The impact of tuning of multiple key system parameters such as penetration rate, interactive traffic rate, and multcasting factor were investigated in detail.

The comparison of the centralized multimedia server system to the P2P pervasive system showed that P2P pervasive networking approach provides better scalability through mesh networks nodes cooperation whenever there is a demand growth. Besides, P2P pervasive networking can result a large saving of the required system bandwidth.

As illustrated by the case study presented in this chapter, the P2P technology for distribution of multimedia streams provides an efficient and common platform for provisioning of pervasive computing Services. It is expected that, in the near future, P2P pervasive networking will have a significant impact across a wide spectrum of multimedia services for business, healthcare, and governmental sectors.

References

1. Thouin, F. and Coates, M. Video-on-Demand Networks: Design Approaches and Future Challenges, *IEEE Network Magazine*, March/April 2007.
2. Hua, K. A., Tantaoui, M., and Tavanapong, W. Video delivery technologies for large-scale deployment of multimedia applications, in *Proceedings of the IEEE*, Vol. 92, September 2004.
3. Souza, L., Ripoll, A., Yang, X. Y., Hernández, P., Suppi, R., Luqu, E., and Cores, F. Designing a video-on-demand system for a Brazilian high speed network, in *Proceedings of the 26th IEEE International Conference on Distributed Computing Systems Workshops*, 2006.
4. Venkat, R. P., Harrick, M., and Srinivas, R. Designing on Demand Multimedia Service, *IEEE Communication Magazine*, Vol. 30, No.7, July 1992.
5. Tsong-Ho, W. and Korpeooglu, I. Distributed Interactive Video System Design and Analysis, *IEEE Communication Magazine*, Vol. 36, No. 1, March 1997, USA, pp. 100–108.

6. Ramanathan, S. and Rangan, P. Adaptive feedback techniques for synchronized multimedia retrieval over integrated networks. *IEEE/ACM Trans Netw* 1993; 1(2): 246–260.

7. Nafaa, A., Murphy, S., and Murphy, L. Analysis of a Large-scale IP/VOD Architecture for Broadband Operators: A P2P-based Solution, *IEEE Communications Magazine*, Vol. 46, No. 12, December 2008, pp. 47–55.

8. He, Y., Lee, I., and Guan, L. Distributed throughput maximization in P2P VoD applications. IEEE *Trans Multimedia* 2009; 11(3): 509–522.

9. Nafaa, A., Murphy, S., and Murphy, L. Analysis of a Large-Scale VOD Architecture for Broadband Operators: A P2P-Based Solution, *IEEE Communications Magazine*, December 2008.

10. Smith, D. E. and Walfham, I. P. TV bandwidth demand: multicast and channel surfing, in *IEEE International Conference on Computer Communications INFOCOM 2007*, Anchorage, AK, May 2007.

11. Agrawal, D., Beigi, M. S., Bisdikian, C., and Lee, K-W. Planning and managing the IPTV service deployment, in *10th IFIP/IEEE International Symposium on Integrated Network Management*, Vol. 25, No. 21, May 2007, pp. 353–362.

12. Annapureddy, S. et al., Exploring VoD in P2P swarming systems, in *Proceedings of INFOCOM '07*, Anchorage, AK, May 2007, pp. 2571–2575.

13. Do, T. T., Hua, K. A., and Tantaoui, M. A., P2vod: Providing fault tolerant video-on-demand streaming in peer-to-peer environment, in: *IEEE International Conference on Communications*, Paris, France, 2004 , pp. 1467–1472

14. H. Chi, Q. Zhang, J. Jia, and X. Shen, http://bbcr.uwaterloo.ca/~xshen/paper/2007/esasip.pdf \t "_blank" Efficient Search and Scheduling in P2P-based Media-on-demand Streaming Service, *IEEE J. Selected Areas of Communications*, Vol. 25, No. 1, pp. 119–130, 2007.

15. Cui, Y., Li, B., and Nahrstedt, K. oStream: Asynchronous streaming multicast. *IEEE J Select Areas Commun* 2004; 22(1): 91–106.

16. Li, J. Peerstreaming: An on-demand peer-to-peer media streaming solution based on a receiver-driven streaming protocol, in *Proceedings of IEEE MMSP*, Shanghai, China, October 2005, pp. 1–4.

17. Liao, X. et al., AnySee: peer-to-peer live streaming, in *Proceedings of IEEE INFOCOM '06*, Barcelona, Spain, April 2006, pp. 1–10.

18. Hefeeda, M. et al., PROMISE: Peer-to-peer media streaming using collectcast, in Proceedings of ACM Multimedia '03, Berkley, CA, November 2003, pp. 45–54.

19. Janardhan, V. and Schulzrinne, H. Peer assisted VoD for set-top box based IP network, in *Proceedings of ACM SIGCOMM '07 Wksp. Peer-to-Peer Streaming IP-TV*, Kyoto, Japan, August 2007, pp. 335–339.

20. Shen, Y., Liu, Z., Panwar, S. S., Ross, K. W., and Wang, Y. Streaming layered encoded video using peers, in *Proceedings of ICME*, Amsterdam, The Netherlands, July 2005. pp. 966–969

21. Xu, X., Wang, Y., Panwar, S. P., and Ross, K. W. A peer-to-peer video-on-demand system using multiple description coding and server diversity, in *Proceedings of IEEE ICIP*, Singapore, October 2004, Vol. 3, pp. 1759–1762.

22. Alwakeel, S. Modeling and provisioning of a P2P-based VOD architecture. *Int J Video Image Process Netw Secur,* 2010; 10(3): 01–14.

23. Hassanien, A-E., Abawajy, J. H., Abraham, A., and Hagras, H. *Pervasive Computing Innovations in Intelligent Multimedia and Application.* Springer Series: Computer Communications and Networks, Springer, Berlin, Heidelberg, New York, 2010.

24. Aarts, E. Ambient intelligence: A multimedia perspective. *IEEE Multimedia* 2004; 11(1): 12–19.

25. Nafaa, A., Murphy, S., and Murphy, L. Analysis of a Large-scale IP/VOD Architecture for Broadband Operators: A P2P-based Solution, *IEEE Communications Magazine,* Vol. 46, No. 12, December 2008, pp. 47–55.

26. Hefeeda, M. M., Bhargava, B. K., and Yau, D. K. Y. A hybrid architecture for cost-effective on-demand media streaming. *Comput Netw* 2004; 44: 353–82.

27. Begen, A., Glazebrook, N., and Steeg, W. Reducing channel-change times with the real-time transport protocol. *IEEE Internet Comput* 2009; 13(3): 40–47.

28. Breslau, L. et al., Web caching and Zipf-like distributions: Evidence and implication, in *Proceedings of IEEE INFOCOM'99,* New York, March 1999, pp.126-34.

29. Smith, D. E. IP TV bandwidth demand: multicast and channel surfing, in *Proceedings of IEEE INFOCOM'07,* Anchorage, Alaska, USA, 2007, pp. 2546–2550.

30. Alwakeel, S. S. Bandwidth provisioning models for a large scale IP-based video-on-demand broadband network. *Egyptian Comput Sci J* 2009; 33(1): 10–25.

14

Using Universal Plug-n-Play for Device Communication in *Ad Hoc* Pervasive Environments

Hamid Mukhtar
National University of Sciences and Technology

Djamel Belaïd
Institut Telecom

14.1 Introduction

As the mobile devices are increasing in number as well as improving in capabilities, mobile services are gaining much attention and popularity among the researchers and developers. The focus is on enabling the ordinary user to access ubiquitous services anywhere and anytime. These services will be provided by peer devices or by the underlying infrastructure. When considering pervasive environments, devices must be able to discover one another and then interact with each other using specific protocols for pervasive communication.

With technologies such as web-enabled smartphones, and due to the ubiquity of Wi-Fi hotspots everywhere around us, IP communications are well on their way to becoming pervasive. However, we believe the future of wireless networks will not follow the traditional base-station to mobile model, but will be peer-to-peer or *ad hoc*. This can be inferred from the fact that Wi-Fi alliance is currently in the process of standardizing the Wi-Fi Direct protocol for peer-to-peer communication using Wi-Fi protocol. This is not possible using traditional Wi-Fi-based approach that uses access points (APs) for communication between two devices. This will make it easier for any two devices to make *ad hoc* Wi-Fi connections between them.

The communication between two devices in a pervasive environment takes place at two levels: network-level communication using protocols such as Bluetooth, Wi-Fi, and Infrared and application-level protocols such as Bonjour, Jini, service location protocol (SLP) and universal plug and play (UPnP), and so on. The network-level communication is established using physical layer and is dependent on the properties of the physical medium. Network-level protocols are mostly used for device discovery, that is, it allows a user to connect different devices with each other. The application-level protocols, on the other hand, are designed to work on top of different network-level protocols. Most of the time, they are used for service discovery, that is, to enable a high-level interaction of a user with different applications on the device.

A typical user would be more interested in the services provided by the device rather than the device itself. For example, consider that a user wants to browse the photos available on his mobile phone using a PC. The mobile phone should be discoverable by the PC and then the user should be able to interact with it for browsing the contents through the PC. The user first discovers the device and then interacts with the services of the devices once it is discovered. In this chapter, we are concerned with the service discovery and user interaction part of the problem.

14.1.1 Service Discovery

Service discovery mechanisms provide means for dynamically discovering available services in a network and for providing the necessary information to use them [1]. From users' point of view, service discovery simplifies the task of: (i) searching and browsing for services, (ii) choosing the right service (with desired characteristics), and (iii) utilizing the service. From administrators' point of view, service discovery allows: (i) building and maintaining a service-oriented network, (ii) introducing new services and devices, and (iii) allowing others to introduce and use these services in a controlled manner.

Service discovery is not only useful in infrastructure-based environment—where mostly fixed PCs offer and utilize services using an existing cable or wireless network—but it can also be used in an *ad hoc* communication system, where no fixed infrastructure is present but the nodes themselves can form the network on the fly.

To understand the importance of service discovery, let us consider a scenario in which a user wants to print a colorful PowerPoint presentation in her laptop at an airport terminal while she is hurrying to catch her flight, which is due in a few minutes. In the absence of a service discovery system, she has to locate a printer nearby, determine whether it can support color printing, setup a network with the printer by obtaining an IP address, and finally download the printer driver to her laptop. As we can see clearly, all these procedures can take up reasonable amount of time as well as help from the surroundings.

Now consider a service discovery system for the same scenario. All the user has to do is to run the service discovery application, specify the parameters of interest and it will automatically come up with the best solution: the nearest color printer with its capabilities and how to connect and invoke a print request on it. As it can be observed, a proper service discovery mechanism facilitates a user in simplifying the complex task involved in configuration of user devices.

There are two possible approaches to carry out service discovery in a pervasive environment: peer-to-peer and centralized.

14.1.1.1 Peer-to-Peer Service Discovery

In peer-to-peer discovery mode, each device on the network acts as a peer. There is no central server and each device can act as a client as well as a server at the same time. A device manages a service repository that contains the description of the services it offers and might contain, additionally, other services discovered in the network. Service request messages are multicast in the network and responses are unicast to only the request sender.

14.1.1.2 Centralized Service Discovery

In this mode, a central lookup server exists in the network that contains the description of all the available services in the network. Any device offering a service registers itself with the server and any device looking for a service contacts the server to inquire about a service location.

In Section 14.2.2.3, we describe some service discovery architectures that use either centralized or peer-to-peer discovery mechanism. In general, for a pervasive environment where *ad hoc* networks may be formed on the fly, a peer-to-peer discovery is more appealing as having a centralized server in such environments is not always possible. It is also possible to combine the best of both worlds; for example, SLP, described below, can work in both centralized and peer-to-peer modes at the same time.

14.1.2 Review of Some Service Discovery Architectures

In this section, we give a brief overview of some of the important existing architectures and platforms for service discovery viz. Jini, Salutation, SLP, Service Discovery Protocol (SDP), UPnP, and Bonjour. A more detailed description and comparison has been provided in [1].

14.1.3 Jini

Jini [2] is Oracle's (previously Sun Microsystems) Java-based approach for service discovery. It consists of an architecture and a programming model. Jini addresses the issue of how devices connect with each other to form an *ad hoc* network and how these devices offer services to other devices in this network. Each Jini network, also called Jini community, maintains a Lookup Table on a lookup server. This lookup table is a database containing references to services.

Devices and applications register with a Jini network using a process called discovery and join. To join a Jini community, a device or application places itself in the lookup table. Apart from pointers to services, the lookup table can also store the Java code of services. Hence, it is possible that services may upload device drivers, interfaces, and other programs that may help the users to access the service.

Although Jini can be implemented in 46K of Java binaries, its dependency on Java requires a device to have a JVM running on it. Moreover, the centralized lookup server is not a good option for *ad hoc* environments. Hence, Jini might not be a susceptible choice for pervasive environments.

14.1.4 Salutation

Service discovery in Salutation [3] is defined on a higher layer, and the transport layer is not specified. Thus, Salutation is independent of the network technology and may run over multiple infrastructures such as TCP/IP or IrDA. It is independent of any programming language.

The Salutation architecture consists of Salutations Managers (SLMs) that have the functionality of service brokers. Services register their capabilities with an SLM, and clients query the SLM when they need a service. After discovering a desired service, clients are able to request the utilization of the service through the SLM.

A lightweight version (Salutation-Lite) has been developed for resource-limited devices. It is based primarily on IrDA to leverage the large number of infrared capable devices. It provides a means to determine the operating system, processor type, device class, the amount of free memory, display capabilities, and other characteristics of a handheld device.

14.1.5 Service Location Protocol

The SLP [4] has been developed by IETF and aims to be vendor-independent standard. It is designed for TCP/IP networks and is scalable up to large enterprise networks. The SLP architecture consists of three main components:

- User Agents (UA) perform service discovery on behalf of the client (application or user).
- Service Agents (SA) advertise the location and the characteristics of services.
- Directory Agents (DA) collect service addresses and information received from SAs in their database and respond to service requests from UAs.

Before a client (UA or SA) is able to contact the DA, it must discover the existence of the DA. There are three different methods for DA discovery:

- *Static Discovery*: the SLP agent obtains the address of the DA using DHCP.
- *Active Discovery*: the UAs and SAs send service requests to a specific SLP multicast group address. A DA listening on this address receives the request and responds directly to the agent.
- *Passive Discovery*: the DA periodically sends out multicast advertising for its services. UAs and SAs learn the DA address from the received advertisements and are now able to contact the DA via unicast.

The DA is not a required part in SLP. In case of absence of a DA, UAs repeatedly send out their service request to the SLP multicast address. All SAs listen for these multicast requests and, if they advertise the requested service, they will send unicast responses to the UA. Also, an SA multicast an announcement of their existence periodically, so that UAs can learn about the existence of new services.

14.1.6 SDP in Bluetooth

The Bluetooth stack contains the SDP, which is used to locate services provided by or available via a Bluetooth device. The SDP supports the following inquiries: search for services by service type; search for services by service attributes; and service browsing without *a priori* knowledge of the services characteristics. Once services are discovered with SDP, they can be selected, accessed, and used by mechanisms out of the scope of SDP, for example, by other SDPs such as SLP and Salutation. SDP can coexist with other SDPs, but it does not require them.

Unfortunately, SDP is limited only to Bluetooth and does not provide a complete solution for a pervasive environment, where heterogeneity of technologies and protocols is expected most of the times.

14.1.7 Universal Plug and Play

UPnP [5] is an architecture based on top of the TCP/IP network. It has been put forward by Microsoft and can be thought of as an extension to Microsofts Plug and Play technology that is already implemented on all Windows platforms. At the time of writing, UPnP forum consists of an industry consortium of about 800 companies.

UPnP is proposed for small offices or home computer networks, where it enables peer-to-peer mechanisms for auto-configuration of devices, service discovery, and control of services. UPnP device model is hierarchical. The Simple SDP (SSDP) [SSDP99] is used in UPnP to discover services.

The fact that UPnP is platform- and programming language-independent makes it a favorite choice as a service discovery technique. Since it uses a peer-to-peer discovery technique, it is also suitable for a pervasive environment. It is for these reasons that we have selected UPnP technology to be a part our service composition platform. We have provided additional information about UPnP in Appendix B. For usage of UPnP in the context of our service composition architecture, the reader is referred to Chapter 5, Service Discovery.

14.1.8 Bonjour

Apple's Bonjour [6] is an open, standard-based networking technology that automatically connects electronic devices on a network; a zero-configuration networking solution. It can be used for dynamic discovery of services over standard, ubiquitous IP networking protocol.

When a new bonjour-enabled computer or device is added to a network, it automatically assigns a link-local address to itself (addresses in the range 169.254.xxx.xxx). To perform name services, Bonjour uses a variant of DNS called Multicast DNS-Service Discovery (mDNS-SD). Bonjour query and response packets contain the information needed for service discovery. To find out about other devices that have services, an application transmits a multicast query and receives responses from the devices running the appropriate services. The query/response transaction follows the standard DNS format for naming and lookup.

Bonjour works on a network subnet, just like UPnP, making it ideal for *ad hoc* local area networks that do not have central DNS servers.

Based on the above comparison, we can identify that UPnP contains most of the useful features as compared to others. In the next few sections, we describe in detail the UPnP technology, its benefits, and usage in a number of different applications and scenarios.

14.2 Universal Plug-n-Play

UPnP technology defines an architecture for pervasive peer-to-peer network connectivity. UPnP is a set of network protocols designed to make it easier to attach devices to computers and networks. UPnP standard specifications have been developed by the UPnP Forum, which currently consists of over 900 companies that work with and comply with the standard. The goal of UPnP is to allow devices to connect seamlessly and to simplify network implementation in the home and corporate environments. Toward this end, UPnP Forum defines and publishes UPnP device control protocols (DCPs) built upon open, Internet-based communication standards.

What is "universal" about UPnP technology is that no device drivers are used for UPnP communication; common protocols are used instead. UPnP networking is media independent. The scope of UPnP is large enough to be suitable for a truly pervasive environment in many existing as well as new and exciting scenarios, including home automation, printing and imaging, audio/video entertainment, and comprising devices such as kitchen appliances, handheld devices like PDA's and mobile phones, as well as the existing computing environment. UPnP has already been supported by various vendors for many home and office devices such as printers, cameras, and media players and has also been introduced in various mobile phones recently by implementation of UPnP API on top of Android, iOS, and Symbian operating systems.

Advantages of UPnP are that it is built on top of TCP/UDP and uses simple object access protocol (SOAP) messages over HTTP for communication. Hence, it is independent of any operating system and programming environment and can be deployed without modifying the existing systems. With UPnP, a device can dynamically join a network, obtain an IP address, convey its capabilities, and learn about the presence and capabilities of other devices. A device can be a UPnP server by offering services, or a UPnP client by requiring services, or both at the same time. Figure 14.1 shows the different standard protocols used in UPnP.

The specification of UPnP is written down in the UPnP Device Architecture (UDA). There are already a lot of stacks and applications for that specification. Most UPnP stacks wrap protocols that are required by UPnP and that are defined by UPnP, respectively. Some of them provide typing by offering classes such as Device, Service, Action, and State Variable. Developers using such stacks do not need to care about details of UPnP

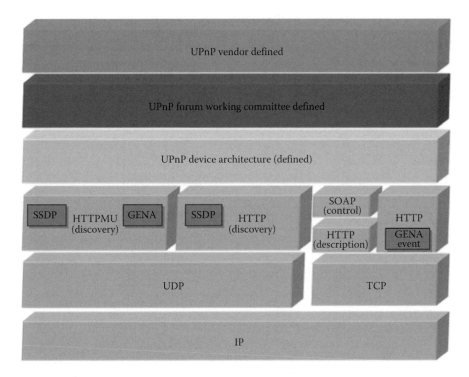

FIGURE 14.1 Standard Protocols in UPnP.

protocol, but they can instead concentrate on building UPnP applications. There are stacks for different programming languages and for different operating systems.

Figure 14.2 shows a network of UPnP-enabled devices. Each device exposes a set of services and may embed other devices exposing yet some other services. Whenever a device is brought into a network, it announces its services using multi-cast packets in the network. The device can also provide a presentation page which can be accessed by a user via a browser application to see the device's properties and services. Apart from this, each device is able to query other devices for their capabilities and to execute actions on it. A UPnP device can subscribe to services provided by another UPnP device.

14.2.1 Benefits of UPnP

The UPnP architecture offers pervasive peer-to-peer network connectivity of PCs of all form factors, intelligent appliances, and wireless devices. Some of the benefits of UPnP are:

- *Internet-Based Technologies.* UPnP technology is built upon IP, TCP, UDP, HTTP, and XML, among others. Thus, UPnP can be embedded in any device, as small as sensor devices, that can support the standard Internet protocols.

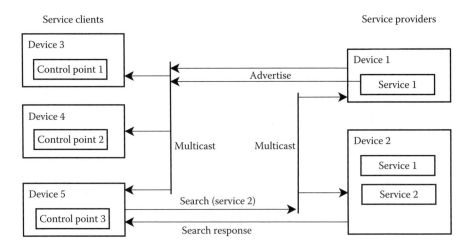

FIGURE 14.2   UPnP devices, control points, and services.

- *Common Base Protocols.* Although UPnP is built on top of open, standard protocols, it is not restricted to only these protocols. Vendors may agree on base protocol
  sets on a per-device basis. For example, audio/video devices may support a number
  of protocols and formats as discussed in Section 14.4.
- *Extendable.* Vendors are not restricted to selected few defined services of the
  UPnP forum. Each UPnP product can have value-added services layered on top of
  the basic device architecture by the individual manufacturers.
- *Media and Device Independence.* Being an application-level protocol and due to its
  adoption of open protocols, UPnP technology can run on any network technology
  including Wi-Fi, coax, phone line, power line, Ethernet, and 1394.
- *Platform Independence.* Vendors can use any operating system and any programming language to build UPnP products. UPnP applications developed using different programming languages such as C/C++, C#, and Java and implemented for
  different platforms such as Windows, Linux, Macintosh, or mobile phone operating systems can interoperate.
- *Standardized DCPs.* The UPnP initiative involves a multi-vendor collaboration for
  establishing standard DCPs. These are based on protocols that are declarative,
  expressed in XML, and communicated via HTTP.
- *Programmatic Control.* UPnP architecture enables the devices and applications to
  be controlled programmatically.
- *User Interface Control.* UPnP allows separation of User Interface from the UPnP
  service architecture. This enables vendor control over device user interface. A
  UPnP device can also offer its service as a Web service, so users can interact with
  UPnP services using a conventional Web browser.
- *Remote Access.* Although UPnP specifications are meant for LANs and WLANs,
  it is possible to use UPnP over WANs. Similarly, the capability of accessing a

UPnP device through a Web browser means the ability to interact with UPnP devices remotely.

- *Zero Configuration.* The UPnP architecture supports zero-configuration and automatic discovery. The users need not configure a UPnP device to join a network. As soon as a UPnP device joins a network, it can announce itself and the offered services. The device can then leave the network smoothly and automatically without leaving any unwanted state information behind.

- *Keep-Alive Messages.* In order to keep devices and control points aware of the presence of one another, they can resend advertisement messages from time to time in the network. If such an advertisement is not noticed from an expected device during a pre-defined interval, then the device has to be queried in order to know its status in the network.

14.3 Steps in UPnP Networking

Addressing is Step 0 of UPnP networking. Through addressing, devices and control points get a network address. Addressing enables *discovery* (Step 1) where control points find interesting device(s), *description* (Step 2) where control points learn about device capabilities, *control* (Step 3) where a control point sends commands to device(s), *eventing* (Step 4) where control points listen to state changes in device(s), and *presentation* (Step 5) where control points display a user interface for device(s).

The foundation for UPnP networking is IP addressing. A UPnP device or control point may support IP version 4-only, or both IP version 4 and IP version 6. If a dynamic host control protocol (DHCP) server is available in the network, that is, the network is managed; the device or control point must use the IP address assigned to it. If no DHCP server is available, that is, the network is unmanaged; the device or control point must use automatic IP addressing (Auto-IP) to obtain an address. Auto-IP is defined inRFC 3927 and defines how a device or control point: (i) determines whether DHCP is unavailable, and (ii) intelligently chooses an IP address from a set of link-local IP addresses. This method of address assignment allows a device or control point to easily move between managed and unmanaged networks.

The next steps in UPnP networking are as following:

1. *Discovery:* When a UPnP device gets an IP address, it is able to advertise itself in the network. Similarly, a control point can also search for related devices and services after it is assigned an IP address. The messages exchanged in both cases contains queries by control points or responses by the devices, containing a few, essential specifics about the device or one of its services, for example, its type, universally unique identifier, a pointer to more detailed information and optionally parameters that identify the current state of the device.

2. *Description:* During the discovery phase, the control point only sense the presence of a device and knows very little about it. In order to learn more about the device and its capabilities, or to interact with the device, the control point must retrieve the device's description from the URL provided by the device in the discovery message. As discussed above, the description will contain information about the

devices, its services, as well as details about the embedded devices and their services inside the root device. The description is XML based and has a standard format. Among other things, it contains information about the device vendor, information like the model name and number, the serial number, the manufacturer name, URLs to vendor-specific web sites, and so on. The description also contains URLs for control, eventing, and presentation. For each service, the description includes a list of the commands, or actions, to which the service responds, and parameters, or arguments for each action; the description for a service also includes a list of variables; these variables model the state of the service at run time, and are described in terms of their data type, range, and event characteristics.

3. *Control:* Given the description of a device, the control point can send actions to a device's services. To do this, a control point sends a suitable control message to the control URL for the service (provided in the device description). Control messages are also expressed in XML using the SOAP. Depending upon the service invoked by the control point the device may return action-specific values in response to the control message. The effects of the action, if any, are modeled by changes in the variables that describe the run-time state of the service.

4. *Eventing:* A UPnP description for a service includes a list of actions the service responds to and a list of variables that model the state of the service at run time. If any of the variables is changed due to external command by a control point or due to internal configuration of the device, the corresponding service publishes the updated value of the variable. This is done by sending event messages in the network. Event messages contain the names of one or more state variables and the current value of those variables. These messages are also expressed in XML. A control point may subscribe to receive this information if it is of interest. More than one control points may subscribe to receive notification about a single variable.

   A variant of eventing is multicasting. Through multicast eventing, control points can listen to state changes in services without subscription. This form of eventing is useful first when events which are not relevant to specific UPnP interactions should be delivered to control points to inform users, and second when multiple controlled devices want to inform multiple other control points.

5. *Presentation:* A device may also provide a presentation URL, which can be accessed through a normal web browser. If a device has a URL for presentation, then the control point can retrieve a page from this URL, load the page into a browser, and depending on the capabilities of the page, allow a user to control the device and/or view device status. The degree to which each of these can be accomplished depends on the specific capabilities of the presentation page and device.

14.4 UPnP A/V Architecture

With the rapid development of home networking technologies and due to increased usage of multimedia-based contents, there is much need to share the multimedia contents between devices. This includes audio, video, images, photos, slideshows, and so on. UPnP forum has already defined some standardized DCPs (Device Control Points) while the vendors can also create specifications for different devices of their own. For example, the UPnP Device Architecture version 1.0 defines a service type known as Audio that 1 provides programmatic control for volume, tone, and spatial balance of a device with audio output. This service type contains a number of state variables such as volume, treble, bass, balance, fade, and so on and a number of actions that permits to query or adjust these variables. It also allows the vendors to include non-standard actions in the service definition according to the needs. UPnP A/V is sitting on top of UDA. A UPnP A/V stack is an implementation of UPnP AV Architecture. In contrast, an A/V application is an executable program that is running over UPnP A/V. Such application could use a UPnP A/V stack but this is not required.

In the remaining of this section, we focus on the UPnP audio/video architecture which is quite comprehensive and has been implemented and used by a number of vendors worldwide. We describe how users can share contents between UPnP devices based on the standardized UPnP A/V architecture.

The UPnP A/V architecture [7] defines three entities: a Media Server, a Media Renderer, and a Control Point, which are used together for controlling and sharing multimedia contents across various devices regardless of the device, content format or transfer protocol. All three entities are considered as if they were independent devices on the network, but the A/V Architecture supports arbitrary combinations of these entities within a single physical device (e.g., a device having both the Media Render and Control Point). The relationship between the devices is shown in Figure 14.3.

14.4.1 Media Server

The media server is used to locate the multimedia content that is available to media renderers. Media servers include a wide variety of devices such as VCRs, DVD players, satellite/cable receivers, TV tuners, radio tuners, CD players, audio tape players, MP3 players, PCs, and so on. The media server contains (i) a ContentDirectory service which

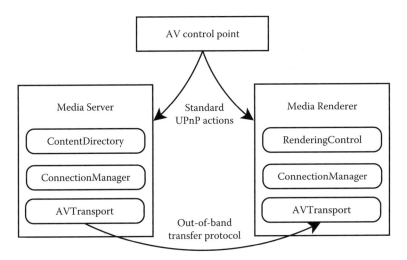

FIGURE 14.3 UPnP audio/video playback architecture.

provides a set of actions that allow the control point to enumerate the content that the server can provide to renderers, (ii) a ConnectionManager service to manage the A/V connections associated with a particular device, and (iii) an optional AVTransport service that is used by the control point to control the playback of the content that is associated with the specified A/V transport.

14.4.2 Media Renderer

The media renderer is used to render (e.g., display and/or listen to) the contents provided by the media server via network. Examples include a wide variety of devices such as TVs, stereos, speakers, hand-held audio players, and so on. The media renderer, like media server, has a ConnectionManager service and an optional AVTransport service to control the flow of the content (e.g., stop, pause, seek, etc.), but instead of the ContentDirectory service it includes a RenderingControl service. This service provides a set of actions that allow the control point to control how the media renderer renders content by modifying characteristics such as brightness, contrast, volume, mute, and so on.

14.4.3 Control Point

The control point is the only component that initiates UPnP actions, usually in response to user interaction with the control point's UI. The control point requests to configure the media server and media renderer so that the desired content flows from the media server to the media renderer, using one of the transfer protocols and data formats that are supported by both the media server and media renderer. The control point is capable of controlling the flow of the content by invoking various AVTransport actions such as stop, pause, FF, REW, skip, scan, and so on. Additionally, the control point is also able to control the various rendering characteristics on the renderer device such as brightness, contrast, volume, balance, and so on. Example of control point is a remote control for TV, VCR, CD Player, and so on. In the next section, we will show how a control point is used to initiate the transfer of session across devices.

While the UPnP A/V architecture specifies UPnP as a protocol for controlling devices, it does not specify any data format or protocol for actual transfer of contents between the devices. This can be any of the standard formats (such as MPEG2, MPEG4, JPEG, MP3, Windows Media Architecture (WMA), Bitmaps (BMP), etc.) and protocols (such as RTP, HTTP GET/PUT/POST, TCP/IP, etc.) or even a vendor-specific format/protocol.

14.4.4 Push versus Pull Transfer Protocols

The UPnP A/V architecture supports both isochronous-push transfer protocols (e.g., IEC61883/ IEEE1394) and asynchronous-pull transfer protocols (e.g., HTTP GET). In the first case, the underlying transfer mechanism provides real-time content transfer between the media server and media renderer. This allows the media renderer to provide the user with smooth rendering of the content without implementing a read-ahead buffer. The pull transfer protocols, on the other hand, do not provide real-time guarantees and a read-ahead buffer is required at media renderer. Algorithms for both these approaches do not differ significantly and can be found in the specifications [7].

14.5 UPnP for Technology Convergence

UPnP, being an open standard, can be used as a basis of network discovery and interaction between devices together with a number of technologies built on top of it. Due to its open, XML-based interfaces, it is easy to incorporate UPnP into other platforms and tools without significant rework. A number of independent studies at different places around the world have led to interesting and innovative concepts with the help of UPnP.

The digital living network alliance (DLNA) [8] is collaboration of the world's leading consumer electronics, PC, and mobile companies. DLNA has created design guidelines for a new generation of DLNA certified products that can work together—no matter the brand. Thus, DLNA is there to avoid and eliminate the compatibility problems between thousands of devices by various manufacturers around the world. DLNA interoperability guidelines allow manufacturers to participate in the growing marketplace of networked devices and are separated into the below sections of key technology components.

1. Network and connectivity
2. Device and service discovery and control
3. Media format and transport model
4. Media management, distribution, and control
5. Digital rights management and content protection
6. Manageability

To allow interoperability between DLNA certified devices, and to provide a common device interaction protocol, manufacturers are inclined to use UPnP as basis for communication among the supported devices. UPnP is the only protocol that supports all of the above-mentioned components required for easier, hassle-free device communication.

Kim et al. [9] have developed a home networking system using UPnP middleware. The proposed Home network system provides network services, maintains the network database, and timetables the home sewer scheduler. The scheduler enables the user to manage all of the home appliances using its services. It consists of client/server programs, a scheduler for home users, appliance emulators, and embedded communication devices using Linux and Window CE platforms. The home server programs are either developed using traditional programming languages or they provide browser-based interaction. The different types of appliances include refrigerators, microwave ovens, toasters, air-conditioners, washing machines, and TVs. They may have graphical or text display interfaces supporting various types of data rates. The implementations mostly are done for Windows CE and Linux operating systems.

Allard et al. [10] present a Jini/UPnP interoperability framework that allows Jini clients to use UPnP services and UPnP clients to use Jini services, without modification to service or client implementations. The architecture introduces service-specific proxies that allow Jini and UPnP clients and services to intermix with each other. Each new service type requires a modest amount of code to be written, but the Jini/UPnP clients and services themselves do not require modification. The framework generates virtual Jini services, one for each instance of a recognized UPnP service. This concept is similarly applied in the other direction (UPnP client using a Jini service). For each supported UPnP service instance, the architecture automatically generates a virtual Jini service instance and registers the instance with available Jini lookup services. For each supported Jini service instance, the architecture generates a virtual UPnP instance which performs UPnP advertisement for the service. The virtual services perform bridging, and translating between Jini method calls and UPnP actions. When a UPnP or Jini service is removed from the network, the architecture automatically destroys the corresponding virtual services.

Shirehjini [11] describes PECo system, a Personal Environment Controller. The aim of PECo is to provide an integrated and intuitive access to the user's personal environment and media repositories. PECo uses an automatically created 3D visualization of the environment. When a user equipped with a PECo enabled device enters a room, PECo discovers the infrastructure and available devices and constructs the integrated user interface. The 3D visualization makes a direct link between physical devices and their virtual representations on the user's device. Then he or she can access identified devices through the 3D interface and directly manipulate them. By interconnecting these two worlds the user can, for example, move a PowerPoint document—which is stored on his notebook—to the beamer by just one drag and drop operation. The fundamental principles behind PECo include device discovery, media management, and standardized device access. All these principles are met by the UPnP technology incorporated in PECo.

Now consider the problems related to audio/video session management when some user mobility is involved. For example, if a user has already started his session on a device when he needs to move nearby where he finds another device which can be used to continue his previously established session. To do this, generally the user has to stop playing the content using a UPnP A/V control point before leaving first location. In the new location, he has to reconfigure the rendering device so that he can continue to play the content from the previous playing position. All these operations are done manually by him. In order to minimize such time-consuming efforts, Hwang et al. [12] have designed a UPnP A/V session manager (USM) for automatically moving a user's A/V session information from one UPnP device to another, by making use of light-weight and low-price RFID readers. First, a user registers new RFID readers and tags using a UPnP control point, and assigns the added RFID readers to the media renderers near them. If a user plays some content on one media renderer and then moves around in the environment such that his presence is detected by the RFID reader near another media renderer, the USM detects his activity by receiving an event message from the previous RFID reader, and then controls the first media renderer to stop the current playing content on it and stores its session information. The USM then controls the new media renderer to play the content from the previous playing position by sequentially invoking actions on it.

While this approach is limited by enforcing the use of RFID cards, another approach proposed in [13] uses user-driven selection of devices of his choice for moving the session between them based on his preferences. Based on the user's preferences, one of the devices in the environment is selected considering factors such as display size, network type, input methods, and so on.

Chintada et al. [14] describe a system architecture consisting of bridging mobile network (SIP) technologies with the home network (UPnP) technologies and hosting the convergence function on the in-home broadband Residential Gateway. The Session Initiation Protocol (SIP) is an IETF specified protocol referred by RFC 3261, and it has various extensions. SIP is an application level protocol that has been designed along the lines of HTTP. The key strength of the SIP protocol is its ability to locate session end-points even when one or both the end-points are mobile. It also has protocol support to enable transfer of session end-points between devices. Their proposed solution can exist in two configurations. First, the bridge function could be internal to the device, when the device supports both the UPnP and the SIP protocols. Second, when the device supports only one of the two protocols, the bridge function is hosted on an edge device like a home gateway or a set-top box.

Integrating a signaling protocol like the SIP with UPnP also enables new usage scenarios. For example, it allows interaction between a videophone and other UPnP compliant devices, like a TV set or a media server. This interaction, for example, makes it feasible to display the caller's image on the television screen, thus allowing the user to relax on the sofa and naturally converse. Another interesting scenario is the one where the user sends to the conversation partner a set of still pictures or short video clips acquired from a UPnP media server already available in the home premises. Such a convergence has been explained in [15].

14.6 Conclusion

This chapter outlined the UPnP technology as an important tool for service discovery and device interaction in pervasive environments. This chapter first introduced the core concepts in service discovery and service-oriented architecture. It then provided an overview of several of the existing SDPs. The in-depth discussion of UPnP along with its benefits and usage scenario as well as different research work related to UPnP was discussed in detail.

References

1. Bettstetter, C. and Renner, C. A comparison of service discovery protocols and implementation of the service location protocol, in *Proceedings of EUNICE Open European Summer School*, Twente, Netherlands, September 2000.
2. Arnold, K., Scheifler, R., Waldo, J., O'Sullivan, B., and Wollrath, A. *The Jini Specification*. Boston, MA, USA: Addison-Wesley Longman Publishing Co., Inc., 1999.
3. Salutation Consortium. White Paper: Salutation Architecture: Overview. http://www.salutation.org/whitepaper/originalwp.pdf, 1998.
4. Guttman, E. Service location protocol: automatic discovery of ip network services. *IEEE Internet Comput* 3(4): 71–80; 1999.
5. Universal Plug and Play Forum. Universal Plug and Play Device Architecture. http://www.upnp.org/, March 2000.

6. Apple. BOnjour. http://www.apple.com/support/bonjour/.

7. UPnP Forum AV Architecture Specification v1.0. http://www.upnp.org/specs/av/UPnP-av-AVArchitecture-v1-20020622.pdf, June 2006.

8. Digital Network Living Alliance. http://www.dlna.org/.

9. Kim, D-S., Lee, J-M., Kwon, W. H., and Yuh, I. K. Design and implementation of home network systems using UPnP middleware for networked appliances. *Consumer Electron IEEE Trans* 2002; 48(4): 963–972.

10. Allard, J., Chinta, V., Gundala S., and Richard III, G. G. Jini meets upnp: an architecture for jini/UPnP interoperability, in *SAINT '03: Proceedings of the 2003 Symposium on Applications and the Internet*, p. 268, Washington, DC, USA, 2003. IEEE Computer Society.

11. Nazari Shirehjini, A. A. A generic upnp architecture for ambient intelligence meeting rooms and a control point allowing for integrated 2d and 3d interaction, in *sOc-EUSAI'05: Proceedings of the 2005 Joint Conference on Smart Objects and Ambient Intelligence*, pp. 207–212, New York, NY, USA, ACM, 2005.

12. Hwang, T., Park, H., and Paik, E. Location-aware upnp av session manager for smart home, in *Networked Digital Technologies, 2009. NDT '09. First International Conference on*, pp. 106–109, 2009.

13. Mukhtar, H., Belaïd, D., and Bernard, G. Quantitative model for user preferences based on qualitative specifications, in *ICPS'09: IEEE International Conference on Pervasive Services 2009*, London, UK, July 2009.

14. Chintada, S., Sethuramalingam, S., and Goffin, G. Converged services for home using a sip/upnp software bridge solution, in *Consumer Communications and Networking Conference, 2008. CCNC 2008. 5th IEEE*, pp. 790–794. IEEE, 2008.

15. Vilei, A., Convertino, G., and Crudo, F. A new upnp architecture for distributed video voice over ip, in *MUM '06: Proceedings of the 5th International Conference on Mobile and Ubiquitous Multimedia*, p. 2, New York, NY, USA, ACM, 2006.

III

Applications

15

Wireless Network Security for Health Applications

Eduardo B. Fernandez
Florida Atlantic University

15.1 Introduction

The Internet, wireless systems, and radio-frequency identification (RFID) sensors are opening a new era in medical care. Most activities in medical care can be integrated and performed remotely and pervasively. This change promises to improve medical care and reduce costs. However, if these systems are not secure, users will lose confidence in them and some of these advances will not be realized. We discuss here some issues and emphasize that, to have secure systems, we must consider security from the beginning, in all phases, and in the whole system.

Medical functions such as assisted living and patient monitoring require the use of computer networks. Often, these networks are built out of wireless sensors. Because of the need to be laid out through long distances and usually operate without supervision, these networks are exposed to a variety of attacks. We look at some medical applications and the corresponding network security issues. The so-called "telehealth" applications include assisted living, pervasive health care, patient monitoring, distance surgery, remote diagnosis, ambulance, and others; they all typically rely in wireless networks.

In particular, we consider here aspects of the so-called "mHealth" (also written as "m-health" or sometimes "mobile health"), a recent term for medical and public health practice supported by mobile devices, such as cellular phones, patient monitoring devices, PDAs, and other wireless devices [1]. Applications of mHealth include the use of mobile devices in collecting clinical health data, delivery of health care information to practitioners, researchers, and patients, real-time monitoring of patient vital signs, and direct provision of care (via telehealth). We consider mHealth to be not a set of independent systems, but just a complementary system to the complete medical system.

Most discussions of the security of wireless networks for health applications consider only the communication aspects of the networks [2,3]. However, as indicated, a health network is part of a complete health application and we need to relate the communication aspects to the medical aspects. We need to understand first what information is needed for medical purposes and how this information is used. This is discussed in Section 15.2. Once we define what information to keep, we present in Section 15.3 a pattern for patient records management. Design patterns were introduced in 1994 and have had an enormous influence in the system design. A pattern is an encapsulated solution to a recurrent problem in a given context. Security patterns encapsulate solutions to security problems [4]. Section 15.4 shows two case studies: Ambient assisted living (AAL) and a sensor-based hospital. Section 15.5 discusses security of wireless devices, threats, and defenses in a wireless network such as the ones used in medical applications. We end with some conclusions.

15.2 Medical Records and Their Regulations

The electronic health care record (EHR) is a lifetime record of an individual with the purpose of supporting continuity of care, and related education and research. It typically includes information about encounters (visits), lab tests, diagnostics, observations, medications, imaging reports, treatments, allergies, and therapies, as well as patient-identifying information and legal permissions [5].

Medical information is very sensitive and must be protected. Most countries have severe restrictions in the use of this information. There are several regulations in the United States about the handling of health information. The best known is the Health Insurance Portability and Accountability Act (HIPAA) [6]. Title II of HIPAA, known as the administrative simplification (AS) provisions, requires the establishment of national standards for electronic health care transactions and national identifiers for providers, health insurance plans, and employers. The AS provisions also address the security and privacy of health data. The standards are meant to improve the efficiency and effectiveness of the nation's health care system by encouraging the widespread use of electronic data interchange.

The HHS has promulgated five rules regarding AS:

- The "Privacy Rule" regulates the use and disclosure of certain information held by covered entities (health care providers, health care clearinghouses, employer sponsored health plans, and health insurers) and their business associates (lawyers, accountants, IT consultants). It establishes regulations for the use and disclosure of protected health information (PHI).

- The "Transactions and Code Sets Rule" defines specific transaction types. For example, the EDI Health Care Claim Transaction set (837) is used to submit health care claim billing information, encounter information, or both, except for retail pharmacy claims. It can be sent from providers of health care services to payers, either directly or via intermediary billers and claims clearinghouses.
- The "Security Rule" complements the "Privacy Rule." While the "Privacy Rule" pertains to all PHI including paper and electronic, the "Security Rule" deals specifically with electronic PHI (EPHI). It lays out three types of security safeguards required for compliance: Administrative, physical, and technical.
- The "Unique Identifiers Rule" establishes that providers must use only the National Provider Identifier (NPI) to identify themselves in standard transactions. The NPI is a unique 10-digit identification number provided by the US Government.
- The "Enforcement Rule" sets civil money penalties for violating HIPAA rules and establishes procedures for investigations and hearings for violations. It seems to be rarely applied, however.

Privacy is the right of individuals or groups to keep their personal information away from public knowledge or their ability to control personal information flow. In the electronic or the real world, people seek privacy, so they can perform their actions without others monitoring them. Individuals should be able to live without being disturbed and users interacting with the Web, navigate without being identified. People providing information to medical institutions or storing their personal records in commercial companies should know what to expect about the privacy of their information. This right is recognized by all civilized societies and is considered a fundamental human right. The first national privacy protection law was the Swedish Data act of 1973. This was followed by the US Privacy Act of 1974. The intent of this act was to protect individuals against invasion of privacy by the Federal Government. This law is complemented by the Computer Security Act of 1987, which defines requirements for federal agencies about the security of their information. In general, privacy laws are more developed in Europe than in the United States.

15.3 Patient Treatment Records Pattern

We present a pattern to describe some of the basic functions to maintain and use patient records in a hospital, the Patient Treatment Records Pattern [7]. A medical record can be thought of as a series of dated treatment instances, or encounters. Each encounter is documented on a patient chart. For our purposes, we will use treatment instance, chart, and encounter synonymously. There is a possibility that a patient chart may contain more than one encounter, however, we will use the term chart or treatment instance to mean one encounter. Each patient encounter is documented and contains dated notes written by physicians, laboratory reports, and letters from consulting physicians. In addition, a patient chart will have vital sign documentation from nurses, imaging reports, specific treatment plans, treatments performed, medications given, assessments of patient condition, and so on. Due to the sensitive nature of the medical information, much of these data need to be organized and categorized to prevent unwanted access to private information by unauthorized personnel.

The Patient Treatment Records Pattern focuses on the private and sensitive nature of medical information and the need for maintaining accurate and organized records. A patient is admitted to a health care facility where all pertinent information is recorded. A physician and other facility assets are assigned to the patient. Following treatment, the patient is discharged. This pattern describes only some of the aspects of patient treatment, which include the creation and maintenance of the patient record and the assignment of the assets for the use by the patient. This pattern describes a general non-emergency treatment situation and does not consider the details of patient diagnosis and treatment.

The Patient Treatment Records Pattern describes the handling of records during the treatment or stay instance of a patient in a hospital. The hospital may be a member of a medical group. Each patient has a primary physician, an employee of the hospital. Upon admission, the patient record is created or information is updated from previous visit(s). Inpatients are assigned a location, nurse team, and consulting doctors.

The Patient Treatment Records Pattern is a *Semantic Analysis Pattern* and corresponds to requirements expressed as a set of use cases [8]. The analysis model is developed from the use cases. We present one of its two component patterns, the patient record. The other two can be found in [7]. Figure 15.1 shows the use case diagram that corresponds to some of the typical needs of patient treatment and which define the structure of the Patient Treatment Records Pattern. There are other use cases such as diagnose, perform patient treatment, and billing that have been left out for simplicity.

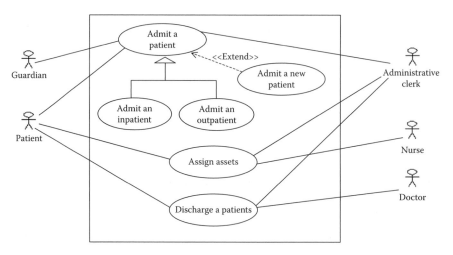

FIGURE 15.1 Use case diagram for patient treatment.

15.3.1 Patient Treatment Record Pattern

Describes the structure of patient records and the process of creating and maintaining them for a stay or treatment in a hospital.

15.3.1.1 Problem

Maintaining accurate records is crucial for patient treatment. A poor record may result in erroneous treatments, loss of insurance, or other problems for the patient or the hospital. How do we keep an accurate picture of what happens during the stay of a patient at a hospital for treatment?

The solution to this problem is affected by the following forces:

- Patient characteristics, for example, age, sex, occupation, race, weight, and others, may have an effect on the diagnosis and treatment of the patient and it is important to keep this information accurate.
- We need a detailed record of what has been done to a patient during a specific stay at the hospital. This is necessary for medical, billing, and legal reasons.
- Patients may return to the hospital and we need to relate new treatments to past treatments.
- There may be different types of patients that we need to classify. Otherwise, the patients or the hospital may incur unnecessary expenses.
- Patients may not be responsible for their decisions or their expenses; we need somebody responsible for the patient.

15.3.1.2 Solution

Maintain a medical history for each patient. This medical history typically contains insurance information and a record of all treatments within the medical group. If the patient is new to the medical group, a patient record and medical history will be created upon admission. If the patient has been treated in any facility within the medical group, there will be an existing patient record and a medical history, which may need to be updated. A treatment instance is created for all patients admitted and updated throughout the patient's stay. The treatment instance will subsequently be added to the patient medical record upon patient discharge. A person or guardian is responsible for each patient. We classify patients into inpatients and outpatients. Use cases realized by this pattern include Assign a Guardian, Modify Medical History, and Admit a New Patient.

Figure 15.2 shows the structure of patient records. A unique stay or treatment instance for every patient is created upon admission to the hospital. The patient may be admitted to the hospital as an inpatient to stay in the hospital, or he may be admitted as an outpatient, in which case he will receive treatment but will not stay at the hospital. The treatment instances are collected into the medical history. A Guardian is responsible for each patient. Guardian can be seen as a role in that a patient may be its own guardian. Additional relevant patient information is recorded into patient info. We can model similarly other aspects of patient treatment. The rest of this pattern and the two other patterns can be found in [7]. A possible structure for medical histories and treatment

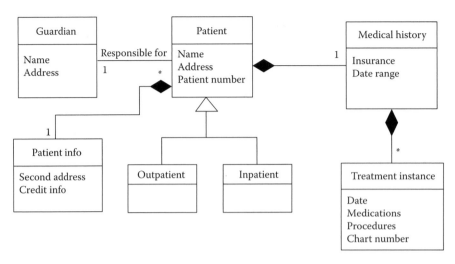

FIGURE 15.2   Class diagram for patient record.

instances is shown in [9]. These use cases do not specify design aspects, some of them, for example, discharge patient, could be performed from wireless devices.

15.4 Two Case Studies

We present two case studies showing the use of wireless networks for health applications. These cases can be expressed as patterns although we have not done so here.

15.4.1 Case 1: AAL

AAL defines architectures for home environments which have devices such as sensors and cameras to support and monitor people with impaired functions or disabilities. Assisted living requires a secure infrastructure of services to be in place at the patient's home or place of care. Many of these services are also valuable for family living and the corresponding architectures differ mostly in the specific types of services they provide [10]. An assisted living system should have the following requirements [11]:

1. *Dependability:* Critical services should be delivered in spite of the failures of useful but noncritical services. Moreover, the system as a whole will have high availability and robustness.
2. *Low Cost and Flexibility:* The general infrastructure will be open with well-defined interfaces, machine checkable quality-of-Service (QoS) assumptions, and support the use of low-cost, third-party devices.
3. *Security and Privacy:* Medical and personal data should be protected with different rights for different roles (health care providers, medical team, relatives, and assisted persons (APs)). In fact, the complete network architecture must be secure.

4. *Quality-of-Service Provisioning:* QoS should be provided at different levels depending on their criticality requirements.

5. *Open Standards:* Any brand or type of device should be able to interoperate with any devices and operating systems.

6. *Light-Weight, Easy-to-Use HCIs:* The user interfaces will be easy-to-use, safe, tolerant of user mistakes, and provide different control levels of information disclosure.

7. *Flexible:* Adaptable and extensible software and hardware architectures.

8. *Interoperable:* Compatible with EHRs, so that the same concepts appear in the information carried along by the devices.

Some possible use cases are shown below [17]; we are converting them into patterns and we show some UML models for them:

1. *Remind of Activity:* The doctor defines the patient schedule. When it is time for the patient (AP) to carry out his time-driven routines, such as taking medicine, taking vital signs, or exercising the system locates active wireless-enabled devices (e.g., TVs, cell phones) in range, and sends reminder messages to one or more devices that are in the proximity of the AP. (The AP can also prioritize the order in which devices will be used.) For example, if the AP is watching TV at the time when the reminder message is scheduled, the TV will be switched to an information channel (with the use of Infrared remote control) and a reminder message will be displayed. In this manner, the AP can be reminded of his/her time-driven routines. Whether or not these routines are followed as advised is detected in a nonintrusive manner by exploiting sensor localization technologies, the prescription bottles can have RFID tags (with unique barcodes) and one or more RFID readers in the environment are activated (by the ALH) to track location changes of these bottles. Figure 15.3 shows a class diagram for this pattern, whereas Figure 15.4 shows a corresponding activity diagram.

2. *Vital Sign Measurement:* In the current practice of glucose monitoring for diabetics patients, a patient measures her glucose level on a daily basis, and brings the

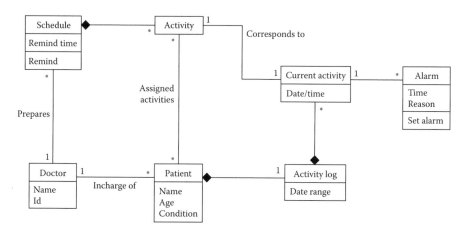

FIGURE 15.3 Class diagram for remind of activity pattern.

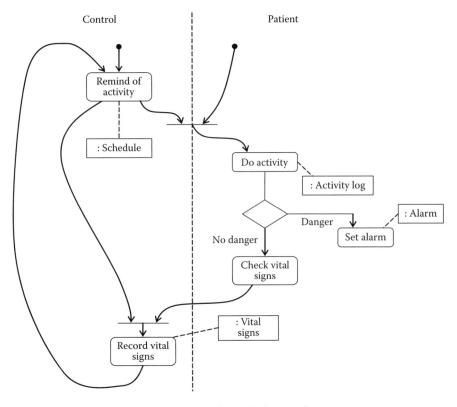

FIGURE 15.4 Activity diagram for use case "remind of activity."

measuring device to her monthly clinic visits where the measurements are retrieved and interpreted by health care providers. With the proposed environment in place, vital signs can be measured and transmitted by Bluetooth-enabled meters to the server. In this fashion, health care providers can monitor various vital signs at convenient time granularity. Should the readings suggest any abnormal health situations, medical instructions can be given before the situations deteriorate.

3. *Personal Belonging Localization:* Personal belongings such as eyeglasses, hearing aids, and key chains can be attached with tags, and located through the use of RFID readers. When a person cannot find her belongings (because of forgetfulness), she can issue a simple vocal command (through, for example, a light-weight, Bluetooth-enabled headset) to the ALH which then schedules the RFID readers to scan the environment and help locate the object.

4. *Personal Behavior Profiling:* With the same set of sensor localization techniques, the assisted living environment can profile the movement of APs in a privacy preserving manner (e.g., without the use of surveillance video cameras) and detect early warning signs for depression (no longer taking medicine regularly, giving up routine activities, or staying in bed for long periods of time) and/or

other chronic diseases such as Parkinson's disease and Alzheimer's disease. The AP wears an RFID tag or an active badge.

5. *Emergency Detection:* In case of the need for emergency attention (e.g., the blood pressure/sugar has been dangerously high/low, and/or the person has been detected via localization techniques to be immobile on the floor for an unreasonably long time), real-time communication channels can be established to notify on-site caregivers (in the case of assisted living), health care providers (in the case of clinical use), and/or designated relatives, and facilitate transmission of electrocardiogram (EKG) and other measures in real-time.

All these approaches require networks and we discuss next some aspects of network security. Later we will discuss the security of sensor networks. These systems are also cyber-physical systems and we should study them as such.

15.4.2 Case 2: Sensor-Based Hospital

A smart (sensor-based) hospital is described in [12]. The goal of that work is to show how technologies of identification using RFID sensors can contribute to build a smart hospital by optimizing business processes, reducing errors, and improving patient safety.

To start with, many assets and actors of the facilities have to be "tagged":

- The medical equipment must embed RFID tags. In the best case the tags should be placed into the devices by the manufacturer and should contain a standardized unique identifier.
- The doctors, nurses, caregivers, and other staff members wear a "smart badge," storing their employee ID number.
- On arrival, each patient receives a wristband with an embedded RFID tag storing a unique identifier, and some information about him (e.g., a digital picture, a unique patient code, etc.)
- Bandages, blood bags, and drug packages all contain RFID labels.

Furthermore, RFID readers are placed at strategic places within the hospital:

- RFID gates are provided at entrances and exits of the hospital.
- Each operating theater contains a least one RFID reader.
- RFID sensors are placed in strategic galleries and important offices. In the best case, every office should contain an RFID reader.
- The staff members (doctors, nurses, caregivers, and other employees) each have a handheld (smart phone, etc.) equipped with an RFID reader and possibly with a wireless connection to the Web.

This setup allows use cases such as:

1. *Perform Patient Identification.* RFID bracelets may include patient ID and some other useful information, for example, blood type.
2. *Blood Tracking.* RFID tags in bags can identify the patient for which they are intended.

3. *Identify Right Patient and Operation.* In the operating theater, avoid confusion of patients, operations to be performed, and place of operations by correlating patient ID with its EHR.

4. *Prevent Drug Counterfeiting.* Can use Electronic Product Codes (EPC) in drug packages. This increases patient safety.

5. *Tracking Equipment, Patients, Staff, and Documents.* All of these carry RFID tags and if we have readers in strategic places we can track all of them. A byproduct is the possibility of an accurate and efficient inventory system.

6. *Avoiding Theft of Medical Equipment.* When something is stolen, we must add to the cost of the lost equipment the time to locate it and to reorder a replacement. We might also need the stolen equipment in an emergency. With RFID tags, we can detect if it leaves the hospital and we can control access to it. RFID tag must be hard to remove.

15.5 Wireless Network Security

We consider some aspects of the security of the wireless networks that support these health applications.

15.5.1 Wireless Devices

When compared to wired networks, there are four generic limitations of all wireless devices: (1) limited power, (2) limited communications bandwidth, (3) limited processing power, and (4) relatively unreliable network connection. The bandwidth available to wireless systems is usually an order of magnitude (or even more) less than that available to a wired device. The processing power is limited due to limited space/cost in case of fixed wireless devices typically used for Wi-Fi networks, and is further limited due to power constraints in other wireless devices. In general, wireless networks are not very reliable. Protocols have been designed to take this lack of reliability into account and to try to improve it. However, in designing these protocols, choices have to be made about the size of the packets and frames to be used. Such decisions can have a profound impact on the effectiveness and efficiency of cryptographic protocols.

To this we must add that security needs for wireless devices are greater than those of regular wired-network devices. This is due to the very nature of their use; they are mobile, they are on the edge of the network, their connections are unreliable, and they tend to get destroyed accidentally or maliciously. These devices can also be stolen, lost, or forgotten. Thus, we need more security processing. Security processing can easily overwhelm the processors in wireless devices. This challenge, which is unique to wireless devices, is sometimes referred to as the "security-processing gap." Non-fixed wireless devices such as cellular handsets and *ad hoc* network devices such as sensors are severely handicapped due to their very low battery power. Even though significant advances are expected in computation and communication speed over the next decade, it is still expected that they will lag behind the power available to fixed computers due to the desire for miniaturization. To make things worse, battery power is only expected to make only modest improvements. The battery limitation in mobile wireless devices is sometimes called "battery gap" and refers to the growing disparity between increasing energy requirements for high-end operations needed on such devices and slow improvements in battery technology. To this we add the fact that there is a large variety of devices using different architectures, several operating systems, and diverse functionality. With increase in functions, the typical problems of larger systems are also appearing in portable devices.

15.5.2 Threats

The analysis of secure systems should start from their possible threats. We can apply our methodology [13] to enumerate the threats to the activities of Figure 15.4. We show the threats for two activities, and the other activities can be analyzed similarly. We then find policies to stop or mitigate the threats. As shown in [14], these policies can be realized by security patterns, which define the system security requirements and can then be converted into design artifacts, for example, secure interfaces. Here, the threats appear as attacker goals. When the details of the design start emerging, we can convert these threats in specific threats to the units of the system, for example, threats to the user interfaces or to the wireless network. This may require adding more security patterns to stop the emergent threats.

Activity 1: Remind of Activity/Task	Activity 2: Do Activity
T11: Control site or patient site is an impostor	T21: Unauthorized reading of activity log
T12: Unauthorized reading of schedule	
T13: Unauthorized writing of schedule	T22: Unauthorized writing of activity log
T14: Denial of service	
Policies to Stop These Threats	
P11: Mutual authentication	P21: Authorization/access control
P12: Authorization/access control	P22: Authorization/access control
P12: Authorization/access control	
P12: Cell phone backup	

15.5.3 General Wi-Fi Threats

The attacks described above and for other applications can be realized through other parts of the application, but the wireless network is a source for many of them.

15.5.3.1 Attacks Related to Access Points (APs)

Detection of access points: This is really an attack preparation. Tools exist for this purpose, for example Netstumbler. These tools can detect for each AP: its MAC address, its location, the transmission channels it uses, and the type of encryption it uses.

Unauthorized (rogue) APs: The attacker sets up its own AP ("malicious association"). Once the thief has gained access, she can steal passwords, launch attacks on the wired network, or plant malware. Fraudulent APs can easily advertise the same network name (SSID) as a legitimate AP, causing nearby Wi-Fi clients to connect to them. One type of man-in-the-middle attack is a "de-authentication attack." This attack forces AP-connected computers to drop their connections and reconnect to the fake AP.

Accidental association: When a user connects to a wireless access point from a neighboring company's overlapping network, it is a security breach in that proprietary company information may be exposed and now there could exist a link from one company to the other.

Direct endpoint Attacks: These take advantage of flaws in Wi-Fi drivers, using buffer overflows to escalate privilege.

15.5.3.2 Denial of Service

This is a common attack in wireless networks because of their frequency sharing with other networks and can be accidental. A reason for performing an intentional DoS attack is to observe the recovery of the wireless network, during which all of the initial authentication information is resent by all devices, providing an opportunity for the attacker to collect this information.

15.5.3.3 Network Injection

The attacker may inject network configuration commands that affect routers and switches.

15.5.3.3.1 Cryptographic Attacks Based on Message Interception

Data sent over Wi-Fi networks, including wireless printer traffic, can be easily captured by eavesdroppers. Weaknesses in encryption protocols such as WEP allows an attacker read messages.

15.5.3.3.2 Operating System Attacks through the Network

Operating systems for wireless devices are becoming more and more complex and attackers can exploit code flaws in them to access their files.

15.5.4 General Wi-Fi Defenses

We have discussed elsewhere how to stop attacks as the ones enumerated for the application above [14]. We consider here general approaches to stop attacks to wireless networks. First, a few security principles that also apply to wireless networks:

- For new systems, use a global-level design, starting from requirements. Define policies, analyze threats, select defense mechanisms. A methodology like the one proposed in [14] is appropriate. Threats can be enumerated as described earlier in this paper. In other words, security must be applied in the whole lifecycle and in all architectural levels of the system.
- Medical records should be integrated with lab, pharmacy, and so on. A global conceptual model should include all entities relevant to patient care.
- Due to the sensitivity of the data, the system should be a closed system, where everything is forbidden unless explicitly authorized.
- System-critical and life-critical functions must have a backup.

- Every access to resources, coming from fixed or mobile devices, should be mediated and checked for validity. There should be no direct access to any data or other resource.
- For existing systems, perform security auditing.

In particular, for the wireless network we should apply these principles:

- The wireless network should be an integrated unit of the whole medical system. All the policies and models that apply to medical records and related information should also apply to the data handled by the wireless network and its devices.
- Each wireless device should be integrated in a system-wide structure. There should be no outsiders that can access health data.
- Wireless device usage policies must be consistent with the total system policies.

Specific policies for wireless access should at least include:

- List devices authorized to access the wireless network.
- List personnel that can access the network.
- Define rules about setting up wireless routers or APs.
- Define rules on the use of Wi-Fi APs or about connecting to home networks with company devices.
- Devices such as network access center (NAC) can be useful to enforce all of the above.
- TrackAP vulnerabilities. It is important to track Wi-Fi endpoint vulnerabilities and keep the Wi-Fi drivers up-to-date.

15.6 Conclusions

We can build patterns for sets of related use cases and add security patterns to them to define a secure unit that can be used by an inexperienced developer to build secure applications. To define the security architecture of the network we need an analysis of the possible threats, the security patterns in the network are introduced to stop or mitigate them. In this way, we can build secure health systems in an integrated way, including health records, pharmacy records, and other related information, not just isolated wireless networks.

Complementary approaches to security focus on specific aspects, for example, intrusion detection [2]. A methodology to build health information systems using agents is proposed in [15], but security is not an objective. An attempt to build an integrated system is described in [16], which has some points in common with our approach. Pazin et al. [17] present a pattern for a rehabilitation clinic using a similar approach to ours but without considering security aspects. A close approach to ours is given in [18], but they do not try to integrate the complete health system; their idea of security by contract can be complementary to our approach. An interesting security architecture for home systems is presented in [19].

A complete design for an integrated system is needed to fully appreciate the value of our ideas and we intend to do so in the near future. We also need specific patterns as building blocks of such a design.

References

1. Jurik, A.D. and Weaver, A., Remote medical monitoring. *Computer*, 2008; 41(4): 96–100.
2. Giani, A., Roosta, T., and Sastry, S., Integrity checker for wireless sensor networks in health care applications, in *Proceedings 2nd International Conference on Pervasive Computing Technologies for Healthcare*, Piscataway, NJ, 2008.
3. Zeng, Z., Yu, S., Shin, W., and Hou, J.C., PAS: A wireless-enabled, cell-phone-incorporated personal assistant system for independent and assisted living, in *Proceedings of 28th International Conference on Distribution Computer Systems*, IEEE 2008, pp. 233–242.
4. Fernandez, E.B., Security patterns and a methodology to apply them, in *Security and Dependability for Ambient Intelligence*, edited by Spanoudakis, G., Maña, A., Berlin, Heidelberg, New York: Springer, 2009.
5. Eichelberg, M., Aden, T., Riesmeier, J., Dogac, A., and Laleci, G.B., A survey and analysis of Electronic Healthcare Record Standards. *ACM Comp. Surveys*, 2005; 37(4): 277–315.
6. Health Insurance Portability and Accountability Act, http://en.wikipedia.org/wiki/Health_Insurance_Portability_and_Accountability_Act.
7. Sorgente, T. and Fernandez, E.B., Analysis patterns for patient treatment, in *Proceedings of PLoP 2004*, http://jerry.cs.uiuc.edu/~plop/plop2004/accepted_submissions
8. Fernandez, E.B. and Yuan, X., Semantic analysis patterns, in *Proceedings of 19th International Conference on Conceptual Modeling*, ER2000, pp. 183–195. Also available from: http://www.cse.fau.edu/~ed/SAPpaper2.pdf.
9. Sorgente, T., Fernandez, E.B., and Larrondo-Petrie, M.M., The SOAP pattern for medical charts, in *Proceedings of the 12th Pattern Languages of Programs Conference (PLoP2005)*, Monticello, Illinois, 7–10 September 2005. http://hillside.net/plop/2005/proceedings/PLoP2005_tsorgente0_1.pdf.
10. Suomalainen, J., Moloney, S., Kolvisto, J., and Keinanen, K., Open House: A secure platform for distributed home services, in *6th Annual Conference on Privacy, Security, and Trust, 2008*, IEEE Press, Fredericton, New Brunswick, Canada, 2008.
11. Wang, Q., Shin, W., Liu, X., Zeng, Z., Oh, C., AlShebli, B.K., Caccamo, M. et al., I-Living: An open system architecture for assisted living. IEEE Trans. Syst Man Cybern.
12. Fuhrer, P. and Guinard, D., Building a smart hospital using RFID technologies, in *Proceedings of 1st European Conference on eHealth (ECEH06)*, 12–13 October 2006, Fribourg, Switzerland, edited by Meier, A., Stormer, H., pp. 131–142, GI-Edition–Lecture Notes in Informatics (LNI).
13. Braz, F., Fernandez, E.B., and Van Hilst, M., Eliciting security requirements through misuse activities, in *Proceedings of the 2nd International Workshop on Secure Systems Methodologies using Patterns (SPattern'07)*. Turin, Italy, September 1–5, 2008, pp. 328–333.

14. Fernandez, E.B., Larrondo-Petrie, M.M., Sorgente, T., and VanHilst, M., A methodology to develop secure systems using patterns, in *Integrating Security and Software Engineering: Advances and Future Vision*, Chapter 5, edited by Mouratidis, H., Giorgini, P. IGI Global, 2006, pp. 107–126.

15. Nguyen, M.T., Fuhrer, P., and Pasquier-Rocha, J., Enhancing E-Health information systems with agent technology. *Int J Telemed Appl* 2009; 2009: 13pp.

16. Evidian, Proteger la confidentialite: le controle d'acces en hopital, White Paper, 2010, http://www.evidian.com

17. Pazin, A., Penteado, R., and Masiero, P., SiGCli: A pattern language for rehabilitation clinics management, in Proceedings of SugarLoafPLoP 2004, http://sugarloaf-plop2004.ufc.br/acceptedPapers/index.html

18. Dragoni, N., Massacci, F., Walter, T., and Schaefer, C., What the heck is this application doing?: A security-by-contract architecture for pervasive services. *Computers and Security* 2009; 28:566–577.

19. Rajavelsamy, R., Lee, J., and Choi, S., Towards security architecture for home (evolved) NodeB: Challenges, requirements, and solutions. *Security Comm. Networks*, 2009.

16

Sensor Networks in Healthcare

Arny Ambrose
Florida Atlantic University

Mihaela Cardei
Florida Atlantic University

16.1 Introduction

A wireless sensor network is composed of sensors that are used to monitor a particular environment. The capabilities of such a network have advanced such that sensor networks can be used in healthcare systems. There has been much progress towards the integration of specialized medical technology with pervasive, wireless networks [1]. The use of sensors can enhance the care provided by a healthcare individual to a patient.

Continuous healthcare monitoring can become extremely expensive for a patient with chronic illness, and the use of sensors would alleviate the physical and financial burden of having a permanent caregiver. This would also give the patient a sense of independence, especially with elderly patients, while still having family and caregivers to monitor them and be alerted to any urgent situations. Wireless networks are the optimal method of providing continuous pervasive monitoring of not only the patient, but also of the environment. This offers a variety of applications for the use of these networks in healthcare.

Medical systems can benefit from wireless sensor networks in a variety of applications. They can be used in disaster-response scenarios to allow for efficient tracking of patients and emergency response personnel [2–4], for continuous monitoring of patients in assisted living facilities [5,6], and for other mechanisms that require wearable sensors and provide efficient delivery of patient data.

16.2 Emergency Response Applications

During emergency response situations, it is important that the responders have a timely and accurate assessment of the health of the patients. Wireless sensor networks can be used to coordinate different teams of rescue personnel and also multiple organizations to create a cohesive and an efficient response effort. These networks are usually comprised of wearable sensors that can be placed on patients for continuous monitoring [4], a way of keeping track of response personnel and the patients, and a means of data collection and storage.

The emergency response software infrastructure CodeBlue [2] integrates wireless devices with a wide range of capabilities into a network that can be used for emergency care or disaster response. The infrastructure consists of wireless vital sign sensors, location beacons, and all the protocols and services required to make the information gathered by them useful to emergency medical technicians, police and fire rescue, and ambulance systems.

It is important that in coordinating the collection and transmission of data there are proper discovery of the wireless devices, so that communication pathways is developed. In this architecture, data collection will be done using some kind of mobile device, for example, personal digital assistants (PDA). It is important that the device name be specific to the application and not to use a low-level network address. This makes it easier for the collection device to request information from only sensors of a specific type or in a certain range. Also the discovery process is decentralized such that there is no concern of one failure making the entire network unusable.

When all sensors have been detected, it is important that communication between the devices is reliable since they may be communicating with other devices that are outside their communication range. The CodeBlue infrastructure (Figure 16.1) also contains *ad hoc* routing techniques that extend the effective communication range of the devices. These devices typically use a multicast communication method to allow one sensor to report its data to multiple receiving nodes.

During a disaster-response situation it is also important that the locations of the rescuers and victims be monitored. CodeBlue uses a radio frequency (RF)-based location tracking system called MoteTrack [7] that operates using the low-power radio transceivers of the sensor nodes. MoteTrack uses beacon nodes which broadcast periodic messages

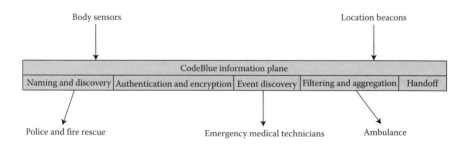

FIGURE 16.1   CodeBlue infrastructure.

that would contain the node's ID and the transmission power that is used to broadcast the message. Each beacon acquires a signature which will be sent to fixed beacons with known locations, serving as reference nodes in determining the location of the mobile nodes based on the power used to transmit the message.

Waterman et al. [4] developed the scalable medical alert response technology (SMART) that is designed to provide health care and patient monitoring. This system was developed to provide continuous monitoring of patients in an overcrowded hospital emergency room, but can also be scaled up for use in disaster scenarios. Sensors are used to observe the patient's electrocardiogram (ECG) signals and oxygen saturation and this information is transmitted to a PDA so that a quickly deteriorating patient can be immediately identified. The location system used is based on MIT Crickets [8], which merges RF and ultrasound to determine indoor locations where the global positioning system (GPS) would be ineffective.

Simulations [6] showed that the SMART system was able to manage 150 individuals. The system was operational after approximately 5 min and each patient was equipped with a monitoring equipment as soon as they arrived. It was found that the use of SMART enabled caregivers to work on other patients without having to continually monitor all patients.

Ko et al. [3] propose the medical emergency detection in sensor networks (MEDiSN) which is also designed for monitoring patients in hospitals and disasters. MEDiSN consists of patient monitors that are custom-built, wearable motes that will collect and secure the data, relay points that will create a multi-hop wireless backbone for transmission of the data and a gateway, as shown in Figure 16.2. This mechanism differs from the others described as it uses a wireless mesh infrastructure of relay points that transmit

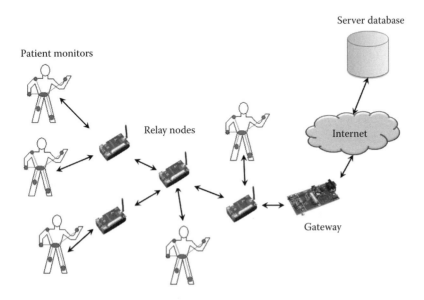

FIGURE 16.2 MEDiSN architecture.

the data from the patient monitors. This increases the scalability of the mechanism so that it can be used in situations with a large number of injured people.

Security and data aggregation are also discussed. MEDiSN secures patient data by performing end-to-end encryption and authentication of the data from the patient monitor. Since it is expected that there will be a large volume of data collected, the authors used a delta compression algorithm to condense the data as much as possible. This algorithm was found to have a very high compression ratio, while providing relatively low implementation complexity. The secured and compressed data are then delivered to the gateway via the collection tree protocol (CTP) [9]. This protocol builds and maintains minimum cost trees to nodes that are designated as tree roots. Data are sent to the base station that requires the least amount of energy with no knowledge of the address of that node. The topology information is kept current using data path validations where data packets are used to query and validate the topology. For MEDiSN, this protocol is enhanced with a mechanism to deliver commands from the gateways to the patient monitors. Data will be retransmitted if there is an error between hops.

Simulations showed that the use of data aggregation combined with CTP cause a 20% increase in the number of messages that reach the gateway. MEDiSN was also compared with CodeBlue in an indoor environment with varying numbers of patient monitors. It was found that there was a significant improvement in the network utilization. This was due to the use of hop-by-hop transmission and retransmission of data. This can be used to mitigate the effects of interference and collisions better than the single hop transmission mechanism used by CodeBlue.

16.3 Smart Home-Care Applications

The wireless network architecture for smart homecare requires that the technology be easily integrated with existing medical practices and technology, should provide real-time, long-term and remote monitoring, the sensor should be small and unobtrusive and should provide assistance to elderly and chronic patients. Using a wireless sensor network for homecare provides doctors with a continuous stream of data which can provide a record of symptoms. The patients can also be monitored in cases where chronic life-threatening conditions may exist so that emergency personnel can be alerted if they require assistance.

Becker et al. [10] propose the SmartDrawer that would observe the medication of chronically ill patients. This drawer contains radio frequency identification (RFID) tags that enable the care provider to monitor the patient to ensure that the medication is being taken as prescribed. The authors expect that the system will have three users: the patient, the caregiver, and the maintainer. The SmartDrawer functionality for the patient is to give an alert when they deviate from the directions. The drawer will contain a record of the patient's medical data and an inventory of the different medication and the specifications on how to take them.

The caregiver will be able to retrieve historical data about the drugs that have been taken and will also be able to add new prescriptions or modify existing ones. The maintainer is described as a system administrator who would be able to add and remove different functionalities and access the data for manipulation so that they may be used in different applications.

The drawer operates only when it is in use. When the drawer is opened it begins to scan the medication that is being removed from it and to record the dosage that the patient is taking. When the drawer is closed it updates the inventory and then stops scanning until it is in use again. This drawer would be especially useful to elderly patients who may have a tendency to forget to take their medication or forget the specific instructions of how to take them.

Wood et al. [5] extend the idea of a single drawer to consist of a wireless sensor network for assisted living and residential monitoring (ALARM-NET). ALARM-NET consists of a body network which is made up of sensors that would be worn by the resident. These sensors would provide physiological sensing that would be customized to the medical needs of the resident. Environmental sensors would be deployed into the living space. These sensors would measure environmental conditions such as air quality, light, temperature, and motion. Motion detection is especially important as it can be used to track the resident's movements. These static sensors will also form a multi-hop network between the mobile body network and the AlarmGate. The AlarmGate manages all the system operations. These nodes enable interaction with the system. It serves as a gateway between the wireless sensors and the rest of the network which includes the back-end database that provides an area of long-term storage for data. This database also contains a circadian activity rhythm analysis program that learns the behavior of the resident and is able to adjust the configuration of the system, so that its operation is optimal based on that specific person. Lastly, the architecture (shown in Figure 16.3) contains a user interface which could be a PDA or a computer that will allow authorized users to view data from the sensors or from a database.

ALARM-NET was designed not only to collect information, but also to alert the healthcare provider if there are any emergency situations based on the patient's medical ailment. A resident, for example, has a condition where he should not remain sedentary for extended periods of time. Accelerometers can be placed in clothes and can alert the caregiver station if there is a prolonged period of inactivity.

Virone et al. [11] designed a network with an architecture that is similar to ALARM-NET. The architecture is multi-tiered, with heterogeneous devices such as light-weight body sensors and other mobile and stationary components. This architecture divides the devices into many layers.

- *Body Network and Subsystems.* This network would include small devices that would contain different sensors such as accelerometers, heart rate, and temperature. These sensors will be necessary to monitor, identify, and locate the patient, as well as to perform a variety of other tasks based on the needs of the care provider. The authors determined that the sensors for this network should use kinetic energy to recharge their batteries. These sensors would also be able to alert the patient during abnormal circumstances. For example, an alert to remind a patient to move around if there are prolonged periods of inactivity.
- *Emplaced Sensor Network.* This network would be made of sensors that have been installed into walls, furniture, etc. so that they can monitor the environment.

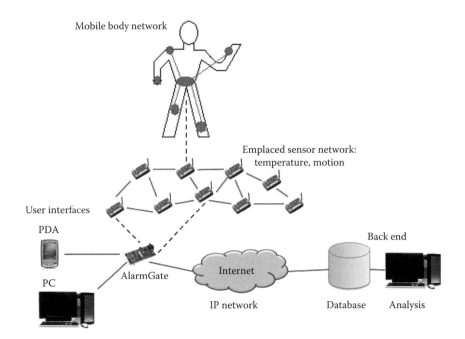

FIGURE 16.3   ALARM-NET architecture.

These sensors are connected wirelessly to a backbone and do not carry out the data calculations or storage. They also serve as relay nodes for the sensors in the body network and the backbone.

- *Backbone.* The backbone connects the stationary sensor network to other devices such as personal computers (PC), PDAs, and databases. It will also connect isolated sensors so that they can participate in more efficient routing.
- *Back-End Databases.* This will be used for long-term storage of patient data. This may include archiving and data mining.
- *Human Interfaces.* These include PDAs, PCs, and laptops that will be used by patients and caregivers to access the system. Access to the systems functions will be provided based on the role of the user which will include data management, queries, and configuration.

This system was designed to be a single-hop structure as the transmission range of the sensors is large enough to cover the entire facility being monitored. The authors recognize that for a multi-floor facility, or in an effort to save energy by reducing the transmission range, a multi-hop protocol would be required.

Stroulia et al. [12] also devise a method of embedding sensors on the patient and in the environment in order to provide a continuous medical monitoring. The Smart Condo Project, monitors the patient, similar to the other mechanisms, but attempts to use a more updated architecture than provided in ALARM-NET.

The architecture, shown in Figure 16.4, begins with a wireless network containing the relevant sensors for the patient. The sensors transmit data to the sink; data include a time

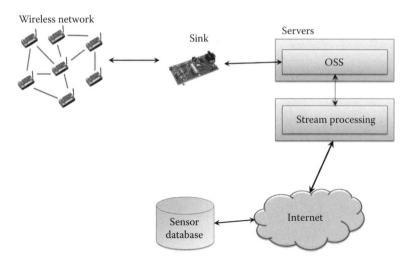

FIGURE 16.4 The Smart Condo software architecture.

stamp, network ID, node ID, sensor ID, sensor type, and event type. The sink transmits the packets it receives to a PC which contains the operation support system (OSS); this helps process the data and determines information about the resident based on the sensor data. The heart of the system is the server that runs the application-specific services; this controls all the functions of the network such as configuring the wireless sensors and interpreting the data received. The sensor database stores raw sensor data and data that have been processed for use by an application.

The authors have not had an opportunity to fully implement The Smart Condo project.

16.4 Other Continuous Monitoring Mechanisms

Otto et al. [13] consider a body area sensor network that is used for health monitoring. This body network was designed such that it can be integrated with a larger telemedicine network for the continuous monitoring of patients. This system consists of individual monitoring networks for users that connect to the Internet where information can be transmitted to healthcare professionals for viewing, processing, and storage.

The architecture of such a system is shown in Figure 16.5 and is comprised of three layers. The top layer is the medical server which is a network of medical personnel, emergency services, and healthcare providers. All persons in this layer are interconnected to enable the medical staff to provide services to thousands of individual users. Each user is equipped with a body area sensor network that will, based on the needs of that user, sample the vital signs and transfer the information to a personal server via a wireless personal network using either Bluetooth or ZigBee. The personal server can be any Internet enabled device including a PDA, cell phone, laptop, or PC which will manage the body network, provide an interface for the user and transmitted the information

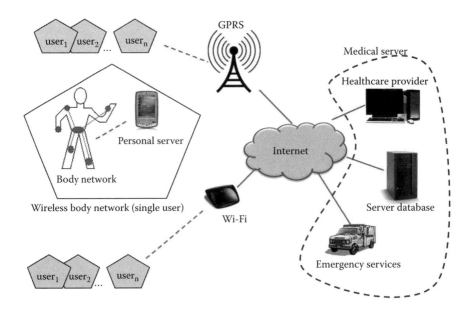

FIGURE 16.5   Telemedical system architecture.

gathered from the sensors using an Internet connection or a cellular network such as general packet radio service (GPRS). The main functions of the medical server are to authenticate users when they connect to the system, download data that is transmitted from the users' personal network of sensors, parse the incoming data and store it in the matching medical records, analyze the data, identify serious irregularities in patient data and alert emergency medical technicians, and forward new care instructions to the user from the healthcare providers.

The personal server includes the authentication information and is configured to connect to the medical server automatically when a communication channel is available. In such cases, the information is uploaded so that it can be stored. If the channel is unavailable, then the personal server will store the data and make continual attempts to connect to the medical server. This ensures that the information being transmitted is as close to real-time as practically possible.

The most critical element of the entire telemedical system is the wireless body area sensor network. This network is made up of sensor nodes that will be able to sense, process, and communicate the physiological signals of the user. These sensors include ECG sensor for monitoring heart activity, an electromyography (EMG) sensor for muscle activity, an electroencephalography (EEG) sensor for brain electrical activity, blood pressure sensors, motion sensors, and breathing sensors which can be used to measure the patient's activity level. Figure 16.6 shows the flow of data between patients and the healthcare providers and emergency services.

Jiang et al. developed CareNet which is an integrated wireless sensor environment for remote healthcare [14]. CareNet is a two-tier wireless network [15] that is built on a heterogeneous networking infrastructure. The lower tier consists of a body network which

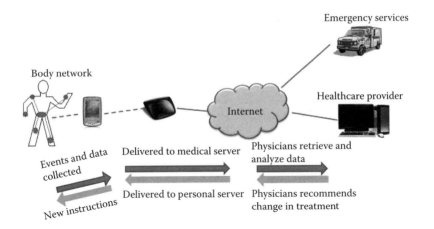

FIGURE 16.6 Data flow between patient and medical services.

employs wearable sensors that will obtain readings from the patient and transmit them via wireless communication to base-station sensors. The upper tier of the network consists of a backbone which would be used for transmitting these data to the healthcare provider. The backbone network offers a prompt and reliable delivery of the medical data. To support this claim the authors designed specific features into the network.

- *Integrated Admission Control and Routing.* This ensures that adding a new patient to the network would not compromise with the service of the existing patients. This is done using an admission control algorithm which would approximate the capacity of the backbone network using its best case routing scenario.
- *Application-Level Routing.* The routing protocol used by the backbone network routers has been implemented at the application level. This was done to make it reusable, that is, it can be used with many different operating systems without any modification to the protocol. The specific method of packet routing is determined based on the patients' activity patterns and so the routing table would be manually configured.
- *Multi-Hop Transmission Control Protocol (TCP) Tunneling.* Packet forwarding is also implemented at the application level. This is done by creating a TCP connection between backbone routers in each hop. This is done so that multiple threads can be used if the packet needs to be sent to more than one router at a time.
- *Mobile Sensor Hand-Off.* Since the sensors are mobile, they would not be able to continuously communicate with the same backbone router at all times. Thus, sensors broadcast their message to all backbone routers within the transmission range. Duplicate packets are then removed based on a timestamp. The first packet received by the router is accepted and all subsequent duplicate packets are discarded.

The network was simulated using 90 patients where the traffic demand and patient location were randomly generated. The results showed that the network is unable to successfully support an increase in the number of patients when all patients are active at the same time. It was found, however, that if the patients maintain a stable mobility pattern, then the system would be able to support 10 more patients without a significant degradation of throughput.

Lupu et al. [16] propose the autonomic management of ubiquitous e-health systems (AMUSE) which uses multiple self-managed cells (SMCs) to implement a monitoring network. Each SMC manages a set of components such as the network of body and environmental sensors. The purpose of an SMC is to make that entire cell independent in that it is able to configure, optimize, and repair itself. The patient SMC should be able to support the addition and removal of sensors and then perform self-reconfigurations. SMCs are then expected to interact with each other and do so via composition and peer-to-peer interaction.

Composition interactions take place when there is a device inside the SMC that can manage the operation of that entire cell. In such a case, the device itself can be considered its own cell as it can function independently of the cell containing it. In such a case that cell will not advertise itself as an independent resource and will allow the cell containing it to govern its interactions with other devices. Peer-to-peer interactions occur between SMCs, for example, a nurse visiting a patient at home would need to interface with the patient SMC in order to interact with it. When the nurse SMC is detected, new policies such as alert behavior can be downloaded to the patient SMC and this will cause the patient cell to recalibrate itself without requiring the entire calibration to be stored in the nurse cell. Presently, the entire AMUSE network has not been simulated.

Zhu et al. [17] offer Vesta which is a security improvement to the AMUSE system. Vesta integrates a variety of security protocols in order to secure the SMC. The AMUSE project was extended and secured in three major areas: secure sensor discovery, authentication module, and access control. In secure sensor discovery, it is ensured that only legitimate sensors can enter the network and that this new sensor will be paired with the intended patient. All wireless communication in the network has been encrypted using a key that is created when the new sensor gains access to the system. When users are authenticated, all communication between them is confidential.

The authentication protocol developed is based on a public key infrastructure where every user would be assigned a public key that identifies them. The nurse controller will broadcast a HELLO message to the patient. When the patient receives this message, the shared authentication process commences, in this process each user is identified and authenticated before any information is shared between them. To guarantee that the process does not fail due to lost or corrupt messages during the exchange, a timer is initiated at the beginning of the authentication process and when the timer goes off the process stops and is reset so that a new process can begin.

Access control is supported using the Ponder2 policy system [18] that manages authorization polices. Services and resources are characterized as objects. If an object is invoked, then access is permitted only if a policy exists that will allow access to this object. Otherwise, the request is denied. This ensures that only authorized users are allowed to view the patient information.

Ng et al. [19] present UbiMon which is a five-component monitoring structure: the body sensor network (BSN) node, the local processing unit (LPU), the central server (CS), the patient database (PD), and the workstation (WS).

- *Body Sensor Network (BSN) Node.* This wireless node was proposed to be a wireless module that either contains a very small battery or none at all. This node would be combined with a physiological sensor such as temperature or ECG. The context-aware sensors can then be used with the BSN to observe the patient activity.
- *Local Processing Unit (LPU).* The data transmitted by the body network are collected in the LPU. This can be a PDA or a cell phone which will communicate with the network via Bluetooth, for short-range transmissions, and GPRS for long range. The LPU will collect data and recognize irregularities in patient data and provide warnings to the patient.
- *Central Server (CS).* From the data received from the LPU, the CS performs trend analysis and derives a pattern of the patient's condition. Based on this analysis, predictions can be made so that life-threatening anomalies can be detected.
- *Patient Database (PD).* This is an area of long-term storage for the patient data, which is obtained from the CS. All data queries for patient data from WSs are satisfied by the PD.
- *Workstation (WS).* This may be a computer, laptop or other handheld device that will be used by healthcare providers to access and analyze the data collected from the BSN. Historical data may also be accessed.

This system also is in its conception phase.

Chakravorty [20] introduces MobiCare that uses an architecture similar to others discussed. On the client side of this network it consists of a BSN which is made up of wearable sensors and actuators that communicate wirelessly and a BSN Manager which will interact with the sensors and the healthcare provider servers. The BSN manager/MobiCare client monitors the sensors in the body network and aggregates the data collected by the sensors. This information is then uploaded to the servers using a secure wireless communication channel.

The authors suggest various devices that would meet the hardware requirements of a BSN manager, which should be user-friendly, energy efficient, and contain wide area wireless connectivity but it was decided that a cell phone would be the most suitable. Using a cell phone actually provides easy mobility for the MobiCare architecture. The MobiCare client will act as the server to connect to a gateway using the cellular 3G (generation) or 4G networks. This form of communication will allow the patient data to be transmitted in real time to the healthcare providers. This becomes especially important during emergency situations.

The healthcare provider side contains the services for the evaluation of the data. Different services will be provided to allow the patient to be continuously monitored and have their data reviewed by specialists. The data can also be stored for long-term health assessments.

Ganti et al. [21] introduced the idea of smart attire. Smart attire consists of sensors that have been embedded into the clothing that will be worn by the patient. These sensors will record the person's activities and location. The information gathered from the sensors can be stored in a database for subsequent analysis. The architecture to be used with this attire is called SATIRE.

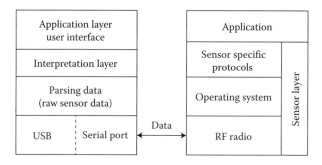

FIGURE 16.7   SATIRE architecture framework.

SATIRE is based on a two-component structure which is a PC or laptop and the mote. A data item is sensed at the sensor layer of the mote and passes to the application layer. The data then moves to the filter layer where data processing would occur. The results of the calculations move to the next layer which contains the operating system and other important system functions. This layer provides an interface for communication with higher level applications. The final layer contains communication devices and drivers which will send the data to the PC. On the PC side, the data are uploaded to the PC and then is passed to the next layer which parses the raw data and stores them in a standard format which may be based on a time stamp. The next layer is used for data processing. It deciphers the data using data mining and signal processing techniques. This prepares the data for use in the application layer which enables the user to perform queries on the information stored. This framework is illustrated in Figure 16.7.

SATIRE works with wearable sensors that are embedded into the users' clothing. When the patients' wearable sensor network is in the range of the rest of the network, the information can be transmitted in real-time to the PC for processing. If the patient leaves this location and is out of range, then the information is stored within the sensors and can be uploaded when the body network is within the communication range of the rest of the network. Because the user is allowed to leave the range of the network and still be able to be monitored, albeit not in real-time, he/she can be tracked over a variety of locations and so a GPS tracker is added to the sensors so that not only the location of the patient can be tracked but also the speed and duration of the activity.

16.5 Algorithms Comparison

All the mechanisms surveyed in this paper have the common objective of improving the healthcare provided to patients by continuously monitoring them so that anomalies can be detected sooner. This premise was used in each article and adapted to different situations such as emergency response and smart home-care.

The algorithms discussed in Section 2 are compared in Table 1. The main objective of these schemes is to facilitate efficient communication among responding emergency services. This enables cooperation and a well organized rescue effort. In Table 2 mechanisms from Section 3 are evaluated. These are methods that enable patients to live in their homes without assistance while retaining the ability to be monitored. Table 3 contains the mechanisms from Section 4 which are other schemes used to monitor patients in miscellaneous circumstances. These mechanisms are compared based on the following criteria:

- *Scalability*: the ability of the system or network to accommodate increasing numbers of users with minimal reduction in the quality of service.
- *Self-configuration*: the network should be able to initialize itself and also adapt to nodes entering and leaving the network without having a loss of service.
- *Data Collection*: this refers to the method used to relay the data to the server for processing. This may be done by a PDA which communicates directly with the server or using a multi-hop network.
- *Security*: The scheme may or may not contain a mechanism that will be used to secure patients' sensor data as it is transmitted through the network.
- *Location Tracking*: some schemes have the ability to keep track of the patient's location through localization protocols.
- *Body Sensor Network*: the patient may be monitored using a BSN that will keep track of their vital information.
- *Environmental Sensors*: smart home-care schemes in particular make use of sensors that keep track of environmental conditions to keep the patient safe.
- *Context Aware Sensors*: these sensors keep track of the patient's body position to determine when a fall has occurred.
- *Architecture*: this is the structure of the network's physical components and their operating organization and configuration.

See Tables 16.1, 16.2, and 16.3.

TABLE 16.1   Emergency Response

Scheme	Scalability	Self-Configuration	Data Collection	Security	Location Tracking
CodeBlue	Large	Yes	PDA	No	RF tracking
SMART	Medium	No	PDA	No	MIT crickets
MEDiSN	Small	No	Multi-hop network	Yes	None

TABLE 16.2   Smart Home-Care

Scheme	Body Sensor Network	Environmental Sensors	Context Aware Sensors	Location Tracking	Data Collection
Alarm-Net	Yes	Yes	Yes	Motion detection	PDA and multi-hop network
Health Monitoring	Yes	Yes	Yes	Context aware	PDA and multi-hop network
Smart Condo	Yes	Yes	Yes	Motion detection	PDA/laptop
SmartDrawer	No	No	Yes	None	RF receiver

TABLE 16.3 Continuous Monitoring

Scheme	Body Sensor Network	Environmental Sensors	Architecture	Data Collection	Security
Telemedical System	Yes	No	Three-tier wireless network	PDA, cell phone, laptop or PC	Yes
Carenet	Yes	No	Two-tier wireless network	Multi-hop network	Yes
AMUSE	Yes	Yes	Self-managed cells	PDA/laptop	Yes (enhanced by Vesta)
Vesta	Yes	No	Enhancement for AMUSE	NA	Yes
UbiMon	Yes	Yes	5-component structure	PDA/cell phone	No
MobiCare	Yes	No	3-component structure	Cell phone	Yes
SATIRE	Yes	Now	2-component structure	PC/laptop	Yes

16.6 Conclusion

Pervasive healthcare monitoring using wireless sensor networks is an important technology for the increasing demography of people with chronic illnesses. This type of monitoring is also extremely important for disaster response to take care and monitor the mass casualties when there are too few medical professionals for each patient to get continuous attention. Sensors can be placed on the body or in the homes of these patients and this can provide vital information that can alert the patient or medical professional when an emergency occurs.

It was shown that these networks can be used in a variety of applications. These include, but are not limited to, the disaster response, assisted living facilities, and smart homecare. These sensors can be used to make the lives of older patients more independent and to provide patients with chronic illnesses the benefit of having their condition monitored without enduring the expense of having a person to monitor them. In smart home care many of the mechanisms employ environmental and context aware sensors to keep track of the residents' location and to determine whether they are safe in their environment.

The architectures of these networks are designed to provide low congestion messages passing between sensors and the data collection unit. Some authors design a network that is only a single hop between the sensors and a cell phone that will use the cellular network to transmit information. Some schemes create multi-hop networks that use a gateway to connect to the Internet so that the information can be sent to a remote server for storage or viewing by the healthcare providers. Others describe an entirely local network where the sensors are able to communicate with the server directly or with the use of relay nodes if the sensors are out of the communication range.

Security is also a major concern when transmitting and storing medical information. Many of the mechanisms discussed, mention some form of securing data including Vesta [17] which is a mechanism that was proposed as a security enhancement for the AMUSE [16] mechanism. There are some mechanisms that do not refer to any mechanism of securing data but this enhancement would be necessary before successful implementation can occur.

Many of the schemes surveyed share a common need for some type of location tracking for the individual being monitored. This is especially important for those such as CodeBlue and MEDiSN that can be used for mass casualties where it would be impossible to locate a patient in need of assistance without some sort of protocol that could be used as a locator. It is advantageous that the mechanisms do not rely on a GPS as the means of tracking as this is not only more costly and significantly increases energy consumption but it has been found that GPS is not very accurate indoors. Most employ RF signals and motion detection to determine the patient position.

As the research continues into providing efficient pervasive healthcare monitoring, it is important that the architecture is low in congestion with a reliable connection to the Internet for remote access. The information should be completely secured and be processed so that it is available in a readable format for both patients and healthcare professionals to access.

References

1. Stankovic, J., Cao, Q., Doan, T., Fang, L., He, Z., Lin, S., Son, S., Stoleru, R., and Wood, A. Wireless sensor networks for in-home healthcare: potential and challenges, in *Proceedings of HCMDSS Workshop*, Philadelphia, PA, 2005.
2. Lorincz, K., Malan, D., Fulford-Jones, T., Nawoj, A., Clavel, A., Shnayder, V., Mainland, G., Welsh, M., and Moulton, S. Sensor networks for emergency response: Challenges and opportunities, *IEEE Pervasive Comput.* 2004; 2(3): 16–23.
3. Ko, J., Musaloiu-E. R., Lim. J. H., Chen, Y., Terzis, A., Gao, T., Destler, W., and Selavo, L. Demo Abstract: MEDISN: Medical emergency detection in sensor networks, in *ACM Conference on Embedded Networked Sensor Systems*, Raleigh, NC, 2008.
4. Waterman, J., Curtis, D., Goraczko, M., Shih, E., Sarin, P., Pino, E., Ohno-Machado, L., Greenes, R., Guttag, J., and Stair, T. Demonstration of SMART (scalable medical alert response technology), in *AMIA 2005 Annual Symposium*, Washington, DC, 2005.
5. Wood, A., Virone, G., Doan, T., Cao, Q., Selavo, L., Wu, Y., Fang, L., He, Z., Lin, S., and Stankovic, J. ALARM-NET: Wireless sensor networks for assisted-living and residential monitoring, Technical Report University of Virginia Computer Science Department, 2006.
6. Curtis, D., Pino, E., Bailey, J., Shih, E., Waterman, J., Vinterbo, S., Stair, T., Guttag, J., Greenes, R., and Ohno-Machado, L. SMART—An Integrated Wireless System for Monitoring Unattended Patients, *J. Amer. Med. Inform. Assoc.* 2008; 15(2): 44–53.
7. Lorincz, K. and Welsh, M. Motetrack: A robust, decentralised approach to RF-based location tracking, *J. Pers. Ubiquitous Comput.*, 2007; 11(6): 489–503.

8. Wang, Y., Goddard, S., and Perez, L. C. A study on the cricket location-support system, *IEEE Int. Conf. Electro/Inform. Technol.* Chicago, IL, 2007: 257–262.

9. Gnawali, O., Fonseca, R., Jamieson, K., Moss, D., and Levis, P. Collection tree protocol, in *Proceedings of the 7th ACM Conference on Embedded Networked Sensor Systems (SenSys)*, Berkeley, CA, 2009.

10. Becker, E., Metsis, V., Arora, R., Vinjumur, J., Xu, Y., and Makedon, F. SmartDrawer: RFID-based smart medicine drawer for assistive environments, in *Proceedings of the 2nd International Conference on Pervasive Technologies Related to Assistive Environments*, Corfu, Greece, June 2009.

11. Virone, G., Wood, A., Selavo, L., Cao, Q., Fang, L., Doan, T., He, Z., Stoleru, R., Lin, S., and Stankovic, J. An advanced wireless sensor network for health monitoring, in *Transdisciplinary Conference on Distributed Diagnosis and Home Heathcare*, Arlington, VA, 2006.

12. Stroulia, E., Chodos, D., Boers, N., Huang, J., Gburzynski, P., and Nikolaidis, I. Software engineering for health education and care delivery systems: The smart condo project, in *SEHC '09, ICSE Workshop on Software Engineering in Health Care*, Vancouver, BC, 2009, pp. 20–28.

13. Otto, C., Milenkovic, A., Sanders, C., and Jovanov, E. System architecture of a wireless body area sensor network for ubiquitous health monitoring, *J. Mobile Multimedia*, 2006; 1(4): 307–326.

14. Jiang, S., Cao, Y., Iyengar, S., Kuryloski, P., Jafari, R., Xue, Y., Bajcsy, R., and Wicker, S. CareNet: An integrated wireless sensor networking environment for remote healthcare, in *3rd International Conference on Body Area Networks*, Tempe, AZ, 2008.

15. Jiang, S., Xue, Y., Giani, A., and Bajcsy, R., Robust medical data delivery for wireless pervasive healthcare, in *IEEE International Conference on Dependable, Autonomic and Secure Computing*, Chengdu, China, 2009.

16. Lupu, E., Dulay, N., Sloman, M., Sventek, J., Heeps, S., Strowes, S., Keoh, S. L., Schaeffer-Filho, A., and Twidle, K. AMUSE: Autonomic Management of Ubiquitous e-Health Systems, *J. Concurrency Comput. Practice and Experience*, 2008; 20(3): 277–295.

17. Zhu, Y., Sloman, M., Lupu, E., and Keoh, S. Vesta: A secure and autonomic system for pervasive healthcare, in *3rd International Conference on Pervasive Computing Technologies for Healthcare (Pervasive Health 09)*, London, UK, 2009.

18. Twidle, K., Dulay, N., Lupu, E., and Sloman, M. Ponder2: A policy system for autonomous pervasive environments, International Conference on Autonomic and Autonomous Systems, Valencia, Spain, 2009.

19. Ng, J., Lo, B., Wells, O., Sloman, M., Toumazou, C., Peters, N., Darzi, A., and Yang, G., *Ubiquitous monitoring environment for wearable and implantable sensors (UbiMon)*, UbiComp, Nottingham, UK, 2004.

20. Chakravorty, R. MobiCare: a programmable service architecture for mobile medical care, in *Pervasive Computing and Communications Workshops*, Pisa, Italy, 2006.

21. Ganti, R., Abdelzaher, T., Jayachandran, P., and Stankovic, J. SATIRE: a software architecture for smart AtTIRE, in *4th ACM Conference on Mobile Systems, Applications, and Services (Mobisys 2006)*, Uppsala, Sweden, 2006.

17

Pervasive Computing for Home Automation and Telecare

Claire Maternaghan
University of Stirling

Kenneth J. Turner
University of Stirling

17.1 Introduction

This section introduces home automation and telecare, and presents the general challenges that they pose.

17.1.1 Home Automation

Home automation has been a goal for many years. For the most part, the current commercial offerings might be better termed home control in that they support control of aspects such as lighting, heating, security, audio, and video. However, they tend to lack flexible management and automation of home functions. Home automation relies on devices in the home being networked, typically with a central hub, with either local or remote control.

Consumers have become increasingly knowledgeable about computer-based capabilities thanks to widespread use of personal computing, mobile phones, media players, and the like. The time is therefore ripe to offer more sophisticated ways of managing the home. This can deal with aspects such as health/social care, comfort, security, entertainment, and energy usage in the home.

17.1.2 Telecare

The global population is aging, with the percentage of older people (over 65) expected to be 19.3% by 2050—and much higher in some developed countries [1]. It is infeasible for society to provide sufficient care homes for this growing segment of the population. It is also socially desirable for older people to remain in their own homes for as long as possible.

Telecare aims to support delivery of care to the home, with a particular emphasis on social care. Telehealth is similar, but focuses on aspects such as health monitoring and support. Remote delivery of care is expensive in manpower, so computer-based support is an attractive and effective solution. Telecare and telehealth can monitor undesirable situations such as night wandering or abnormal medical readings. However, they can also identify potential problems in daily life such as poor sleeping patterns or reduced meal preparation.

Telecare resembles home automation in requiring management of how the home reacts. Although the two applications have some overlap, telecare makes use of specialized devices such as medication monitors, fall detectors, and enuresis (bed wetting) sensors. Telecare is usually linked with a call center for handling alerts. Advantage may also be taken of a wide-area link to upload home care data to a remote facility (e.g., a social work office or a health center). Not only are telecare systems proprietary, they usually require specialized technical expertise and reprogramming to modify the services they offer.

17.1.3 Challenges in Pervasive Computing for the Home

Computing facilities in the home need to be usable, acceptable, interoperable, and automated. Home computing must be appropriate for ordinary householders. Despite increased understanding of computer-based capabilities, consumers will have little understanding of or interest in the technical details of home equipment. The concepts and interfaces therefore need to be readily understood. Home equipment also needs to be acceptable: devices that look out of place in the home are unlikely to be welcome, and devices that need disruptive installations are unlikely to be accepted.

An interesting issue with acceptability is having to satisfy multiple stakeholders [2]. For home automation, for example, the occupants of a household might have different views on aspects like comfort levels and what TV programs should be recorded. For telecare, the range of stakeholders is very much larger: besides the end user, it includes family caregivers and professional carers. These may have differing opinions on how care should be supported in the home, for example, what constitutes an alert condition or whether a less capable person should be discouraged from certain activities (like cooking).

Interoperability remains a challenge. Although a number of standards are available for home automation (see Section 17.2.1), these are often proprietary, low-level, and do not guarantee interworking across different commercial solutions. Since telecare is in its infancy, there is little standardization of telecare equipment interfaces. User-visible interfaces to home equipment are also proprietary and unlikely to be standardized.

As will be seen in Section 17.2, attempts have been made to introduce flexible home automation through user-defined rules or mappings. However, these usually require technical expertise that ordinary householders are unlikely to have or to learn. Telecare management is highly specialized, being performed by technicians or care workers rather than end users. Nonetheless, it would be desirable to open this up to less technical users (e.g., end users themselves or their family caregivers).

17.1.4 Chapter Overview

Section 17.2 describes related work in the areas of standards and interoperability, home system architectures and platforms, automated home management, and user interfaces for the home.

In order to give a concrete illustration, Section 17.3 describes a typical approach to computer-based home support. The Homer project offers a generic solution for both home automation and telecare. The philosophy and high-level architecture of the system are described. The types of components are described, and how they are flexibly integrated into the system. Above the component level, an event server achieves sensor fusion and actuator fusion. Above event level, policies are automated rules for how the home should react to various situations. Above policy level, goals allow users to manage the home through high-level objectives. Finally, user-friendly interfaces at the top level aim to make it easy for non-technical users to manage their homes.

Section 17.4 summarizes the chapter, and evaluates the overall state of the art. Future trends in home automation and telecare are identified.

17.2 Background to Home Systems

This section discusses the work toward challenges in home computing in the areas of standards, platforms, automation, and interfaces.

17.2.1 Standards for Home Automation and Telecare

Standards for home automation are mostly at a low (device) level; compatibility at the higher level of home services remains to be achieved. Examples of device standards include the following:

Infrared: Infrared control of domestic appliances is one of the commonest techniques (*www.irda.org*). However, implementations of infrared control vary widely, so that uniform control is hard to achieve.

HAVi: This standard for home audio–video interfaces (*www.havi.org*) is particularly intended for interconnection of entertainment equipment. However, it has so far seen only limited use.

KNX: This standard from the Konnex Association (*www.knx.org*) has evolved from earlier work on the EIB (European Installation Bus). KNX is widely used in building management and home automation for control and interconnection of lighting, devices, and media.

Lonworks: This proprietary standard from the Echelon Corporation (*www.echelon.com*) has been widely used for device interconnection in building management, home automation, and transportation.

UPnP: Universal Plug and Play (*www.upnp.org*) takes the plug-and-play idea from personal computers into networked devices. Although UPnP offers sophisticated capabilities, it has not yet seen widespread use in the home.

X10: This standard (*www.x10europe.com*) is common for controlling appliances using existing mains wiring, but with some support for wireless. However, the degree of control is very limited (on, off, dim to some level).

Since telecare is still an evolving approach, standards are not yet complete and widespread. At present, proprietary implementations are usual, from companies such as Cisco, General Electric, Intel, Philips, Siemens, and Tunstall. Ongoing work on interoperability of telecare systems includes:

Continua: The Continua Health Alliance (*www.continuaalliance.org*) is developing standards for interoperability of healthcare systems, including home telehealth systems. However, telecare continues to be a somewhat separate approach that is not yet well supported by standards.

ETSI: The European Telecommunications Standards Institute (*www.etsi.org*) has been working on higher level standards for telecare. However, much work remains to be done on interoperability at the service level.

TSA: The Telecare Services Association in the UK (*www.telecare.org.uk*) has been developing telecare support standards. However, these are focused on procedures (e.g., how a call center should operate) rather than on technical interoperability of equipment or services.

17.2.2 Home System Platforms

Several general architectures are relevant to home systems. Major examples include:

Jini: This is an SOA (Service Oriented Architecture) that extends Java for distributed and federated systems (*www.jini.org*). The Jini architecture supports service components, service registration and lookup, access control, distributed events, and transactions. Although a general-purpose approach, Jini has found use in the design of home systems [3].

OSGi: Originally "Open Services Gateway initiative" (*www.osgi.org*), this is a widely adopted platform for the home (and vehicles) that follows the principles of SOA. Several projects have used OSGi for home automation and healthcare, for example, Atlas [4], e-HealthCare (*ehealth.sourceforge.net*), Match (Mobilising Advanced Technologies for Care at Home [5]), and Saphire [6]. OSGi offers an infrastructure that allows effective interworking of components ("bundles"). These communicate by registering events they are interested in and services they offer.

A number of commercial solutions have platforms for home automation, such as:

Control4: This package (*www.control4.com*) is a leading approach that offers a framework for third-party devices to interoperate with Control4. Companies can embed the "Control4 operating system" into their own devices, resulting in tightly integrated solutions. This resembles the "app store" concept that has proven popular with mobile device manufacturers, suggesting that this could work just as well with applications for the home.

Cortexa: This package (*www.cortexa.com*) offers sophisticated home control. Cortexa is integrated with some of the most popular home automation technologies such as HAI and Insteon (*www.insteon.net*). Control can be exercised from a touch-screen in the home, a web interface or an iPhone. Customers can use only Cortexa user interfaces, and must purchase Cortexa-compatible hardware.

Girder: This package from Promixis (*www.promixis.com*) supports a variety of home devices. Programmability is achieved through the ability to map input events to output events in a flexible way. This, however, is designed for those with specialized technical knowledge.

HAI: This system from Home Automation Inc. (*www.homeauto.com*) deals with home control, safety, entertainment, and energy usage. However, this package is intended more for system installers than for programming by end users.

HomeSeer: This is widely used to control a variety of home devices (*www.home-seer.com*). Its advantages include remote control from devices such as PCs and mobile phones.

IQare: OmniQare (*www.omniqare.com*) offer a range of services to users in areas such as safety, communication, shopping, services, contact, and well-being. Developers can also integrate their own services using a custom XML-based protocol. However, the focus is more on telecare and older users rather than on home automation. The system also acts as a gateway to services, rather than managing these as a whole.

Many research projects have worked on smart homes, with an emphasis on either home automation or telecare/telehealth. Examples from a wide range include the Gator Tech smart house [7], the Gloucester smart house [8], House_n [9], Safe at Home [10], Saphe (Smart and Aware Pervasive Healthcare Environment, *http://ubimon.doc.ic.ac.uk/saphe*), and Saphire [6]. Dewsbury et al. [11] are unusual in supporting simulations to design smart houses for individual user needs.

17.2.3 Automated Home Management

Section 17.2.2 mentioned commercial solutions such as Cortexa and Girder that support home management through some basic form of rules. However, greater flexibility can be offered by policy-based management: user-defined rules that automatically determine how the home system should react. This is a general approach that has been used in applications such as access control, network management, and system management.

Drools (*www.jboss.org/drools*) is a well-known example of a rule-based system that implements the Java standard for a rules engine (JSR 94). Drools is typically used to enforce business rules rather than to control systems. Unlike other policy-based approaches, it does not make a sharp distinction between events and conditions.

Ponder [12] is a popular example of a policy-based system. This offers a mature approach for managing systems through policies. Ponder has found applications in areas such as system management and body sensor networks. It supports policy domains, policy conflicts, and policy refinement.

The inspiration for home management in this chapter came from the project Accent (Advanced Component Control Enhancing Network Technologies, *www.cs.stir.ac.uk/accent*). Originally developed for controlling Internet telephony [13], Accent and its accompanying policy language Appel (Adaptable and Programmable Policy Environment and Language, *www.cs.stir.ac.uk/appel*) have now been extended for new applications such as home management.

Goals are high-level objectives for how a system should behave. They are often broken down into sub-goals and ultimately into concrete actions to achieve goals. Kaos ("Knowledge Acquisition in Automated Specification" [14]) is an example of goal refinement in requirements engineering. Other approaches to goal refinement such as [15] are typically based on logic. In the context of policy-based management, the idea is to refine goals into policies that can realize them (or alternatively to combine policies that achieve goals). As an example, Bandara et al. [16] use event calculus for formal refinement of goals into policies. Ache [17] is specifically for goals in a home context. However, the kinds of goals supported by Ache are restricted to comfort and cost, and the approach emphasizes how the system can learn to meet these goals.

17.2.4 Home System User Interfaces

Interface design techniques of relevance to home systems include programming by demonstration, tangible programming, and visual programming.

Programming by demonstration (e.g., a CAPpella [18], Alfred [19]) allows the user to set up a situation and then demonstrate how the system should respond to it. However, it can take significant effort to demonstrate all the situations that might arise. It can also be difficult to demonstrate rare events.

Tangible programming uses real-world analogues to define system behavior. Rodden et al. [20] define rules by assembling jigsaw pieces. Media Cubes [21] are similarly used to define rules by placing action requests next to devices. As noted by the designers of [22], approaches like these require users to think in unnatural, device-oriented terms. Instead [22], focuses on requirements expressed using words from a "magnetic poetry" set. However, what can be expressed is deliberately restricted to avoid complex natural language processing.

Visual programming is an attractive option for home control. The approach of [23] allows end users to define rules graphically in ubiquitous computing environments, although what may be stated is very restrictive. iCAP [24] allows new devices to be defined by drawing icons. These are then dragged onto a situation window (for rule conditions) or an action window (for system response). Newman et al. [25] allow components to be interconnected visually, but are almost entirely focused on home media.

17.3 Homer System Approach

To make the ideas of this chapter concrete, this section discusses the Homer system developed at the University of Stirling to support home automation and telecare. The overall architecture, its components, and internal framework are discussed. An event server supports sensor and actuator fusion by mapping between lower level component events and higher level policy events. User-defined policies define automated rules for how the system should react to events. High-level user goals are dynamically realized through policies. Finally, user-friendly interfaces to the home system are offered.

17.3.1 Philosophy

Homer aims to act as a middle layer between users and components (hardware or software), hiding complexity from both developers and users. Components register themselves with Homer, which exposes their functionality to the home user. Instances of components can then be created and placed within a model of the home environment. The user can view and control the state of all component instances. Home management policies can also be written to make use of the defined component instances.

A common limitation of current home automation systems is that the home logic is hidden, so that it cannot be controlled or changed by the user. This can result in the home behaving in ways which the user does not understand and cannot discover without contacting the system installer. It is also common to find that changes to the home logic have to be made by the system installer (at cost). Some commercial systems do allow the user to create rules for the home. However, the user interfaces for this are often complex and hard to use, meaning that the average householder is unlikely to attempt any changes. Homer includes an integrated policy server that is made available through an HTTP interface. This makes it easy to create new extensions and interfaces.

17.3.2 System Architecture

A component within Homer simply wraps and exposes a particular type of device or software service. Each component is implemented as an OSGi bundle, which allows components to be added at run time. As examples of components, there is support for X10 devices (mains-controlled), Tunstall devices (telecare), SMS (messaging), and weather forecasting.

Services within Homer support common tasks on behalf of components. These services are also implemented as OSGi bundles and can be used by any other bundle within Homer. Examples of services include communications port support, logging, and email. Services ensure that components do not need to communicate directly with each other. This means that components can simply advertise what they do, with control being exercised by the user.

The third element in Homer is the framework that manages the collaboration between components and the user. The overall system architecture appears in Figure 17.1; the internal details of the framework are explained in Section 17.3.4.

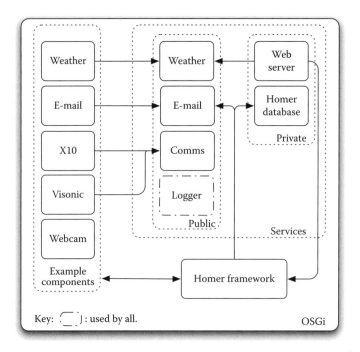

FIGURE 17.1   Homer system architecture.

17.3.3 Home System Components

A component within Homer offers some kind of capability to the user. This functionality is associated with a particular system device type, and is split into three different aspects: triggers (events), conditions (states), and actions. A component acts as a proxy between Homer and the underlying hardware or software service. Components are categorized according to their underlying technology. However, users view the system using different categories that *they* create and which make sense to them.

As an example, X10 supports a variety of hardware that uses a common protocol and is therefore one Homer component. X10 devices are either appliance modules (on, off) or lamp modules (on, off, dim), which are treated as two different system device types. The user has access to a simple device management application which allows them to create new instances (user devices) of hardware within the home. Users create their own categories (user device types) instead of having to refer to the system device-type names (here, X10 appliance modules or X10 lamp modules). This allows users to name and group devices in ways meaningful to them, without having to be aware of the underlying technology. For example, user-defined categories for X10 devices might include heating and lighting.

As a further example, Visonic make a range of wireless sensors for the home. Being supported by a common protocol, these are managed by one Homer component. The different kinds of sensors (movement, opening, etc.) are treated as different system device types. As the user installs these sensors, they can be allocated to categories meaningful to the user (security, doors, etc.). Decoupling the user view of devices from the underlying technology makes the system more usable. Users can refer to devices within the home as they wish. Devices are associated with particular technologies only at installation time.

All components offer services in the form of triggers, conditions, and actions; these are specific to each system device type. For example, an X10 appliance module system has actions "turn on" and "turn off," whereas an X10 lamp module also supports "dim." Components report their advertised triggers, check their advertised conditions on request, and perform their advertised actions.

Multiple underlying technologies can support one user device. For example, the lounge TV might be controlled by both an X10 appliance module for on/off as well as infrared for changing channels and volume. For any user device, the actions that can be carried out are those of the parent system device types. Suppose the kitchen TV made use of only an X10 appliance module. Both TVs could be powered on or off, while the lounge TV could also have its channels and volume changed.

Device support has been created for use in both home automation and telecare. Examples in various categories which Homer currently supports are as follows:

Appliance Control: Appliances controlled via the mains include lighting, fans, and TVs. Appliances controlled via infrared include TVs, audio-visual systems, and DVD recorders.

Communication: Communications services include email, SMS (Short Message Service), Facebook, Twitter, message display on a digital photo frame, and speech input/output (using code from the University of Edinburgh).

Energy Consumption: Energy usage is monitored per appliance. This allows Homer to react to how much energy is being used and helps reduce energy consumption. For example, clothes washing might be delayed until other energy demands are lower.

Environment: Oregon Scientific sensors (*www.oregonscientific.com*) are used for humidity, temperature, and so on. The Google Weather API is used to obtain the current weather or a forecast for chosen locations.

Home Automation: Sensors from companies like Tunstall (*www.tunstallhealth. com*) and Visonic (*www.visonic.com*) include movement detectors, pressure mats, reed switches (cupboard, door, window), and RFID readers (active badges). Homer is capable of handling a variety of home actuators (though the current support is limited). Future support will include curtain/blind controllers, garage door controllers, and remote door locking and unlocking.

Telecare: Telecare sensors from Tunstall and Visonic include alarms (pendant, wrist), hazard detectors (flood, gas, smoke), medicine dispensers, and pressure mats (bed, chair). Specialized sensors include detectors for enuresis, falls, and seizures.

User Interfaces: Various "Internet buddies" are supported as they appeal to ordinary users. Examples of these are the i-Buddy "angel" (*www.unioncreations. com*), the Nabaztag "rabbit" (*www.nabaztag.com*), and the Tux Droid "penguin" (*www.ksyoh.com*). The WiiMote (*www.nintendo.com*) can be used to communicate using gestures and tactile output. Using code from the University of Glasgow, similar functions are available from the Shake (Sensing Hardware Accessory for Kinaesthetic Expression, *www.dcs.gla.ac.uk/research/shake*). The Homer interface described in Section 17.8 supports iPad (*www.apple.com/ipad*), iPhone (*www.apple.com/iphone*), and web interfaces. These offer different services and means of control to the home user.

17.3.4 Homer Framework

Figure 17.2 shows the internal structure of the Homer framework. The component gateway is used for communication between components and Homer. Similarly, the system gateway is used for communication between Homer and private services such as the database and web server. The event server is discussed in Section 17.3.5, while the policy server is covered in Section 17.3.7. The events hub acts as the central communicator for all interested parties. For example, the component gateway listens for any condition checks or action requests for each system device type, so as to then contact the relevant component. As a means of distributing messages, the component gateway also tells the event hub about any triggers from its components.

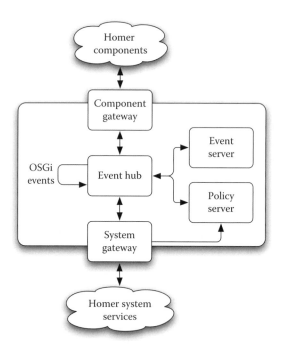

FIGURE 17.2   Homer internal architecture.

Events carry unique identifiers for various pieces of information about each device. This includes system device information (e.g., type "X10 Appliance Module," instance "B3" as the X10 module address), user device information (e.g., type "lamp," instance "bedside light"), location information (e.g., context "home," location "bedroom"), and event details (e.g., trigger "turned on," condition "is off," action "dim" with parameters "50"). This allows different system elements to listen for different kinds of information. For example, a Twitter service could listen for any kind of trigger event and report this via a tweet. As another example, a kitchen display could listen for events in the kitchen to keep the device statuses up-to-date.

17.3.5 Event Server

Event logic defines complex mappings between component events and higher level policy events. This supports the policy server by listening for relevant events and transforming them into new ones. Combining multiple, device-level triggers into a higher-level trigger is commonly known as sensor fusion. By analogy, mapping an action request (from the user or policy server) to multiple component-level outputs is called actuator fusion. The following categories of event mappings illustrate the range of possibilities:

in → in: One or more component triggers are mapped to one or more higher level triggers (most commonly taking a combination of component triggers and posting a higher level trigger). In other work, this is called sensor (data) fusion, where raw input from several sensors is combined to produce higher level, more meaningful events. For example, a more accurate prediction of falls might be obtained by combining fall detector data with movement detector data. This allows for synthetic input events, for example, a macro trigger that reports someone has entered the house. This might build on raw sensor inputs that the front door was opened and that there is a movement in the porch.

out → out: One or more high-level actions are mapped to one or more component actions. This allows for synthetic actions, for example, a macro action for contacting someone. This might first trying calling the user's cell phone. If the call is not answered within 10 s, a text message might be sent.

in → out: One or more triggers are mapped to one or more actions. This is what policies are normally used for, where given triggers and conditions result in actions. Policies are designed to be formulated by users, so the approach is intentionally simple. However, there are times where the policies supported by Homer may not be flexible enough. Event logic can then be used by more technically minded users to define more complex behaviors.

out → in: One or more actions are mapped to one or more triggers. This allows a policy action to trigger the execution of further policies. For example, suppose the lounge light is turned on under policy control because it has become dark outside. This action can result in a trigger that the lounge is now brighter. Other policies might react to this by increasing TV brightness and closing the curtains.

Event logic could be coded in a conventional programming language. However, the aim is to allow changes in the home system without requiring specialized technical expertise. Event logic is therefore described in a visual design language (as a simple form of programming). This is an application of Cress: Communication Representation Employing Systematic Specification (*www.cs.stir.ac.uk/ kjt/research/cress.html*).

Cress [26] is a graphical notation and a toolset for designing service flows, for example, in grid, voice, or web services. A root diagram describes a basic service. This may be extended by feature diagrams that automatically add capabilities to the basic service. In a home context, the diagrams describe event logic. A compiler automatically converts these diagrams into BPEL (Business Process Execution Language [27]) and deploys them into a BPEL engine (ActiveBPEL, *www.activebpel.org*).

Triggers, conditions, and actions are communicated as OSGi events, whereas BPEL processes have a web service interface. The event server therefore maps bidirectionally between OSGi events and web service calls. The logic consists of BPEL processes created from diagrams.

For space reasons, event logic is illustrated with only two small examples (and simplified syntax); see [28] for a more extensive set of examples. A diagram has numbered activities in ellipses. These are linked by arcs that can be governed by value conditions or event conditions. An assignment in a node or along an arc is preceded by "/." A rule box (rounded rectangle) declares types, variables, etc. Input/output actions refer to *component.direction.operation*. Such actions include *Device* (component output with optional response), *Receive* (component input), and *Respond* (response to component input). Although not illustrated in this chapter, Cress offers a complete methodology for service creation, including automated specification, verification, validation, implementation, and performance analysis.

Figure 17.3 shows the logic for a sample in → in mapping. This can generate a fall alert by combining fall and movement inputs. It starts on reception of a fall detector message (node 1). After 30 s (node 2), a motion input is awaited (node 3). If this occurs, it is assumed that the user has not had a serious fall so the logic terminates (node 4). If there is no motion input after 30 s, a fall alert event is generated (node 5).

Figure 17.4 shows the logic for a sample out → in mapping. This produces a frost alert if heating is turned off when the temperature is freezing. It starts on reception of a message to turn the heating off (node 1). The component instance of interest is then set to "outside" (arc from node 1 to node 2), and the outdoor temperature status is read (node 2). The status response will be a numeric string that is converted to a temperature (node 2). If the temperature is 0 or below (arc from node 2 to node 3), a frost alert is initiated (node 3). Finally, the logic terminates (node 4).

17.3.6 Policies for Home Management

Policies define how the home should react to events. A wide variety of policies can be defined for both home automation and telecare. These policies cover aspects such as appliance control (e.g., lighting control), communication (e.g., how to be contacted), comfort (e.g., room temperature), entertainment (e.g., favorite programs), modalities

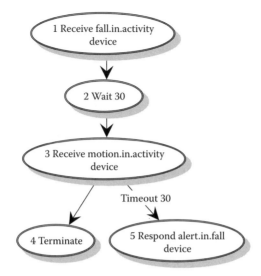

FIGURE 17.3 Fall detection logic.

(e.g., use of speech), reminders (e.g., appointments), security (e.g., intruder detection), system aspects (e.g., access control), and telecare (e.g., medication alerts).

Homer policies have a "when-do" format. The *when* clause comprises triggers and conditions that can be combined with *and, or,* and *then.* The *do* clause can contain multiple actions and conditional groups. A policy can be represented as a tree: a hierarchy of

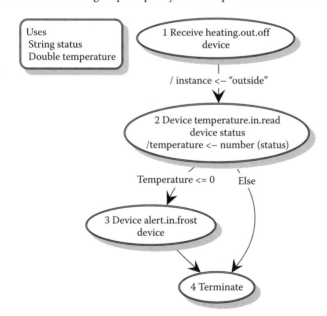

FIGURE 17.4 Frost alert logic.

terms combined with explicit precedence. How a particular user interface implementation decides to display this to the user is irrelevant to Homer (see Section 17.8). The following examples illustrate various policy formats:

```
when the house is unoccupied and it is dark do turn on the hall light

when John arrives home or Mary arrives home do play music and
(if the outside temperature < 10 then do turn on the heating)

when Mary leaves the house and (it is raining or it is forecast to rain)
do say 'remember your umbrella as it might rain today'

when someone gets up at night then the front door is opened
do say 'go back to bed' and illuminate a path back to the bedroom
```

Users can find it difficult to differentiate between triggers (e.g., the door opens) and conditions (e.g., the door is open). This confusion is eliminated by treating triggers and conditions similarly. For each element in a *when* clause, the policy server listens for triggers as well as state changes that affect conditions. State changes imply that something has happened to trigger the change of state, allowing the rest of the *when* clause to be evaluated. Conditions are also evaluated when an event triggers a policy.

It is very rare for triggers to occur at the exact same time, therefore policies can have a time interval in which all triggers occur and all conditions are met. The time interval depends on the particular policy being defined. For example, take the following two *when* clauses, the first trying to determine if the house owner has arrived home and the second determining if a visitor arrived at the home may:

```
when the garage door opens then the garage door closes then the
front door opens (within 5 minutes)

when movement is detected on the porch and the front door bell is
pressed (within 1 minute)
```

A policy is represented as a tree structure, wrapped using JSON (JavaScript Object Notation, *www.json.org*). When Homer is given a new policy, it is saved to a database and passed to the policy server. This then loads the policy into a custom tree structure. The *when* part of the tree is represented with a root node (*and*, *or*, *then* or *event*). Each node stores any child nodes or, in the case of an event node, stores the details of the particular trigger or condition.

A policy is represented by such a tree, along with other relevant properties such as its name. A policy can be enabled or disabled by the user; this dictates whether the policy should be listening for events. Multiple policies can exist at run time. Currently, the policy server has been tested with 50 enabled policies, and is expected to scale comfortably to 200–500 policies. If a policy is enabled, then it can be either waiting or activated. A waiting policy is simply awaiting one of its child nodes to report that an event has happened. If a policy is activated, one of its child nodes has reported that something has happened. The policy therefore starts a countdown according to its time interval (the default time being 60 s). When future events occur, the policy's *when* tree is evaluated to determine if it is now triggered. In that case, the policy is executed and its child nodes are reset. If the time interval passes before the *when* is satisfied, then the policy does not run and its child nodes are reset.

In the case of an *and* or *or* node, a listener is registered for each child node, whereas a *then* node registers a listener only for its first child. If an *or* node has a child whose trigger/condition becomes true, its child nodes will stop listening until the policy is reset. If a child of a *then* becomes satisfied, it will stop listening for that child and instead start listening for the next child. By listening only for what is relevant, the policy server can run efficiently and reduce the number of event listeners required.

Conflicts are almost inevitable with policies. Typically, conflicts arise because the policies of different people are inconsistent. For home automation, the householder may wish to reduce heating levels at night, but a family member may wish more heat if they are not in bed. For telecare, a resident of sheltered housing may wish to watch TV at any hour, but the warden may require TVs to be switched off after 11 PM. Even the policies defined by one individual might be inconsistent, especially if the user defines many policies over time. For example, the user might have a policy of saving energy but also wish the house to be warm in winter.

Resolution policies are defined to detect and resolve conflicts among policy actions. These are specialized policies that a system designer or administrator rather than a householder is expected to define. A library of resolution policies has been created for dealing with likely home conflicts, so little or no extra work may be needed. The tool Recap (Rigorously Evaluated Conflicts Among Policies [29]) is used to statically detect conflicts among policies. It also automatically generates outline resolution policies for use dynamically.

There may be conflict if the particular user device and desired action are the same (e.g., both actions wish to dim the hall light). If the parameter values are different, then there is indeed conflict (e.g., the dim levels are different). In such a case, the more recent policy might be preferred as an example of a generic action. Other generic actions include choosing a higher priority policy or one defined in a higher level domain (e.g., all houses operated by *stirling.org*, as opposed to one particular house). A resolution may also use regular policy actions (such as turning an appliance on). If appropriate, the householder can be told of a conflict and be asked what to do.

For home systems, many conflicts are obvious (e.g., trying to turn the heating on and off at the same time). However, conflict detection can be quite subtle. Suppose that two policies wish to send different text messages to the same person. Superficially, this is the same kind of conflict as the dimming example above, but in fact both actions should almost certainly be allowed (e.g., one is reporting that the house is too cold, while the other is reporting that a favorite TV program is starting soon). In practice, what this means is that the resolution policy library needs to handle specific situations (dimming a light, turning an appliance on or off, sending a message, etc.).

17.3.7 Goals for Home Management

The approach to goal definition and refinement extends earlier work on Internet telephony [30]. The tool Ogre (Optimising Goal Refinement Engine [31]) performs static and dynamic analysis of goals, resulting in execution of the most appropriate policies.

A goal is a user objective for the home such as making it comfortable or complying with medication. It can be easier for users to identify their goals (e.g., "I wish to be secure") than it is for them to define a comprehensive set of policies (e.g., "alert me if I leave the house and a window is open," "inform a neighbor if the house is entered while I am on holiday").

Goals are defined in terms of subgoals, for example, comfort might include aspects such as lighting level, ambient noise level, and room temperature. These subgoals are called goal measures, and are the means of assessing how well a goal is achieved. Most approaches to goal refinement take a logic-based approach (see Section 17.2.3). However, this is impractical in view of its technical difficulty and run-time inefficiency. Instead, Homer takes a numerical approach. This allows (sub)goals to be given appropriate weights, allows goal refinement to take current circumstances into account, and allows goals to be achieved as far as possible (and not necessarily in some absolute sense).

A goal measure is a formula over relevant system variables. These variables are classified as controlled (managed by the home system), uncontrolled (e.g., environmental factors), or derived (defined in terms of (un)controlled variables). Some variables have a natural measurement (e.g., temperature in degrees, additive intake in grams). Other variables are placed on a numerical scale that ranks them (e.g., security, chill risk).

Syntactically, a goal is a simplified form of policy. There is no trigger because goals always apply. A goal may have a (compound) condition that uses information like time of day or an environment value. Unlike a policy, a goal has a single action that maximizes or minimizes some measure.

Suppose the goal is to minimize household disturbance at night. This might be expressed as follows:

```
when the hour is in 23:00..07:00
do minimise household disturbance
```

The measure of household disturbance then needs to be broken down into factors such as the ambient noise level and other residents being active at night. These uncontrolled system variables are used to define a measure of household disturbance, for example

$$1.5 \times \text{threshold (noise\_level, 70)} + 0.5 \times \text{night\_activity}$$

A goal measure is usually a linear weighted sum whose weights are automatically inferred from typical values of the variables. Defining a goal measure then requires only a choice of the relevant variables. The *threshold* function is used when a factor should be counted only over a certain level (here, 70 dB).

In both home automation and telecare, there are often multiple stakeholders with their own goals. As examples, the following cover goals for home automation as well as telecare:

Doctor: minimize allergen exposure (e.g., food additives, pollen); maximize medication compliance (e.g., taking the correct dose of medicine on time).

Family: maximize user activity (e.g., avoid over-sleeping or watching too much TV).

Householder: maximize household security (e.g., detect intruders, keep doors, and windows locked); minimize home discomfort (e.g., ambient noise level, room temperature).

Social Worker: maximize social contact (e.g., going out, phoning friends).

Warden: minimize household disturbance (e.g., noise, night-time activity).

Goal refinement into policies is treated as an optimization problem: choose the set of policies that maximizes the overall evaluation function. There are normally multiple goals. It is therefore necessary to combine their individual measures into an overall evaluation function that is usually a linear weighted sum of individual goals. In fact, goal achievement does not critically depend on the choice of weights [30].

Policies can define an effect that specifies how they modify one or more system variables, and thus how they contribute to goal measures. The effect of a policy is an abstraction of the actions it can perform. More specifically, an effect is defined in the same terms as goal measures. At definition time, this identifies the relationships between goals and policies. At run time, this is used to determine the policies that optimally satisfy the goals.

Static analysis is performed when a goal or policy is created, modified, or deleted. Whether a policy contributes to a goal is determined by comparing its effects with how the goal measure is defined (i.e., which system variables it uses). A policy is considered to contribute to a goal if it affects one or more system variables involved in the goal measure. A policy effect may modify an arbitrarily complex goal measure. The sense of the effect is therefore not known until run time, when it may worsen or improve the evaluation of including the policy.

Dynamic analysis is performed when a trigger selects policies that contribute to goals. The selection of policies is then dynamically optimized. This may also include finding the best choice of variables appearing in a policy (e.g., the length of time that heating is turned on). By taking current circumstances into account, dynamic analysis ensures that the system response is always optimal.

17.3.8 Home User Interfaces

Home automation companies generally offer a range of applications, from mobile phones to wall-mounted touch screens and web-based applications. However, most of these applications offer only basic control over the home. System configuration and programming are rarely offered to the user. Instead, these aspects are often fixed at installation time (and have to be changed by the supplier on user request). Even where the home system allows users to make changes, the user interface is often hard to use and requires technical knowledge.

Homer, however, aims to satisfy a range of users from novices (perhaps even technophobes) to experts. The authors have carried out a survey of what users wish to be able to manage in the home [32]. It was found that users would like to control many aspects of the home, but they are reluctant to program it. Users like the idea of managing the home through policies, but need these to be easily modified and adapted. They are hesitant to use an approach that looks like programming, mostly due to fear that it will be complicated, confusing, and designed for the technically minded.

Homer provides a web-based API that allows external entities to manage the home system. This supports any application that can make HTTP calls with JSON objects. For security, authentication uses an application key and a secret key. This openness makes it easy to write new applications for the home, using different styles and technical approaches. For example, Homer has web-based, iPhone, and iPad applications.

The web-based application uses the Google Web Toolkit (*http://code.google.com/webtoolkit*) to expose devices activities within the home, browsable by location, or device type.

The iPhone application shown in Figure 17.5 makes it possible to browse devices by location or type. The current state of a device can then be viewed, its history of past events is available, and the device can be asked to perform selected actions. A live Twitter feed is also available for all events within the home.

The iPad application is the main user interface. It offers full control over the home as well as integrating the ability to view, write, and edit policies. It has therefore been designed for a range of user capabilities. It aims to make simple tasks easy, and more complex tasks possible. It combines controlling and programming the home in one sleek and simple interface. Blending programming with control means that the user is less likely to notice, or fear, the programming aspects. The other benefit of combining these is that the user does not have to make a conscious decision to program the home or to use advanced features of the home interface. Instead, the user is able to

FIGURE 17.5   Homer iPhone application.

apply policies to any object (person, room, device, etc.) that they are viewing or controlling.

Users think about the home in different ways, so they must be allowed to define logically equivalent policies in ways that suit them. For example, a device-oriented policy might say "turn on the coffee machine when I get up," while a situation-oriented view might say "when I get up turn on the coffee machine." Alternative perspectives like these can also be applied to viewing policies. Policies are tightly integrated with the home management user interface. The interface reflects perspectives such as people, time, rooms, and devices. Each perspective can be viewed, controlled, and programmed through the same screen. Users can therefore easily view existing policies. They can also define new policies for whatever aspects of the home they wish, and in whatever way they think about these.

Since users vary in their technical abilities, they can choose different capability levels. These expose or hide various aspects of the underlying policy language. The three levels are as follows:

Simple: This offers basic capabilities such that triggers and conditions can be combined only with *and*, and actions are simple lists. An example would be: *when* trigger1 *and* trigger2 *do* action1 *and* action2.

Medium ("I'm a little scared"): This adds the capability to use *or* with triggers and conditions. An example would be: *when* condition1 *or* trigger1 *do* action1.

Advanced ("let me do everything"): This adds the capability to combine triggers and conditions with *then*, includes an associated time interval, and uses conditions in actions. An example would be: *when* trigger1 *then* (condition1 *or* trigger2) (within 2 minutes) *do* action1 *and* (*if* condition1 *then do* action 2).

Figure 17.6 gives an example of editing a policy at the medium level. The user gives the policy a name for future reference, specifying when the policy is triggered, and what to do. Each element can be edited, reordered, or removed within the current section, and can also be moved in and out of subsections for grouping. Choosing new elements can be done in multiple ways (from different perspectives) to support different user views. For example, an element that describes when the front door opens can be found through Locations (Home > Hall > Front Door > opens), Devices (Doors > Front Door > opens), or People (Someone > opens front door). These distinctions are irrelevant when saving and displaying policies, so the user can easily view the same logic from different perspectives.

For advanced users, two buttons allow saving the current set of triggers and conditions (the *when* clause) and the current set of actions (the *do* clause). Saving a *when* clause supports a form of high-level sensor fusion. This gives a name to a high-level trigger as a set of triggers/conditions associated with a person, device, or location. For example, in the sample policy of Figure 17.6, it would be possible to save the last three terms of the *when* clause as "someone left the house." This would then simplify the current policy and enable that macro trigger to be used in another policy (accessed as People > Someone > left the house). Saving the action list similarly offers a form of high-level actuator fusion. This time, a group of actions associated with an entity can be given a friendly name for use in other policies.

FIGURE 17.6   Homer iPad application.

17.4 Conclusion

17.4.1 Summary

Home automation and telecare have been introduced. It has been argued that a common technical approach can support both of these, the main differences being in the specific components and services. The challenges to be met include achieving acceptability, usability, interoperability, and automated support. The background to home systems has been discussed in the areas of standard, platforms, automation, and interfaces.

The Homer system has been described as a concrete example of a system that supports both home automation and telecare. The Homer philosophy is to make home control visible to users and manageable in simple ways. A Homer component embodies some underlying device technology. Homer services provide common capabilities to components. A distinction is made between system device types (that reflect particular technologies) and user device types (that reflect how the user wishes to treat devices in the home). Devices are supported in categories such as communication, environment, home automation, telecare, and user interface. The Homer framework mediates between components and the rest of the system. Component events can be mapped by the event server to/from higher level events used at the policy level. Users can define policies for how they wish the home to react to different situations. At a higher level, goals define high-level objectives that are optimally realized through policies. User-friendly interfaces allow the home to be managed through interfaces such as the iPad.

17.4.2 Evaluation

The Homer system represents mature work on a number of aspects:

- Service platforms are not new. Homer is based on the widely accepted OSGi framework, and so can take advantage of the stability and maturity of this as an infrastructure. To that extent, Homer has a similar basis to several other approaches such as Atlas and Saphire described in Section 17.2.2.
- Component architectures are also not new. However, Homer is unusual in offering an architecture that is well integrated with policy-based management (which has previously seen little use in the home). In particular, Homer components support the kinds of capabilities that make them easy to use in policies.
- Event logic in Homer is a generalization of sensor fusion studied in other work. It allows visual definition of how component events should be combined, and supports a flexible mapping among inputs and outputs.
- The Homer policy language is broadly similar to other rule-based approaches such as Drools and Ponder discussed in Section 17.2.3. A key difference is that policies for the home need to be usable by ordinary users, and need to reflect the kinds of control required for home automation and telecare. Homer thus supports distinctive forms of policies that are not found in other rule-based languages.
- Goals for home management have been studied by only a few researchers; Section 17.3.7 describes one of few examples. Although general approaches have been developed for goal refinement, their techniques such as event calculus or formal logic are not suitable for the real-time, fuzzy demands of the home. In contrast, Homer is able to realize goals at run time to optimally satisfy changing circumstances.

Each of these aspects has been developed and evaluated over a number of years. For example, the policy server derives from work on the project that started in 2001, while the component architecture and goal server derive from work on the project that started in 2005. Only the Homer interfaces are relatively recent. The elements of Homer have been evaluated in a lab setting and with actual users.

17.4.3 The Future

Although a number of low-level device standards exist, interoperability across manufacturers and compatibility across service providers remain a challenge. Some standards for home automation like KNX and X10 have been widely adopted, though many proprietary standards continue to exist. Standards for telecare and telehealth are only now emerging, and still have a long way to go. The major future challenge is achieving interoperability at a higher (service) level.

Some generic architectures of use in the home exist, but commercial solutions are usually individual and offer very limited interworking across manufacturers. OSGi has shown itself to be an effective service-oriented platform for use in applications such as the home and healthcare. The authors and others have demonstrated that it is relatively straightforward to take third-party solutions (e.g., new types of devices and services) and wrap them for use in OSGi. There is therefore hope of an industry consensus on a service-oriented platform for the home.

The design of usable and acceptable home systems remains a challenge. While consumers have gradually become accustomed to more sophisticated devices around the home, the additional complexity of home automation or telecare poses new challenges. The traditional keyboard–screen–mouse interface for computing is unsuitable for the home. Instead, new interaction modalities are likely to become the norm, such as speech, touch, and gesture. Stakeholder conflict, especially in telecare, will need further study. This will require better techniques for identifying conflicting viewpoints, and better automation for handling these.

End-user programming has been a goal for many years. For the home, a variety of interesting techniques have been developed using, for example, tangible interfaces. At present, these approaches tend to be restrictive in what the user can express. They are also often device-oriented in their focus, and are thus less suitable for ordinary users (who would prefer to manage the home on their own terms). It will therefore be desirable to raise the level of user interfaces to something akin to the policies and goals discussed in this chapter.

Various improvements are planned with respect to Homer. The Homer iPad application will be integrated more fully with the event server and policy conflict handling. Although these aspects of Homer may be too complicated for the typical user, they could still be beneficial. It may also be desirable for raw device data to be made visible to the user instead of some of it being hidden at a lower level.

It is difficult to make policy definition simple for users. However, additional help is planned to make this even simpler. Users could be allowed to learn what is possible as they start to write policies. This might slowly introduce new features and teach users about them when they have shown they are comfortable with the current set of features. Template policies could be included within the application so that users simply have to fill in the gaps; this could help users to understand policies better. Policies could be shared with other users through a public "gallery" that could be rated and commented on. This could help inspire users to be more creative and could show them what is possible. A final idea is to "reward" users as they progress in their use of policies.

Although various parts of Homer have been tested and evaluated, Homer as a whole will shortly be put through extended trials in end user homes.

References

1. Gavrilov, L. A. and Heuveline, P. Aging of population. In *The Encyclopedia of Population*, edited by Demeny, P., McNicoll, G. pp. 27–50. London, UK: MacMillan, 2003.
2. McGee-Lennon, M. (Ed.) *Including Stakeholders in the Design of Home Care Technology*. Computing Science, University of Glasgow: Glasgow, 2007.
3. Rigole, P., Holvoet, T., and Berbers, Y. Using Jini to integrate home automation in a distributed software system. In *Distributed Communities on The Web*, edited by Plaice, J., Kropf, P. G, Schulthess, P., and Slonim, J. number 2468 in Lecture Notes in Computer Science, pp. 185–232, Springer, Berlin, Germany, April 2002.
4. Kind, J., Bose, R., Yang, H-I., Pickles, S., and Helal, A. Atlas: A service-oriented sensor platform, in *Proceedings of Workshop on Practical Issues in Building Sensor Network Applications*, Institution of Electrical and Electronic Engineers Press, New York, USA, 2006.
5. Turner, K. J. A home-based system to support delivery of health and social care. In *Smart Healthcare Applications and Services*, edited by Röcker, C. and Zieffle, M. Pennsylvania, USA: IGI Global, Hershey, 2011.
6. Hein, A., Nee, O., Willemsen, D., Scheffold, T., Dogac, A., and Laleci, G. B. SAPHIRE— intelligent healthcare monitoring based on semantic interoperability platform—The homecare scenario, in *Proceedings of 1st European Conference on eHealth*, edited by Meier, A. pp. 15–21, Gesellschaft für Informatik, Bonn, Germany, 2006.
7. Helal, A., Mann, W., Elzabadani, H., King, J., Kaddourah, Y., and Jansen, E. Gator Tech smart house: A programmable pervasive space. *IEEE Comput*, 2005; 38: 50–60.
8. Orpwood, R. The Gloucester smart house for people with dementia, in *Proceedings of Workshop on Technology for Aging, Disability and Independence*, Engineering and Physical Sciences Research Council, Swindon, UK, June 2003.
9. Intille, S. S. The goal: Smart people, not smart homes, in *Proceedings of 4th International Conference on Smart Homes and Health Telematics*, edited by Nugent, C., and Augusto, J. C., pp. 3–6, IOS Press, Amsterdam, Netherlands, June 2006.
10. Woolham, J., Frisby, B., Quinn, S., Moore, A., and Smart, W. *The Safe at Home Project*. London: Hawker Publications, 2002.
11. Dewsbury, G., Taylor, B., and Edge, M. Designing safe smart home systems for vulnerable people, in *Proceedings of Dependability in Healthcare Informatics*, edited by Procter, R. N. and Rouncefield, M. pp. 65–70, University of Lancaster, UK, 2001.
12. Damianou, N., Lupu, E. C., and Sloman, M. The Ponder policy specification language, in *Policy Workshop* 2001, number 1995 in Lecture Notes in Computer Science. Springer, Berlin, Germany, 2001.

13. Turner, K. J., Reiff-Marganiec, S., Blair, L., Pang J, Gray, T., Perry, P., and Ireland J. Policy support for call control. *Comput Standards Interf*, 2006; 28: 635–649.
14. van Lamsweerde, A. and Letier, E. From object orientation to goal orientation: A paradigm shift for requirements engineering, in *Proceedings of Radical Innovations of Software and Systems Engineering in The Future*, number 2941 in Lecture Notes in Computer Science, pp. 153–166, Springer, Berlin, Germany, March 2003.
15. Rubio-Loyola, J., Serrat, J., Charalambides, M., Flegkas, P., Pavlou, G., and Lafuente, A. L. Using linear temporal model checking for goal-oriented policy refinement frameworks, in *Proceedings of Workshop on Policies for Distributed Systems and Networks*, pp. 181–190, IEEE Computer Society, Los Alamitos, CA, USA, 2005.
16. Bandara, A. K., Lupu, E. C., Moffett, J. D., and Russo, A. A goal-based approach to policy refinement, in *Proceedings of Workshop on Policies for Distributed Systems and Networks*, pp. 229–239. IEEE Computer Society, Los Alamitos, CA, USA, 2004.
17. Mozer, M. C. The neural network house: An environment that adapts to its inhabitants, in *Proceedings of AAAI Symposium on Intelligent Environments*, edited by Coen, M., pp. 110–114, AAAI Press, Menlo Park, CA, USA, 1998.
18. Dey, A. K., Hamid, R., Beckmann, C., Li, I., and Hsu, D. A CAPpella: Programming by demonstration of context-aware applications, in *Proceedings of Conference on Human Factors in Computing Systems*, pp. 33–40, ACM Press, New York, USA, 2004.
19. Gajos, K., Fox, H., and Shrobe, H. End user empowerment in human centered pervasive computing, in *Proceedings of 1st International Conference on Pervasive Computing*, edited by Mattern, F., and Naghshineh, M., number 2414 in Lecture Notes in Computer Science, pp. 134–140, Springer, Berlin, Germany, 2002.
20. Rodden, T., Crabtree, A., Hemmings, T., Humble, B. K. J., Åkesson, K-P., and Hansson, P. Configuring the ubiquitous home, in *Proceedings of 6th International Conference on The Design of Cooperative Systems*, pp. 215–230, IOS Press, Amsterdam, Netherlands, May 2004.
21. Blackwell, A. F. and Hague, R. AutoHAN: An architecture for programming the home, in *Proceedings of Symposium on Human Centric Computing Languages and Environments*, pp. 150–157, ACM Press, New York, USA, 2001.
22. Truong, K. N., Huang, E. M., and Abowd, G. D. Camp: A magnetic poetry interface for end-user programming of capture applications for the home, in *Proceedings of Ubiquitous Computing*, edited by Davies, N., Mynatt, E., and Siio, I., number 3205 in Lecture Notes in Computer Science, pp. 143–160, Springer, Berlin, Germany, September 2004.
23. Knoll, M., Weis, T., Ulbrich, A., and Brändle, A., Scripting your home, in *Proceedings of Symposium on Human Centric Computing Languages and Environments*, number 3987 in Lecture Notes in Computer Science, pp. 274–288, Springer, Berlin, Germany, May 2006.
24. Sohn, T. and Dey A. K. iCAP: An informal tool for interactive prototyping of context-aware applications, in *Proceedings of International Conference on Human Factors in Computing Systems*, pp. 974–975, ACM Press, New York, USA, April 2003.

25. Newman, M. W., Elliott, A., and Smith T. F. Providing an integrated user experience of networked media, devices, and services through end-user composition, in *Proceedings of Symposium on Human Centric Computing Languages and Environments*, number 5013 in Lecture Notes in Computer Science, pp. 213–227, Springer, Berlin, Germany, May 2008.

26. Turner, K. J. and Tan, K. L. L. A rigorous methodology for composing services, in *Proceedings Formal Methods for Industrial Critical Systems 14*, edited by Alpuente, M., Cook, B., and Joubert, C., number 5825 in Lecture Notes in Computer Science, pp. 165–180, Springer, Berlin, Germany, November 2009.

27. Arkin, A., Askary, S., Bloch, B., Curbera, F., Goland, Y., Kartha, N., Lie, C. K. Thatte, S., Yendluri, and P., Yiu, A. eds. *Web Services Business Process Execution Language. Version 2.0.* Organization for The Advancement of Structured Information Standards, Billerica, MA, USA, 2007.

28. Turner, K. J. Device services for the home, in *Proceedings of 10th International Conference on New Technologies for Distributed Systems*, edited by Drira, K., Kacem, A. H., and Jmaiel, M. pp. 41–48, Institution of Electrical and Electronic Engineers Press, New York, USA, May 2010.

29. Campbell, G. A. and Turner, K. J. Policy conflict filtering for call control, in *Proceedings of 9th International Conference on Feature Interactions in Software and Communications Systems*, edited by du Bousquet, L., and Richier, J-L., pp. 83–98, IOS Press, Amsterdam, Netherlands, 2008.

30. Turner, K. J. and Campbell, G. A. Goals and conflicts in telephony, in *Proceedings of 10th International Conference on Feature Interactions in Software and Communications Systems*, edited by Nakamura, M., and Reiff-Marganiec, S., pp. 3–18, IOS Press, Amsterdam, Netherlands, June 2009.

31. Campbell, G. A. and Turner, K. J. Goals and policies for sensor network management, in *Proceedings of 2nd International Conference on Sensor Technologies and Applications*, edited by Benveniste, M., Braem, B., Dini, C., Fortino, G., Karnapke, R., Mauri, J. L., Monsi, M. S. H., pp. 354–359. Institution of Electrical and Electronic Engineers Press, New York, USA, 2008.

32. Maternaghan, C. How do people want to control their home? Technical Report CSM-185, Department of Computing Science and Mathematics, University of Stirling, UK, December 2010.

18

Online Social Networks and Social Network Services: A Technical Survey

Huangmao Quan
Temple University

Jie Wu
Temple University

Yuan Shi
Temple University

18.1 Introduction

Social functions are natural consequences of human societies. Before communication technologies, social functions tend to evolve within cultural boundaries, such as location and families. Communication technologies, from mountain top signaling to Voice-over-IP, have broken those boundaries more or less and enabled multi-culture social functions.

Empowered by low-cost, high-power personal computing devices, the combined computing and networking capabilities have created a fertile ground for innovative forms of social activities. *Online social network* (OSN) serves as a means of social activity and has become a mainstream information media in the industry and in the public. Both government and entrepreneurs recognized the value of OSNs and have put forth efforts to capitalize on them. On the other hand, social networks also appear in public discourse. One significant example is social network analysis [1–3].

This article surveys OSNs. We provide our commentary mainly using technology factors, as well as their psychological and social backgrounds. Our article begins with a background of taxonomy of OSNs by comparing similar counterparts. Our survey attempts to explain OSN II in Section 4, and proposes a taxonomy and retrospectively compares it with other online communities, and even pre-internet social networking communities. In Sections 3–5, we attempt to plot the possible evolution directions and future technology challenges. In general, this article states and attempts to answer the following questions:

- What are OSNs in comparison with conventional Web services?
- What different forms of infrastructure and application services are available in OSNs?
- Who are the main players in OSNs and what are they doing?
- What are the current developing trends and how successful will they be?
- What new services and technologies of OSNs are expected to appear in the future?

18.2 Background

It took the radio 38 years and TV 13 years to reach 50 million users. However, Facebook has added 100 million users in less than 9 months* and OSNs have become a part of our lives and changed us. On the other hand, we have changed OSNs as well. From its early form, which simply provides identity and relationship services to hundreds of services that associate various applications with personal data. Social network services (SNSs) have gained their popularity and serve more and more granular target markets, such as medication, science, education, and so on. However, despite there being extensive amount of work on research of either online communities or off line social networks, they are generally not applicable in the context of OSNs. To make things even worse, with regard to the meaning of social network, sociology and computer science is lost in translation. In this section, our survey will reflect the development of OSNs in the past and then explicit the definition of OSNs in application and platform levels.

18.2.1 History

Generally speaking, in information technology, an OSN is basically a type of website that provides social identity and social relationship services: who are you, who do you connect with. To broaden our conceptual view, social networking, the concept of social phenomena per se, emerged before the Internet. Retrospectively, the first virtual community without propinquity, which we prefer to be called "off-line social network," appeared in the 17th century: the Royal Society of London formed a community through letter exchanging. Since then, various virtual communities became less and less

---

* www.socialnomics.net.

geographically binding and more and more based on common interests and activities [4]. Moreover, social networking*, or relationship initiation [5], as a common phenomenon for human beings in many aspects, originally existed in online virtual communities [6] before the appearance of OSNs. For example, dating sites and community sites supported lists of friends. Although most of those websites help strangers connect based on shared interests or activities—networking with others—it was not until the turn of the last century that new type of websites became recognizable, which extents and maintains pre-existing social networks by encouraging users to create a profile and affiliate with friends. They are called OSNs or social network sites interchangeably. Since the turn of the last century, some famous OSNs, such as Myspace and Facebook, were growing in popularity and proliferating. And more importantly, all the trends of human society have driven the growth of OSNs: internet capacities, hardware and software features, mobile communication, business model of Web 2.0, and so on. As a result, OSNs did not only hit the mainstream, but also became a global phenomenon [5].

18.2.2 What are OSNs?

Today, OSNs are used extensively as public social interactive and collaboration tools. The OSN distinguished itself in structure and behavior patterns from other relationship-initialized information systems, such as business relationship management system and collaboration software. According to Weyer's definition [7], interaction with OSNs is "an autonomous form of coordination of interactions whose essence is the trust cooperation of autonomous, but interdependent agents who cooperate for a limited time, considering their partners" interests, because they can thus fulfill their individual goals better than through non-coordinated activities'. Based on the definition, OSNs have four notable characters:

- No propinquity
- No persistent connection
- Trust based on interdependence
- Autonomous collaboration

No propinquity means an actor in OSN has no or little knowledge about the other actor in the other end of the tier (Latent or Weak ties [8]). No persistent connection means actors keep only temporary connections, unlike relationships in the real world, which are more stable. Trust, based on interdependent and autonomous collaboration means actors on OSNs have no obligation to serve others, their motivation of collaboration is from the awareness of interdependence.

Despite us believing that the above characters should be representative of most OSNs, we would also point out that the diversity of OSNs may fuzzy those characters. Some OSNs actually focus on strong ties, such as LinkedIn, which serves close communities that share more real-world connections. Whereas, OSNs vary widely by application and their key technological features are fairly consistent [5]. For example, most OSNs allow

---

* Social networking is activity initialed [7]. It a type of activity. While social network is an abstract Web connecting people by relationships.

people to articulate friends and publicly display connections, and their own profiles. Additionally, research in various fields developed its own taxonomies. Therefore, for better internal consistency, it is elastic to set a hard line between OSNs and general social networks. In our paper, OSNs are defined as Web-based services that allow individuals to:

- Construct a public or semi-public profiles within a bounded system.
- Articulate a list of other users with whom they share a connection.
- View and traverse their list of connections and those made by others within the system. The nature and nomenclature of these connections may vary from one site to other [5].
- Provide users' online presence to describe their current state and activities [9].

18.2.3 What is SNS?

Except structure and behavior patterns, OSNs also distinguish themselves from traditional Websites by providing various SNSs. Generally speaking, there are two types of SNSs. The first type is "organic" SNS, which is people-focused and embeds social network features within. For example, Twitter provides Microblogging [10], which has a core value of connecting friends and transferring ideas throughout a group of people. The second type is "hybrid" SNS, which is content-focused and combines traditional Internet services and social networking by integrating social features. For example, Flickr-Yahoo's photo-sharing Website combines photo repository and social networking features together. The first type aims towards the maintenance of pre-existing social networks and helps connect people based on common language, while the second type caters to diverse audiences.

All the functions of OSNs are delivered by SNSs. Some of them provides fundamental infrastructures that allow other services to build onto. We call them infrastructure SNSs. Another category only serves a specific purpose or application. We call them application level services. Despite that the SNSs may vary in their forms, they share the same goal, which is to fulfill human needs (Figure 18.1). It is human nature that drives us to socially connect with each other [11].

18.3 Social Network Services

While working on this survey, we found a dilemma while presenting all OSNs in a uniform classification. This is partly because there are two concepts—SNSs and social network platforms (OSNPs). These two concepts are logically hierarchic as applications and platforms, but normally used interchangeably. For example, despite most users agreeing that Twitter is an OSN, Twitter may not be a proper name for an SNS; microblogging would be the more appropriate one. Therefore, we intend to use OSNPs to describe Websites that host SNSs. To further clarify concepts of the OSN, this section will focus on SNSs, and the next section will talk about OSNPs.

In our paper, the term, SNSs, means Internet services provided by OSNs to *end-users*. Similar to other information systems, an OSN congregates a set of services, such as

	Expressing identity	Status & self-esteem	Giving & getting help	Affiliation and belonging	Sense of community
Blogs	✔	✔	✔		
Video, content sharing, tagging sites (e.g., YouTube, deli.ciou.us)	✔				
Self-forming groups (e.g., Yahoo or Google groups				✔	✔
Profile-driven social networks (e.g., MySpace, LinkedIn, Facebook)	✔	✔		✔	
Rating, review sites (e.g., epinions, TripAdvisor)		✔	✔		
Purpose-driven social network (e.g., SparkPeople, Slashdot, Serma, Communispace)	✔	✔	✔	✔	✔

FIGURE 18.1 Human needs vs. social network services.

email, instant messaging, multimedia sharing, and so on. However, compared with traditional online services, the services provided by OSNs are more user-driven: giving more social context of users, SNS is usually described as an individual-centered service compared with traditional online communities where services are content-centered or group-centered [12]. Moreover, as we discussed in Section 2, SNSs may be "hybrid" products of traditional services. Because of the above reasons, we start by grouping SNSs according to their purposes. Then, we propose a map to explicit a taxonomy. We go on to survey specific examples of SNSs according to their classifications (Figure 18.2).

Broadly speaking, SNSs can be classified into infrastructure services and application services (Figure 18.3). Among them, infrastructure services provide the most basic and essential information about a social actor's identity, personal information, and his/her relationship connections.

- *Social profile:* Social actors' personal information of characteristics, such as name, gender, age, and so on.
- *Social identification:* The unique proof or evidence of identity, which is usable by other SNSs.
- *Social graph:* A relationship graph mapping of actors' friends and how they are related.

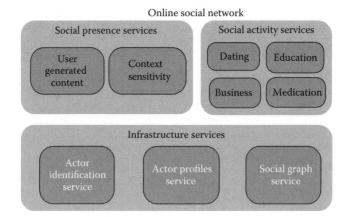

FIGURE 18.2 Classification of social network services.

FIGURE 18.3 Example of social network services.

Supported by infrastructure services, OSNs provide various applications to facilitate social interactions and impress other people by social presence [13]. The term, "social presence," was derived from social present theory [14]. It is originally used to present awareness of the other person in a communication interaction. The interaction is effective if only social presence provides a meaningful indication of one person [15]. Thus, application level services can be classified according to their purposes:

- *Social interaction:* The services which include online communication of any sort, such as comment, instant message, and feedback.
- *Social presence:* Personal stamps created by users for others to indicate their state, log of activities, and performance.

Features of SNSs are different to different people depending on their activities. Therefore, our survey will concentrate efforts mostly in discussing services of infrastructure and social presence. In the next section, we will overview each SNSs, one by one.

18.3.1 Identification and Profile Service

The identity service, as one of the core components of OSNs, is critical to users because it affects others' awareness as well as bolsters users' self-esteem, sense of belonging, role, and hierarchy within online communities [16]. Additionally, persistent identity is required to build a stable connection of friends.

Identities are a set of characteristics that separate self from others. Although most OSNs require users to represent themselves authentically [5], most OSNs provide loose identity, for example, in twitter, no identification is required. Loose identity may not reflect users' authentic personality and lead to identity theft or impersonation (see Section 5 for details). In spite of coarse nature, in most conditions, identities of OSNs are reliable because other users can refer signals of profiles and public friend lists to validate them [5]. Moreover, different OSNs may select different strategies to rigidify their users. Some of them even encourage users to articulate fake identities or avatars [17].

The profile service of OSNs presents individual's choosing identity information. It can be updated in a specific time-frame and with a particular understanding of audience.

Firstly, the profile service is responsible for creating and maintaining profiles. For example, most OSNs ask users to fill out forms with a series of questions with regard to their personal information, which normally includes descriptors, such as name, sex, age, and interests. The profile is generated by the answers to those questions. Most OSNs also allow uploading profile photos.

Secondly, the profile service controls the visibility of a profile–profile closure. It provides an individual's fragmented profile according to audiences' discretion. For example, by default, basic profiles on LinkedIn and facebook are visible to anyone and even crawled by search engines. Therefore, they are publicly visible regardless of whether or not the audience has a connection with the user. Alternatively, the full profile, which includes sensitive information, such as marriage status and religious views, are available only to either paid accounts (LinkedIn) or connected friends (Facebook). Facebook and Myspace also allow users to choose whether they want their profile to be public or "Friends only." Facebook implied more complex settings for profile closure, for example, users can grant permission to a certain network [18]. The counteraction of profile fragmentation is studied in Backstrom et al. [19] and Liu and Terzi [20], as a technical challenge. This will be discussed in Section 5.

Finally, based on this current business model [21], most OSNs' profile services have little or no interoperability with one another. Our paper will discuss this issue in detail in Section 5.

18.3.2 Social Graph Service

Social Graph Service is responsible for building, maintaining, and retrieving ties based on shared affinities. For SGSs; relationships, reputations, and searchs are three key components of the SGS: How to explicit connections—not only their connection status, but also how to quantitatively describe users feelings about connections, such as Ebay's reputation system.

For SGSs, one typical usage scenario is searching and extending connected people who share affinities or complementary capabilities [22]. For example, Alice and Bob belong to the same OSN and both are interested in politics. They are familiar with one another and trust each other. Alice has to work on a book that requires illustrations, and is looking for a freelance artist. By searching with SGS, Alice could be able to find Bob who is connected to Sarah, a student of art. Alice could then approach Sara for the project by sending a connecting request. Since Sarah knows that Jim and Bob trust each other and Bob trusts Sarah, it could be infered that possible trust between Alice and Sarah is high. Then, Sarah would confirm Alice's request with no hesitation. Finally, they will complete a happy and safe business transaction.

1. *Connection component:* Considering that a sense of affiliation is not equivalent to a true sense of community, connections in SGSs can either be bidirectional or unidirectional [5] to present relationships toward either individuals or groups, such as "became a fan" on facebook. To broaden our view, Brzozowski et al. [23] describes multiple relationship types in online communities. Contrary to popular belief, this research distinguishes allies and enemies in types of connections semantically. They argued that a better social network can be achieved by employing multi-typed connections.

   Compared with off-line social networks, people mostly use OSNs to reveal hidden relationships, which results in connecting people within their extended social networks [8]. For example, Bob and Alice know each other in reality. In other words, they share some offline connections. Alice has a friend, Ted, who works at BIG company as HR. All three have profiles on Facebook. Presumably, Bob needs to find a job. By tracking Alice's friends list on Facebook, Bob will have a good chance of tracking down Ted and contacting him with regards to a job. In this case, Facebook plays a bigger role more than just being an information media exchanging Website. It also extends and maintains relationships to benefit Bob's social activities—seeking a job in our example.

2. *Reputation component:* To avoid issues with regards to divergent reputation definition [7], in this paper, we simply define reputation as general opinions toward

individuals. There is extensive research on mathematical framework [24–27] for modeling caculatable reputations and the way reputations propagate [28].

Despite inferring reputations in OSNs being theoretically possible, applications that infer affinity and trusted third parties are limited. This greatly roots in structures of OSNs, which are not fully compatible with trust metrics. For example, Eigen Trust [29], a variation of the PagRank algorithm [30], provides a globally accepted trust rating as reputation. It was originally designed for P2P systems without considering the limitation of OSNs [31]. Despite lots of metrics trying to combine personal trust opinions and global reputations by converging former ones into a single measurement from the whole group, personal aspects are far more complex to express in quantifiable ways than in multi-agent systems. Because users themselves bear some responsibility in contributing to reputation management, most of them are either technically sufficient or would rather spend dedicated time doing so. A negative example is Orkut,* which is used to allow people to express feelings about fellow friends through a rating system known as "karma points," but was finally abandoned due to lack of popularity. Therefore, most live reputation systems of OSNs are based on simple models. Testimonials are still the most popular method for providing member reputation, for example, the number of successful transactions on Ebay and customers' reviews on Amazon.

3. *Search component:* The search refering social graph will leverage performance [32]. Based on the development of Web search techniques [33], searching information in OSNs obeys small world search strategies [34]: using local information about their close contacts [35], for example, when users search jobs on LinkedIn, the results will be ordered by degrees that represent social distances to HRs.

Milgram [34] also explicits a greedy algorithm for small world searching: i will select its neighbor j who is closer to the target t in terms of social distance Y; that is, $Y_{j,t}$ is minimized over all j in i is Web of connections. The problem is developed by Kleinberg [36]. His often cited paper proposed a decentralized search algorithm to solve searching in small worlds with partial information. Kleinberg [37] also provides a theoretical foundation by proving efficient searchability in social networks: a simple algorithm that combines information of network connections, therefore social identities can succeed in efficient searching.

Another principal character of searching in OSNs is based on tagging. The content is semantically annotated for better understanding and searchability, for example, tag cloud has recently been utilized by most OSNs. Our paper will discuss this issue in the semantic web part of Section 18.5.

18.3.3 Social Presence Services

Most Social Presence Services in OSNs are created by users to enhance their impression, for example, adding multimedia content, or modifying the look and feel of their page.

---

* www.orkut.com.

Some OSNs, such as Facebook and Orkut, allow users to add modules or applications that publish various content and interact with others. All that data describe the nature of individuals' presence in the OSNs. The data also enhances their sense of self-worth and stimulates actors to maintain passive interaction within them.

Apart from interactions with their friends on SNs, actors of OSNs tend to present them by current state and activities. Therefore, OSNs offer different mechanisms to support social presence, such as custom messages, online status, the listening of music, watched movies, and so on. By assembling all that data, other actors can form an overall impression of his/her presence in the OSNs and even some clue on what he/she is like in the real world.

There are two types of mechanisms for Social Presence Services. The first type provides social presence by user-generated content, for example, Bob posts a microblogging entry saying "I am feeling good." The second type, context sensitivity, automatically obtains social presences from individuals' real-world context, such as location and time. A case in point is Google Latitude*, which enables a user to allow connected people to track their location. The main difference between context sensitivity and other interactions is that the attention paradigm is reduced to shorter time periods [13]: Compared with interactions such as a direct chat or e-mail, context sensitivity aware friends with no cost of time. In the following paragraphs, we will give detailed explanations and examples of both types.

1. *User-Generated Content:* One principal character of OSNs is that most content is user-generated. User-generated Content (UGC) refers to online media content that is produced by end-users. For example, Wikipedia, a web encyclopedia which has 14 million articles†, is written and edited by its users, who can be anyone with access to the site. Admittedly, UGC services are not necessarily SNSs. Considering that OSNs rely on content created by its users who update profiles, communicate with friends, and participate in communities. Integrated UGC services could boost the usage of OSNs through improved user engagement. On the other hand, utilizing OSNs to support online collaboration could improve the quality of UGC [38]. Although UGCs can be created with little or no restriction, monitoring and administration are also necessary to avoid offensive content, copyright issues, and so on.

   Microblogging is a brief text blogging that allows users to send blogs in limited length. It is inspired by cell phone SMS. It can also update multimedia, called micromedia, such as photos or audio. Its most distinguished character is a variety of means of submitting: web, text messaging, instant messaging, email, and so on. Another character is brief: a single sentence or a very short video. By congregating short entities, a logs of the daily events are presented.

   Social news is a team that refers to web services, in which users can submit and vote on news stories. Compared with formal news, which is published by a media agency, social news is collected and edited by end-users. Two of the most popular

---

* www.google.com/latitude/
† en.wikipedia.org/wiki/Wikipedia.

social news websites are Slashdot and Digg. Today, even media giants such as, CNN start to adapt to social sews.*

Social bookmarking is a service used to share, organize, search, and manage bookmarks of Web resources, typically a Web page. Social bookmark services also encourage users to add, modify, and remove annotations on Web pages. Annotated Web pages will also be visible to other users who share similar annotations which indicate their similar interests. Clouds of tags is a team used to describe clusters of tags or bookmarks provided to users dynamically according to their interests. Social bookmarking services also provide feeds for their lists of bookmarks to allow subscribers to become aware of new bookmarks [39]. For example, users of Google Reader† can either share and tag interested Web pages or subscribe to friends' feeds. Some extra features, such as ratings, may be added to social bookmarking services, for example, research in semantic Web [40,41] propose social bookmarking systems, which are embedded by more semantic means. By referring ontology knowledges in the real world, such as hierarchical relationships.

Wiki is a collaborative encyclopedia service that allows any user to contribute content by creating and editing Web pages. Wikis may serve different purposes, such as learning, collaboration, and knowledge-sharing. Wikis can facilitate social processes [42] in the sense of rewarding contribution [43]. Wiki softwares are also used in corporate intranets, mostly as knowledge-management systems, but in terms of online services, Wikipedia is the most famous and among one of the most typical examples of Web 2.0 services. Wikipedia reached three million english articles in August 2009 and enjoyed the title of being "the largest encyclopedia in the world."

2. *Context Sensitivity*: Context sensitivity is a set of SNSs that provide services considering sensed information of users. It became popular greatly because of the rise in using mobile devices for social networking. Contrary to manual implications of personal and contextual information, mobile devices can automatically obtains users' information by various sensors, such as GPS. With the mention of coupling gathered data automatically, such as time coupled with location, or personal information coupled with location, a new service could be created. For example, location sensitivity can provide localized information according to users' location. Another example is personal information coupled with health information [44].

18.3.4 Social Interaction Services

Unlike social presence services, which are content-focused, social interaction services exist mainly for helping other people and for carrying activities to increase the senses of affiliation, belonging, and community.

---

* CNN iReport, www.ireport.com.
† reader.google.com.

1. *Multiplayer Online Games:* MOGs, such as Second Life and World of Warcraft, enact social networking processes by providing an avatar for game players. They can communicate and live along-side other players. Some OSNs also provide light-weight MOGs as part of their services.

2. *Reviews and Opinions:* Online feedback services mainly provide two functions. First, they publish reviews and opinions to beware public users, which affect customers' decision to buy a certain product. Second, they collect reviews and options from end-users by providing them facilities for writing their own personal views, such as satisfactions or dissatisfactions. Combining social networking with ROS will allow customers to compare products based on reviews from their own connections.

3. *Finance:* Finance services, such as micropayment, provide a system for members to pay for tangible or virtual goods. It is capable of handling arbitrarily small amounts of money. Micropayments have to be suitable for the sale of non-tangible goods over the Internet. This imposes requirements on speed and cost of processing the payments: delivery occurs nearly instantaneously on the Internet, and often, in arbitrarily small pieces. On the other hand, OSNs can easily obtain reputation references to support micropayment.

4. *Groups:* A group is a loosely coupled system of mutually interacting interdependent members. But, a group is more than just the collection of members. Groups can be defined by psychological and temporal boundaries, interact with each other, and task and goals [45]. Most online communities grow slowly at first, due in part to the fact that the strength of motivation for contributing is usually proportional to the size of the community. As the size of the potential audience increases, so does the attraction of writing and contributing. This, coupled with the fact that organizational culture does not change overnight, means creators can expect slow progress at first with a new virtual community. As more people begin to participate, however, the aforementioned motivations will increase, creating a virtuous cycle in which more participation begets more participation.

18.4 Online Social Network Platforms

This section introduces and compares several popular OSNPs. Despite there being many main players on our candidate list, we select the platform that is distinguished either in its features or ability to represent its own class. One exception may be Facebook, with more than 250 million active users and nearly half of them logging on more than once per day. Despite Facebook's popularity, the potential of OSNs is still far beyond reach and greatly relies on its divided types and separated target markets. As we will discuss later, such diversity promises us that OSNs play or will play in almost every aspect of our lives.

Generally, OSNPs attempt to provide various SNSs in one platform to achieve diverse requirements. Namely, we categorize them as general purpose OSNPs, such as Myspace and Facebook, and other OSNPs that serve narrower target markets called Niche Communities [5], such as LinkedIn. Their different business models end up becoming divided structures and technical specifications. The rise of OSNs shifts the online community from websites dedicated to interests, into Websites organized around people. According to social sciences, the web became an ego-network [46] composed of a person (social actor), friends and family members (social ties), and other people a person knows without personal emotion (social alter). Based on different social network components, the second classification perspective is based on various focuses. We propose this via classified OSNPs into three categories:

Social media is a type of OSNP mainly congregated around SNSs that aim for mass communication, like other news media. Notably, traditional news media, such as CNN and BBC, relay through broadcasting (one to many), whereas social media are more decentralized and rely on dialogues between users (many to many).

Social interaction includes various types of combinations between friends or well-known people. It forms a clique of people who use OSNs as an extension of their social interaction platform, such as LinkedIn or alumni sites. In entertainment, the virtual worlds can be classified into this kind, such as Second Life and the Sims Online.

Social networking: OSNs focus on developing social ties or maintaining the existing ones. In other words, making friends out of strangers. Here, *Networking* emphasizes relationship initiation, unlike social interaction, which is normally between close friends. Social networking focuses on relationships built with strangers [5]. The most common types are dating sites, such as eHarmony and a business relationship site such as LinkedIn.

Admittedly, the above classifications are somehow fuzzy and some platforms may be hybrids of two or more types. One example is LinkedIn, which is for both social interaction and social networking: although its main function is seeking a job—a social interaction platform. LinkedIn also plays another important role: networking HR and former colleagues. Next, we will present several live OSNPs and further explain our points on common or unique characters of OSNPs.

LinkedIn is a business-oriented social networking site. Founded in December 2002 and launched in May 2003, it is mainly used for professional networking and has become a powerful business service. As of October 2009, LinkedIn had more than 50 million registered users, spanning more than 200 countries and territories worldwide. LinkedIn controls what a viewer may see based on whether the viewer has a paid account. LinkedIn allows users to opt out of displaying their network. Compared with other OSNPs, LinkedIn's business model is unique, in which it charges the user for accessing personal information.

Flickr [47] is an image and video-hosting website, web services suite, and an OSN. Flickr provides both private and public image storage. A user uploading an image can set privacy controls that determine who can view the image. A photo can be flagged as either public or private. Private images are visible by default only to the uploader, but they can also be marked as viewable by friends and/or family. Privacy settings can also be decided by adding photographs from a user's photo stream to a "group pool." If a group is private, all the members of that group can see the photo. If a group is public, the photo becomes public as well. Flickr also provides a "contact list" that can be used to control image access for a specific set of users in a way similar to social tier tools of other OSNs.

Facebook is the world's largest social network, with over 350 million active users and half of them visiting the site once per day*. It basically provides a platform to share a common interest, idea, task, or goal within its users, where they are able to develop or maintain personal relationships. Moreover, it also provides applications of various services, such as social bookmarking and instant messaging. Like other social networks, the site allows its users to create a profile page and forge online links with friends and acquaintances. Facebook launched its API in 2007, providing a framework for software developers to create applications that interact with core Facebook features. But, its API placed several restrictions on having complete access to an individual's social graph.

Ning was launched in October 2005. It is an OSNP for people who want to create their own social networks. Ning competes with social sites like MySpace and Facebook by appealing to people who want to create their own social networks around specific interests with their own visual design, choice of features, and member data [5]. The unique feature of Ning is that anyone can create their own social network for a particular topic or need, catering to specific membership bases. Ning has both free and paid options to fully eliminate advertisements. When someone creates a social network on Ning, it is free by default and runs ads that Ning controls. If the person creating the social network chooses, they can pay to control the ads (or lack thereof) in exchange for a monthly fee. A few other premium services, such as extra storage and bandwidth and non-Ning URLs, are also available for additional monthly fees. However, Ning does allow developers to have some source level control of their social networks, enabling them to change features and underlying logic.

Realtravel tries to solve the problem: how to extract information from data. In a collective knowledge system, the aggregate content must be more useful: create aggregate values by integrating user contributions of unstructured content with structured data. RealTravel attracts people to write about their travels to share stories and photos with semantic annotation. Travel researchers enjoy the benefit of all experiences relevant to their target destinations.

18.5 Research Topics and Challenges

18.5.1 Key Technologies

Broadly speaking, there are two types of concerns in OSN research: data access issues and data publication-related issues. In this section, we will discuss the following types of research topics as well as challenges:

- Distributed architecture
- Fragmented user identity
- Contextual information associated users and possible abuse
- Identity and trust
- Policies within network and web of trust (*dilemma: usability vs. privacy)
- Deeper adaptive user experiences

---

* http://www.facebook.com/press/info.php?statistics.

We will also cover some challenges of different social aspects: bridging online and offline social networks, positive interactions, and so on.

Distributed Architecture: One key question pertaining to architecture of OSNs is whether a decentralized architecture is sustainable, profitable, and usable, and consequently, what do we stand to lose if we adopt a decentralized architecture. Considering fragmentation of web capabilities, how to avoid overhead in processing information in OSNs? What is the minimum set of new functionalities that the future web should incorporate?

In contrast to the increasingly sophisticated capabilities of services, the fundamental architecture of the web has not changed much over the past 10 years. Existing social networks usually employ a "hub and spoke" model, where the website is the hub of all activity within the network, and where there is a *client* and a *server*. Since all traffic must pass through the hub, that site may become a bottleneck. Furthermore, each transaction must pass up one spoke to the hub, and then down another spoke, when the people interacting may be much closer to each other, in network terms, than either is to the hub site.

Services and applications in OSNs have become quite sophisticated in the features they provide. There is an opportunity to create an architecture that distributes the load. Such an architecture would require better interoperability between OSNs, more-so than what we have available, and should remove any dependence on an "always-on" network connection.

However, the hurdles for distributed OSNs are great, some being fundamental, such as incompatible assemblies, different data access APIs, and the entity data model. Despite difficulties, some prototypes of distributed OSNs have developed. The appleseed project is an open source OSN framework which is based on a distributed model. For instance, a profile on one Appleseed website could "friend" a profile on another Appleseed website, and the two profiles could interact with each other.

Privacy and Trust: For OSNs, identities and links are more important than content. The privacy issue on OSN can be classified into three types: identity disclosure, link disclosure, and content disclosure.

A report* finds that over half (52%) of social network users post risky information online. For example, the report states that 73% of adult Facebook users shared content only with friends, but only 42% of users state that they customized their privacy settings.

When using web-based social networks to refer trust values, most of the information that sociology considers important is not available (e.g., we do not know the history between people, the user's own background, and how likely they are to trust, in general, the familial/business/friend relationship between users). Thus, we must understand trust only from the available information. Privacy is also implicated in users' ability to control impressions and manage social contexts. Boyd (in press-a) asserted that Facebook's introduction of the "News Feed" feature disrupted students' sense of control, even though data exposed through the feed was previously accessible. Some research argued that the privacy options offered by OSNs do not provide users with the flexibility they need to handle conflicts with Friends who have different conceptions of privacy;

---

* State of the Net 2010-Consumer Reports.

they suggest a framework for privacy in SNSs that they believe would help resolve these conflicts.

Identity and Profile: Most research in this subject focuses on enhancing user security without compromising usability. The question left is, how to mirror OSNs with users' real identities?

To answer this question, we need to dive into reality first—identity mapping and wall barriers. Since each user has many registrations or accounts, the attention is dispersed. Identity in different OSNs exists as separate, isolated islands of discourse, unable to exchange meaningful information, leverage their accumulated knowledge, or connect with other communities that share their concerns. As the user takes a more active role in the production of content, and even services, and becomes a "prosumer," this situation leads to a somehow chaotic scenario where the same user is present in an uncountable number of different platforms, taking the best-of-breed for any aspect of social interaction or simply following or joining their friends. This situation creates an increasingly inconvenient and uncomfortable situation where users not only own different accounts, each one with a specific set of credentials, but also deal with an increasing amount of personal information scattered throughout several sites, each with different data usage policies and privacy protection conditions. Finally, how can we allow users who may want to deliberately fragment their online identity to do so?

There are also several independent initiatives focusing on how to break the wall by providing persistent identity. They first appeared as liberty alliances, such as Microsoft's .Net identity system named .Net Passport originally, and changed into Live Passport. Microsoft had accumulated various services, such as the Hotmail and MSN Spaces. But such effort faced significant resistance from other companies and users. There is great concern that online identity might become the property of a single corporation. Such centralized control would be devastating. As a result, vender-neutral identity services emerged, such as OAuth and OpenID. They both provide an open protocol to allow secured API authorization in a simple and standard method. Similarly, OAuth allows using anonymous tokens instead of usernames and passwords as identity. The granularity of permission can be either site level or application level, even a defined duration. OAuth can also grant a third party site access to their information stored with another service provider, without sharing their access permissions or the full extent of their data.

Structured Data: According to the collective intelligence theory from Doug Engelbar: The grand challenge is to boost the collective IQ of the organizations and of the society. To achieve this, the information on the web has to be structured. Semantic Web is used to define information and services on the web, making them possible for the web to "understand" and satisfy the requests of people and machines that use the web content. According to Tim Berner-Lee, "The Semantic Web is not a separate web, but an extension of the current one, in which information is given a well-defined meaning, better enabling computers and people to work in cooperation."

One attempt is FOAF + SSL. It is a machine-readable ontology describing persons, their activities, and their relations with other people and objects. Anyone can use FOAF to describe him or herself. FOAF allows groups of people to describe social networks without the need for centralized databases. FOAF is a descriptive vocabulary team expressed using the Resource Description Framework (RDF) and the Web Ontology Language (OWL). Computers may use these FOAF profiles to find, for example, all people living in Europe, or to list all people both you and a friend of yours know. This is accomplished by defining relationships between people. Each profile has a unique identifier (such as the person's e-mail addresses, a Jabber ID, or an URI of the homepage or weblog of the person) which is used when defining these relationships.

Other efforts include microformats, such as XFN and hCard. It is a Web-based approach to semantic markup that seeks to re-use the existing XHTML and HTML tags to convey metadata and other attributes. This approach allows information intended for end-users, such as contact information, geographic coordinates, calendar events, and the like to also be automatically processed by the software. Unlike the formal semantic Web, which is more complex, the microformat is light-weight and easy to implement in even today's web markup languages, for example, HTML5 adapts to several microformats.

Mobile social networking: Mobile social networking is a concept combining mobile communication and social networking. To illustrate the scale of mobile social networking, the number of unique visitors to the Facebook mobile site increased fivefold from 5 million per month in January 2008 to 25 million per month in February 2009. The latter figure represents 18% of Facebook's 120 million users (February 2009), a proportion that has gradually increased over time, and it will continue to do so in the coming years. Social networks with an established presence on the fixed line Internet are clearly benefiting from extending their services over mobile channels.

One obvious advantage of mobile social networking is context sensitivity, which means, in terms of places, time, and people makes services more information-sensitive. Mobile devices can collect more personal information than normal PCs, such as locations and contacts. By adding various sensors into mobile devices, new types of applications can go beyond the existing domains. Location-based services (LBSs) are among the most popular ones. An LBS is an information and entertainment service, accessible using mobile devices through the mobile networking, which utilizes the ability to make use of the geographical position of the mobile device. It can be used in a variety of contexts, such as health, work, personal life, and so on. LBS services include services of identifying a location of a person or object, such as discovering the nearest ATM or the whereabouts of a friend or employee. LBS services include parcel-tracking and vehicle-tracking services. LBS can include mobile commerce when taking the form of coupons or advertising directed at customers, based on their current location. They include personalized weather services and even location-based games.

Accessibility and user experience: How to deepen and adapt user experiences are also important practices of making OSNs useful. Human factors can be explicated in two levels: the general user experience and especially UI guidelines for accessibility. Accessibility means how the information in the OSNs can be correctly built and maintained, so all of these users can be accommodated while not impacting the usability of the site for non-disabled users. User experience covers a wider context of how to capture and better support social activities. For example, an OSN focus or hobby focus can be treated as a self-organized system, in which global patterns emerge from local actions and structured subsequent local actions.

The challenge in this research is in the mapping of quantitative measurements of interactions based on network traffic to qualitative analyses of social relationships. It is easy to know what people are doing in the network, but it is harder to know why. Most research is empirical and their fundamental theories are beyond the scope of our survey.

18.6 Related Concepts

As the increase in the popularity of OSNs constantly rises, academic research is emerging from diverse disciplinary and methodological information systems that can take advantage of the users' social and personal data, address a range of topics, and build on a large body of social network research. Broadly speaking, research on social networking can be divided into: how to effectively and positively communicate in OSNs,

After Milgram's study revealed small world property in social networks, research also found that social networking shared common characters, such as weak ties, power-law, and fuzzy boundaries. As a sociology concept, social networking is a social structure composted by individuals and relationships within them. An OSN or virtual community, however, is an internet-based community and information system of social networking. The idea of social networking is both old and new. Although it is a common phenomenon existing in every human interaction, when we talk about Social Web, in this paper, we intend to focus on online social networking sites, which are also called "online social networks" or "virtual communities." But the theories may build on each other.

18.6.1 Social Web

The concept of Social Web that research expected, is the web of people. It shares the same features of real social relationships, such as six degrees of separation phenomenon, scale-free, and so on. Unlike OSNs, which maintain weak ties, strength, and latent tie, social web has its limitations, for example, one factor is Dunbar's number, which points out the "theoretical cognitive limit to the number of people with whom one can maintain stable social relationships," is generally accepted to be about 150.

The "Six Degrees of Separation" phenomenon was first investigated by Stanley Milgram [48] in 1960, where he addressed letters to a particular stockbroker in New York and gave them to people, randomly picked at locations in the United States, far away from that of the final receiver. The condition for passing the letter, so that it reaches the addressee, was that one could post it only to people they knew personally by first name. Eventually, most of the letters reached the destination and the average number of hops was six. Since then, there have been various studies demonstrating how this effect may help people conduct their everyday lives.

This effect, also known as *Small Worlds* or *Scale-Free Networks*, has been revisited with analytical techniques starting with the seminal work of Barabasi [4,8,28]. Barabasi studied many natural and man-made networks and found that they all exhibit degrees of clustering with hub and spoke topologies and remote links between clusters. These real networks are fundamentally defined by a few highly connected nodes, but even a very small number of remote links (weak ties) are sufficient to dramatically decrease the average separation between nodes. Analytically, Barabasi measured this clustering effect with power-law distributions, showing varying power law exponents for networks, such as movies (by their actors especially Kevin Bacon), members of an audience (through auditory cues), social systems (family ties, school ties, friendships, etc.), biological organisms (biochemical signals), the brain (neural interconnections), and especially, the Internet. In fact, there have been a series of studies of the structure and topology of Internet-based networks best summarized in [25], including the web, email, instant messaging, virus/worm propagation, and P2P networks. Before this work, identifying and quantifying the scale-free nature of the Internet, every new algorithm proposed by researchers for improving network performance was typically tested on random networks generated by consensus tools (such as the Waxman Network Topology Generator*), which in retrospect, resulted in incorrect solutions, which should now be re-examined.

One set of concepts related to OSNs are strong ties, weak ties, and latent ties [49]. Some research explains why relationships in OSNs are weak ties. The positive effect of it is: "communities of interest are defined by their worldviews, and whenever a community of interest rigorously exposes its worldview in a fashion that permits its knowledge to be federated with the worldviews and knowledge of other communities, the whole human family is enriched"—Steven Newcomb. The research designed for positive social change also found that OSNs may differ in purpose, but their architectures and interactive patterns share a lot in common [50]. Such topics have been extensively studied in theoretical works, such as complex network theory [51].

18.6.2 ERP

Finally, we will explicit some systems, which are not OSNs, but share some characteristics. ERP systems, such as customer relationship management system and human resources management system, are basically role-driven. An user gets a role and a responsibility for the quality of the data in the process. There is no consistency in relationships. They are initiated by an individual sending a request for participation in a narrowly defined project, and would be forwarded based on expressed affinities and the recommendations of trusted third parties. The resulting ad hoc community would dissolve with the completion of the stated objective.

Another similar business system is groupware, which is a software systems for collaborating within a group, such as email, calendaring, text chat, and wiki. Despite the notion of collaborative work systems, which are conceived as any form of human organization that emerges any time collaboration takes place, whether it is formal or informal, intentional or unintentional. In normal terms, it is business software and not public accessible OSNs.

18.7 Conclusion

The work described above is an ongoing dialogue for both practitioners and researchers. New social network services emerge every day. The platforms we analyzed adjust themselves continuously. Methodologically, we can only make causal claims, is limited by a

---

* http://www.math.uu.se/research/telecom/software/stgraphs.html\.

snapshot of the development of OSNs. Our work surveys the web services combining social networks and information system, and leverages the advantages of each type of system. We noticed that most current ONSs implemented only very simple models of social networking and cannot mirror the richness of real-world complexity. On the other hand, due to either technology restrictions or business concerns, the big players in the market cannot, or would not, open their platform to achieve the full potential of OSNs. We hope our survey can advocate a future research agenda to melt the gap.

References

1. Cross, R., Parker, A., Borgatti, S. P. A bird's-eye view: Using social network analysis to improve knowledge creation and sharing. *Knowledge Direct* 2000; 2(1): 48–61.
2. Loscalzo, S., Yu, L. Social network analysis: Tasks and tools, 2008; 151–159.
3. Scott, J. P. *Social Network Analysis: A Handbook*, Sage Publications Ltd, May 2000.
4. Granovetter, M. S. The strength of weak ties. *Am J Sociol* 1973; 78(6): 1360–1380.
5. Boyd, D., Ellison, N. Social network sites: Definition, history, and scholarship. *J Comput-Mediat Commun* 2008; 13(1): 210–230.
6. Smith, M., Kollock, P. *Communities in Cyberspace*, 1 ed., Routledge, February 1999.
7. Ziegler, C-N., Lausen, G. Propagation models for trust and distrust in social networks. *Information Syst Frontiers* 2005; 7(4-5): 337–358.
8. Haythornthwaite, C. Strong, weak, and latent ties and the impact of new media. *The Information Soc* 2002; 18: 385–401.
9. Biocca, F., Harms, C., Burgoon, J. K. Toward a more robust theory and measure of social presence: Review and suggested criteria. *Presence: Teleoperators Virtual Environ* 2003; *12(5)*: 456–480.
10. Java, A., Song, X., Finin, T., Tseng, B. Why we twitter: understanding microblogging usage and communities, in WebKDD/SNAKDD'07: *Proceedings of the 9th WebKDD and 1st SNA-KDD 2007 Workshop on Web Mining and Social Network Analysis*, New York, NY, USA, 2007, ACM, pp. 56–65.
11. Cacioppo, J. T., William, P. *Loneliness: Human Nature and the Need for Social Connection*. Norton Press, New York, 2008.
12. Oreilly, T. What is web 2.0: Design patterns and business models for the next generation of software. Social Science Research Network Working Paper Series, August 2007.
13. Bentley, F., Metcalf, C. J. The use of mobile social presence. *IEEE Perv Comput* 2009; 8(4): 35–41.
14. Short, J., Williams, E., Christie, B. *The Social Psychology of Telecommunications*. John Wiley and Sons Ltd, 1976.
15. Chung, D., Debuys, B., Nam, C. Influence of avatar creation on attitude, empathy, presence, and para-social interaction, 2007; 711–720.
16. Stets, J. E., Burke, P. J. Identity theory and social identity theory. *Social Psychol Q* 2000; 63(3): 224–237.
17. Boyd, D. M. Friendster and publicly articulated social networking, in *CHI '04: CHI '04 Extended Abstracts on Human factors in Computing Systems*, New York, NY, USA, 2004, ACM Press, pp. 1279–1282.

18. Baatarjav, E-A., Dantu, R., Phithakkitnukoon, S. *Privacy Management for Facebook*, 2008, pp. 273–286.

19. Backstrom, L., Dwork, C., Kleinberg, J. Wherefore art thour3579x?: anonymized social networks, hidden patterns, and structural steganography, in *WWW '07: Proceedings of the 16th International Conference on World Wide Web*, New York, NY, USA, 2007, ACM, pp. 181–190.

20. Liu, K., Terzi, E. Towards identity anonymization on graphs, in *SIGMOD '08: Proceedings of the 2008 ACM SIGMOD Internationa Conference on Management of Data*, New York, NY, USA, 2008, ACM, pp. 93–106.

21. McCown, F., Nelson, M. L. What happens when facebook is gone? in *JCDL '09: Proceedings of the 9th ACM/IEEE-CS Joint Conference on Digital Libraries*, New York, NY, USA, June 2009, ACM, pp. 251–254.

22. Jordan, K., Hauser, J., Foster S. The augmented social network.

23. Brzozowski, M. J., Hogg, T., Szabo, G. Friends and foes: ideological social networking, in *CHI '08: Proceeding of the Twentysixth Annual SIGCHI Conference on Human Factors in Computing Systems*, New York, NY, USA, 2008, ACM, pp. 817–820.

24. Levien, R., Attack-resistant trust metrics, 2009, 121–132.

25. O'Donovan, J. Capturing trust in social web applications, 2009, 213–257.

26. Pitsilis, G., Marshall, L. Modeling trust for recommender systems using similarity metrics, 2008, 103–118.

27. Ruohomaa, S., Kutvonen, L. *Trust Management Survey*, vol. 3477, 2005, pp. 77–92.

28. Guha, R., Kumar, R., Raghavan, P., Tomkins, A. Propagation of trust and distrust, in *WWW '04: Proceedings of the 13th International Conference on World Wide Web*, New York, NY, USA, 2004, ACM, pp. 403–412.

29. Kamvar, S. D., Schlosser, M. T., Garcia-Molina, H. The eigentrust algorithm for reputation management in p2p networks, in *Proceedings of the 12th International Conference on World Wide Web (WWW '03)* (New York, NY, USA, 2003), ACM, pp. 640–651.

30. Page, L., Brin, S., Motwani, R., Winograd, T. The pagerank citation ranking: Bringing order to the web, Technical report, Stanford Digital Library Technologies Project, 1998.

31. Massa, P., Avesani, P. Controversial users demand local trust metrics: an experimental study on epinions.com community, 2005.

32. Chi, E. H. Information seeking can be social. *Computer* 2009; 42(3): 42–46.

33. Chakrabarti, S. *Mining the Web: Discovering Knowledge from Hypertext Data*, 1st ed., Data Management Systems, Morgan Kaufmann, August 2002.

34. Travers, J., Milgram, S. An experimental study of the small world problem. *Sociometry* 1969; 32(4): 425–443.

35. Adamic, L. A., Adar, E. How to search a social network. 2004.

36. Kleinberg, J. The small-world phenomenon: An algorithmic perspective. Technical report, Ithaca, NY, USA, 1999.

37. Kleinberg, J. Complex networks and decentralized search algorithms, in *International Congress of Mathematicians (ICM)*, 2006.

38. Hu, M., Lim, E-P., Sun, A., Lauw, H. W., Vuong, B-Q. Measuring article quality in wikipedia: models and evaluation, in *CIKM'07: Proceedings of the 16th ACM Conference on Conference on Information and Knowledge Management* (New York, NY, USA, 2007), ACM, pp. 243–252.

39. Chen, F., Scripps, J., Tan, P-N. Link mining for a social bookmarking website, in *WI-IAT '08: Proceedings of the 2008 IEEE/WIC/ACM International Conference on Web Intelligence and Intelligent Agent Technology* (Washington, DC, USA, 2008), IEEE Computer Society, pp. 169–175.

40. Kiryakov, A., Popov, B., Terziev, I., Manov, D., Ognyanoff, D. Semantic annotation, indexing, and retrieval. *Web Semantics: Sci, Serv Agents on the World Wide Web* 2004; 2(1): 49–79.

41. Wu, X., Zhang, L., Yu, Y. Exploring social annotations for the semantic web, in *WWW '06: Proceedings of the 15th International Conference on World Wide Web*, New York, NY, USA, 2006, ACM, pp. 417–426.

42. Cress, U., Kimmerle, J. A systemic and cognitive view on collaborative knowledge building with wikis. *Int J Computer-Supported Collab Learning* 2008; 3(2): 105–122.

43. Hoisl, B., Aigner, W., Miksch, S. Social rewarding in wiki systems motivating the community, 2007; 362–371.

44. Morris, M. E. Social networks as health feedback displays. *IEEE Internet Comput* 2005; 9(5): 29–37.

45. Arrow, H., Mcgrath, J. E., Berdahl, J. L. *Small Groups as Complex Systems: Formation, Coordination, Development, and Adaptation*, 1st ed., Sage Publications, Inc, March 2000.

46. *Lin, N. Social Capital: A Theory of Social Structure and Action, Cambridge University Press, 2001.*

47. Mislove, A., Koppula, H. S., Gummadi, K. P., Druschel, P., Bhattacharjee, B. Growth of the flickr social network, in *WOSP '08: Proceedings of the first workshop on Online social networks*, New York, NY, USA, 2008, ACM, pp. 25–30.

48. Milgram, S. The small world problem. *Psychol Today* 1967; 2: 60–67.

49. Viswanath, B., Mislove, A., Cha, M., Gummadi, K. P. On the evolution of user interaction in facebook, in *WOSN '09: Proceedings of the 2nd ACM workshop on Online social networks*, New York, NY, USA, 2009, ACM, pp. 37–42.

50. Kumar, R., Novak, J., Tomkins, A. Structure and evolution of online social networks, in *KDD '06: Proceedings of the 12th ACM SIGKDD International Conference on Knowledge Discovery and Data Mining*, New York, NY, USA, 2006, ACM, pp. 611–617.

51. Mislove, A., Marcon, M., Gummadi, K. P., Druschel, P., Bhattacharjee, B. Measurement and analysis of online social networks, in *IMC '07: Proceedings of the 7th ACM SIGCOMM Conference on Internet Measurement*, New York, NY, USA, 2007, ACM, pp. 29–42.

19

Pervasive Application Development Approaches and Pitfalls

Guruduth Banavar
*IBM India Research
Laboratory*

Norman Cohen
*Thomas J. Watson Research
Center*

Danny Soroker
*Thomas J. Watson Research
Center*

19.1 What are Pervasive Applications?

The vision of pervasive computing has been written about extensively. In a nutshell, pervasive computing is about enabling users to gain access to the relevant applications and data at any location and on any device, in a manner that is customized to the user and the task at hand. This fundamentally takes computing off the desktop and into the spaces that we live in everyday. Mark Weiser [1] called it "invisible" computing. This vision of pervasive computing leads to two fundamental characteristics of pervasive applications—*mobility* and *context-awareness*. Both of these characteristics are a result of the extremely dynamic nature of pervasive computing environments.

Mobility has three implications. First, applications must run on a wide variety of devices, including the devices embedded in various environments and devices carried by users. Second, because devices may be transported to locations were a high-bandwidth network connection is not available, applications must work (perhaps in a degraded mode) with low-bandwidth network connections or in the absence of any network connection. Third, applications that make use of a user's location must account for the possibility that the location will change.

The need for context-aware applications arises because pervasive computing makes applications available in contexts other than a computer workstation with a keyboard, mouse, and screen. The users of a pervasive computing application will typically be focused upon some task other than the use of a computing device and may even be unaware that they are using a computing device. Applications must customize themselves to interact with a user in a manner appropriate to the user's current context and activities, exploiting locally available devices, without distracting the user from the task at hand.

In the simplest case, a mobile application is any stand-alone application that can execute on a mobile device. However, the more interesting and useful case is an application that is networked to other software components executing at different points in the network infrastructure. Ideally, an application is hosted on the network and is able to execute on any device. In this case, the application must be written in such a way so as to be able to execute on multiple software platform architectures and in a manner that exploits the user-interface characteristics of multiple device platforms. Although we have made great strides in network connectivity, not all devices and locations support continuous network connectivity. Thus, applications should support disconnected and weakly connected operation. In summary, supporting mobility implies two major technological requirements—supporting device platform heterogeneity and supporting network heterogeneity.

A context-aware application is one that is sensitive to the environment in which it is being used (e.g., the location or the particular user of the application). The application can use this information to customize itself to the particular location or the user. This implies the following technological requirements:

- Identifying and binding to data sources that provide the right information.
- Composing the information from these sources to create information that is useful for an application.
- Using that information in meaningful ways within the application itself.

As a simple example, a pervasive calendar application will have the following features. First, the application will be able to run on multiple device platforms, from a networked phone (with a limited user interface and limited bandwidth, but always connected) to a smart personal digital assistant (PDA) (with a richer user interface and higher bandwidth, but not always connected) to a conference room computer (with a very rich user interface and very high bandwidth and always connected). Furthermore, I (as the user) should be able to interact with this application using multiple user-interface modalities, such as a graphical user interface (GUI), a voice interface, or a combination of the two. Second, the application will be sensitive to the environment in which it is running; for example, if I bring up the calendar at home, the application might bring up my family calendar by default. If I bring up my calendar in my office when I am almost late for my next meeting, the application might bring up my work calendar with the information about my next meeting highlighted.

In this chapter, we discuss the software engineering challenges and approaches to building pervasive applications with the characteristics mentioned above. This chapter considers the application developer's point of view (as opposed to the infrastructure developer's) and discusses the programming models and tools that can support pervasive application development. The purpose of this chapter is not to propose new techniques for addressing development issues, but rather to summarize some of the promising approaches already being developed and to point out some of the pitfalls that these approaches are trying to avoid. The software infrastructure elements for supporting the execution of such applications are discussed elsewhere [2–5].

19.1.1 Basic Concepts and Terms

A *multidevice application* is one that is able to execute on devices with different capabilities. A *multimodal application* is one that supports multiple user-interface modalities such as GUI, voice, and a combination of both.

In this chapter, we consider an application model in which the application is partitioned according to the well-known MVC application structure [6]. The *view* represents the presentation, and the *controller* represents the application flow, including the navigation, validation, error handling, and event handling. The view and the controller together deal with the user interaction of the application. The *model* component includes the application logic as well as the data underlying the application logic.

In this chapter, we consider only networked applications, because they represent the bulk of interesting and useful pervasive applications. In these networked applications, the application components described above are distributed across two or more physical computers with a network connection between them. A *device platform* is the distributed software platform to which a pervasive application is targeted. A *thinclient application* is a networked application in which the user-interface rendering component is executing on the user's device, whereas the rest of the application is executing on a networked computer. A *thickclient application*, on the other hand, has significant application components executing on the user's device. A *disconnectable application* is one that is able to continue to execute when there are different levels of connectivity between the different components of the application.

The attributes of the environment of an application are referred to as the *context* of the application. The context of an application includes some of the user's significant attributes, such as location, destination, the identities of other people in the vicinity, and the attributes of the task being performed, such as the objective and the artifacts necessary for the task. A *contextaware application* is one that is able to sense some aspects of the environment in which it is executing and adapt its behavior to the sensed environment.

19.2 Why is it Difficult to Develop Pervasive Applications?

There are fundamental reasons why pervasive application development is more difficult than conventional application development. One reason is the heterogeneity of environments in which a pervasive application must be able to execute. The other reason is the need for applications to adapt to dynamic environments. These reasons are discussed in more detail in this section, as are the software engineering issues that arise from them.

19.2.1 Heterogeneity of Device Platforms

End-user devices, such as smart phones and PDAs, come in many varieties and have widely varying capabilities, both hardware (form factor, user-interface hardware, processor, memory, and network bandwidth) and software (operating system, user-interface software, services, and applications). These capabilities are so varied and broad that there are industry standards (e.g., Composite Capabilities/Preferences Profiles [CC/PP] and User Agent Profile [UAProf], by the W3C Consortium [7]) being developed for describing the capabilities of individual devices. There are commercial offerings that support and maintain several hundred device profiles, with new devices being introduced at the rate of more than one every week at the current time. Furthermore, the number of applications that need to support a non-trivial number of these devices is on the rise.

The impact of device heterogeneity on application developers is that applications need to be developed (or ported) to each device and maintained separately for each device. In terms of the MVC Application Model described before, the following sections describe the specific impacts of device heterogeneity.

19.2.1.1 User Interface

The capabilities of the user interface include the output capabilities, such as the screen characteristics (e.g., size and color); the input capabilities, such as the number of hard buttons, rollers, and other controls; and the software toolkit available to manipulate these input and output capabilities. Because of the differences in these capabilities from one device to another, the view component of an application will have to be rewritten for each device. In some cases, the structure of the view will also impact the structure of the controller.

19.2.1.2 Interaction Modalities

Informally, an interaction modality is a significant method of user interaction that leverages a user's natural or learned ability. Examples are keyboard or mouse, speech, pen, and tactile interfaces. (In this chapter, we consider primarily keyboard or mouse and speech.) The view and controller portions of applications may need to be significantly rewritten to enable each modality. For example, a speech-based application could have a different structure from a GUI-based application.

Furthermore, *multimodal* interfaces can use multiple modalities within a single application. For example, a single application may use GUI and speech modalities to reap the benefits of both modalities—GUI for rapid interaction and speech for eyes-free and hands-free operation. Writing such an application requires synchronizing the two modalities, so that when a particular utterance is played, the corresponding elements are displayed on the screen. This synchronization requires careful attention by the application developer.

19.2.1.3 Platform Capabilities

The *software platform* for a device is the distributed software infrastructure on which an application executes, including the device software infrastructure and the server software infrastructure. In many cases, the programming models on the device and the server are different, for example, a Java™-based web programming model on the server and a C-based Application Processing Interface (API) on the device. Even if the programming models are the same on the device and the server, an application may need to be partitioned differently between the device and the server depending on the processor, memory, and network capabilities of a device.

19.2.1.4 Connectivity

If an application needs to execute in a dynamic environment that supports multiple levels of connectivity, the application developer needs to worry about dynamically varying the partitioning of the application between the various connectivity scenarios and resynchronizing partitioned components after re-establishing connectivity. This adds a significant amount of complexity to the application development process.

19.2.1.5 Development and Maintenance Complexity

To summarize the above discussion, there are several software engineering challenges in writing pervasive applications. Consider the development scenario for a pervasive application targeted to N device platforms. In the worst case, this requires one to build N different versions of the application. If the application is targeted to O devices that support multimodal interaction, there will be further complexity in developing versions of the application for separate modalities and for synchronizing the application across those modalities. There may be P different partitions of the application to support various platform and connectivity characteristics. Thus, the worst case development complexity for a single application is a factor of $(N + O) \times P$ times the complexity of developing the application for a single platform. This results in a significant increase in the developer time, which is the costliest resource in a software development organization. Maintenance of the application (i.e., fixing bugs and making enhancements) has a similar complexity.

This complexity is fundamentally a scalability issue. Conventional application development methodologies do not scale for the large numbers of devices and platforms that are in existence today. To address this issue, new methods of reusing application components are being developed. These will be discussed later in this chapter.

19.2.2 Dynamics of Application Environments

In describing our vision earlier, we stated that pervasive applications should be customized to the user and task at hand—also referred to as the context of the application. The context can be highly dynamic. The data sources that provide information about the application's environment are called *context sources*. Consider the complexities of application development in the face of dynamic and heterogeneous context sources.

The context data from different context sources could have different schemas and formats. For example, location data from a cell tower is different from the location data from an IEEE® 802.11 base station. If each pervasive application that uses context data was responsible for collecting and normalizing context data from different sources, applications would indeed be quite complex.

The context information from any one source could be of very low level to be useful for an application. For example, if an application is interested in knowing whether Jane is at lunch, it is not enough to know Jane's exact latitude or longitude, but also how that latitude or longitude corresponds to a building's map (also known as *geocoding*). If Jane's exact status is not available, it may be possible to determine whether she is at lunch from other context sources, such as the time of day, her calendar, the lights in her office, the activity on her computer, and knowledge of her normal habits. Combining these lower level forms of context into a higher level notion of "Is Jane at lunch?" should not be the responsibility of the application that requires that information.

The actual context sources themselves could be highly dynamic. For example, the location of a person can be obtained by a multitude of sources, including a cell tower, a telematics gateway, a wireless local area network (LAN) hub, and an activebadge access point. Each of these sources of location may have a different API and may be more or less applicable to different locations. Applications should not be responsible for discovering these context sources and explicitly binding to them.

In summary, the complexity of using dynamic context information boils down to the question of division of responsibility between the application and a reusable infrastructure. The reusable functions of mediation [8] (including normalization, composition, and binding) should be supported by the infrastructure. The application should only be responsible for implementing the business logic, given the high-level context event.

19.3　Approaches for Developing Pervasive Applications

19.3.1　Developing Mobile Applications

Mobile applications may be stand-alone applications that run on mobile devices; they may be networked applications executing partly on the mobile device and partly in a networked server environment (which, by the way, does not imply that network connectivity is always available). This chapter focuses on the latter variety, because it is more relevant to realizing the pervasive computing vision. Web-based applications, whether they are browser-based or use stand-alone renderers who access web services on a network, are examples of this kind of application. To understand the most common approaches to developing such mobile applications, let us keep in mind the MVC decomposition of an application.

As described earlier, the basic problem of mobile application development to multiple devices, modalities, and connectivity environments is that of complexity, because the same application may have to be rewritten multiple times. The following is a discussion of the approaches that are being used to address this problem.

19.3.2 Presentation Transcoding

An early approach to making web applications accessible via multiple devices was transcoding. The basic idea behind transcoding is to repurpose the existing content written
for one device, say a desktop personal computer (PC), to different devices, via an automated runtime transcoder, typically on a server. This might involve parsing the presentation, typically represented in Hypertext Markup Language (HTML) and converting it
into a markup language that is understood by a web browser on the device, such as
Website META Language (WML) [9] or compact HTML (cHTML) [10]. In this process,
images and other multimedia content may also be transformed into a format that can be
handled by the target device.

This approach works to a limited extent, but has not been widely adopted in the industry. There are several reasons for the limited success of this approach:

- The input does not convey the full semantics of the content, but only the presentation, so transcoders can do no more than reformat the content in ways that are
  usable and pleasing to the end-user on different devices.
- Content authors have little to no control on how a web page is displayed on a
  device.
- Content providers are usually protective of their content and do not want runtime
  intermediaries to alter the carefully tailored presentation that was originally
  designed.

Enhancements to the basic idea of transcoding included the ability for the developer
to *annotate* the content with some of the semantics behind the content. Although this
may be reasonable in some cases where there is static content, this notion breaks down
when there is dynamic content. Transcoding was not widely adopted because it fundamentally does not handle the deeper structure and semantics of applications.

19.3.3 Device-Independent View Component

A more widely used approach evolved, in which the view aspects of an application are
conveyed in a device-independent representation. This device-independent representation describes the intent behind the user interaction within a view component (such as a
page), rather than the actual physical representation of a user-interface control. For
example, the fact that an application requires users to input their ages is represented by
a generic INPUT element with a range constraint; an adaptation engine determines,
based on the target device characteristics, usability considerations, or user preferences,
whether the INPUT element should be realized as a text field, a selection list, or even
voice input. Several device-independent view representations have evolved over the
years, including User-Interface Markup Language (UIML) [11], Abstract User-Interface
Markup Language (AUIML, previously known as Druid) [12], XForms [13], and
Microsoft® ASP.NET Mobile Controls [14].

19.3.3.1 Runtime Adaptation

This device-independent representation is typically converted to a device-specific representation via some kind of automatic runtime adaptation. The runtime adaptation

engine gets the device identifier via the request header of a web application (specifically, the user agent field) and maps that to a database record containing detailed device information. The information in this database record guides the adaptation of the device-independent representation to device-specific representations. Microsoft, Oracle, and Volantis have commercial products using some variation of runtime adaptation.

One of the pitfalls of this approach is to rely entirely on automatic runtime adaptation of the device-independent representation. Fully automatic adaptation can work in certain cases: when the content is simple or when the device variations are not too great. However, experience shows that it is extremely difficult for fully automatic adaptation to produce highly customized and usable interfaces that are comparable with handcrafted user interfaces. This is especially true in modern, highly interactive applications. As a result, most successful systems that use this technique provide a way for developers to provide additional information to guide or augment the runtime adaptation process. The extra information can take several forms:

- Meta-information (e.g., where to split content into multiple pages)
- Style information (e.g., templates and style attributes to use for different devices or classes of devices)
- Code modules that plug into the runtime adaptation engine and alter its behavior for particular target devices

19.3.3.2 Design Time Adaptation

Design time adaptation is a technique that converts the device-independent representation to device-specific representations before the application is deployed to the runtime. The result of design time adaptation is a set of target-specific artifacts that can be viewed and manipulated by the developer. At the end of this process, the developer ends up with a set of target-specific view components, similar to the components that a developer would have built by hand [15]. There are two major advantages to this approach:

- The developer has full control over the adaptation process and the generated artifacts. If the developer is not satisfied with the output, the process can be rerun with different parameters, until the result is satisfactory. The generated artifacts can also be manipulated to add device-specific capabilities for particular devices.
- There is no runtime performance overhead for translating applications, because the translations have occurred at design time.

In the design time adaptation technique, applications are converted from a higher level to a lower level representation and the generated representation can be manipulated by the developer. In this scenario, if the developer modifies the higher level representation and regenerates the application, it is critical that the changes made previously to the lower level representation be preserved. This preservation of changes to a generated artifact after the artifact is regenerated is called "roundtrip" [16]. Failure to enable roundtrip is a potential pitfall of the design time adaptation technique. There are multiple ways to support roundtrip. One way is to provide markers in the generated artifacts that indicate where the developer can make modifications. The developer modifications made within these markers are left untouched by the generation process. The other approach is to capture the history of changes to a generated artifact and to provide the capability to reapply these changes selectively to regenerated artifacts.

Design time adaptation alone cannot be relied upon, for two reasons:

1. Design time adaptation supports only devices that were known at design time. If there are new devices that need to be supported after an application has been deployed, it may not be reasonable to depend on the application provider to target those devices via the design time tool.
2. For dynamic content (again, that will be unknown at design time), it is necessary to have some level of runtime adaptation.

For these reasons, some systems, such as Multi-Device Authoring Tool (MDAT) [17] support a hybrid of design time and runtime adaptations. Design time adaptation results in one or more device-specific application versions that can be deployed to a web application server. Additionally, devices can be classified into a hierarchy of device categories (e.g., PDAs, phones, color phones, and so on) and the application can be adapted at the design time according to this classification. When a device requests the application, the runtime web application dispatcher determines if the request can be satisfied by an existing device-specific application version or whether it falls into a category that has been defined. If not, a device-specific version of the application is generated on the fly and delivered to the device. Thus, runtime adaptation allows MDAT to service requests from devices that do not have a predefined device-specific application version.

19.3.3.3 Visual Tools for Constructing Device-Independent Views

Regardless of the adaptation technique used, systems supporting device-independent views also provide a number of integrated development environment tools for authoring the device-independent content and for specifying the additional kinds of information described above. Consider a visual design tool for developing device-independent content. Typically, visual design tools for developing concrete device-specific content support the well-known "What You See Is What You Get (WYSIWYG)" paradigm. One pitfall that a visual design tool for device-independent content can fall into is to attempt to support WYSIWYG capability. In a device-independent content tool, what the user sees is not what the user is going to get in general, because only a single device can be emulated in the interface. The user may be tempted to customize the design for one particular device, rather than thinking about the overall intent that is appropriate for all devices. A device-independent representation should thus be editable in an editor that displays a generic logical representation that conveys the relationships among elements, such as order, grouping, and any layout hints that may be specified. These issues are discussed in detail by Lawrence and coworkers [18].

19.3.4 Platform-Independent Controller Component

The section above discussed adaptation of the view component of an application to multiple devices. As described earlier, the controller of an application represents the control flow, including data validation and error handling, typically via event-handlers. To address the full range of applications, it is necessary to consider the role of the controller in modern interactive applications. There are several reasons why the controller of an application needs to be targeted to multiple devices:

- Different devices may have different input hardware, ranging from a keyboard, tracking device, and microphone on a PC to a pair of buttons and a scrolling wheel on a wristwatch.
- The flow of an application may be different on different devices. For example, an application that contains a secure transaction may not support this transaction on a device that does not have the appropriate level of security infrastructure. Similarly, an application that supports rich content may choose to skip those pages on devices that are not capable of presenting rich content.
- When a device-independent page is adapted and rendered on multiple devices, the page may be split into multiple device-specific pages for any device that is too small to contain the entire page.
- The controller execution framework may be different for different device platforms. Recall that a device platform is the end-to-end distributed platform that supports the execution of all components of the application. One device platform may support a Java-based Apache Struts™ framework, whereas another may support a different framework such as the base servlet framework, or a different language altogether, such as PHP or C#.

As a result, a complete solution for targeting multiple devices must include the application controller. One approach [17] is to represent the controller in a declarative way using a generic graph representation, where the nodes are device-independent pages and the arcs are control flow transitions from one page to another. This representation addresses the three requirements above as follows:

1. Developers can modify the flow of the application for particular target devices. These are represented as incremental changes to the generic controller.
2. When a device-independent page is split into multiple pages, the appropriate controller elements to navigate among those pages are also automatically generated.
3. The concrete controller code for specific controller platforms (e.g., Apache Struts) is automatically generated from the declarative controller representation. The specific controller framework can be changed as necessary.

19.3.5 Host-Independent Model Component

The above sections discussed approaches for targeting the view and controller components of an application to multiple devices. In this section, we discuss how to deal with the heterogeneity of connectivity environments.

Networked mobile applications vary in the distribution of logic and data between the mobile device and the server, as illustrated in Figure 19.1.

In a thinclient application, views are generated on the server and then rendered on the client device by a component such as a web browser. Controller logic, model logic, and model data all reside on the server, so disconnected operation is impossible. In a thickclient application, the model still resides on a server, perhaps accessed through web

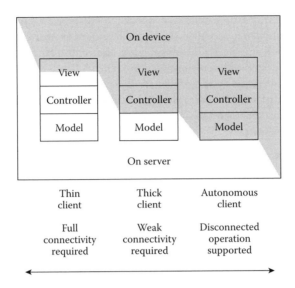

FIGURE 19.1   Distribution of logic and data between the mobile device and the server.

services, but the rest of the application resides on the client device. Caching of data before connection and queuing of updates to be performed upon reconnection enable limited forms of offline operation in a weakly connected environment. The operations allowed are those that can proceed sensibly in the absence of a complete and current model. An autonomousclient application resides entirely on the client device. It maintains its own fully functional model, which may be synchronized from time to time with replicas of the model on a server. As the arrow at the bottom of the diagram suggests, thinclient, thickclient, and autonomousclient applications represent points on a continuum rather than three clearly delineated categories. For example, some nearly autonomous applications have a disconnected mode that closely resembles the connected mode, except that updates made to the model are considered tentative until the model is synchronized with a server-based replica.

Thickclient applications have been supported with varying levels of success. The main drawback of the thickclient approach is that it may not support all the needed functions to support rich interactions in disconnected mode, because the model component is missing on the mobile device. The autonomousclient approach, on the other hand, is the most general technique, because it can support varying levels of connectivity. The remainder of this section is concerned with autonomousclient applications.

A key consideration here is the programming model used for supporting disconnectable applications. We need a programming model that allows the model components of an application (like the view and the controller components) to be shared by multiple versions of a disconnectable application. In this case, we are concerned with connected and disconnected versions of the application.

In the ideal scenario, the logic and the data for the model component is specified once and the tools and infrastructure supporting the programming model extract the right subset of the logic and data for the disconnected mode on each supported device. In reality, this extraction process will likely need to be guided extensively by the developer. The developer will likely specify the model, view, and controller in a generic way (view and controller as described in previous sections). The tools will enable the developer to incrementally refine this generic representation to particular target environments. This is an ongoing area of work and there are significant issues that need to be resolved.

It should be noted that there is a significant level of runtime infrastructure needed for disconnectable applications:

- An application hosting and execution environment is needed on the mobile device.
- If application code is to be downloaded from the server to clients upon demand, a code migration component is needed on the server and device sides to coordinate the partitioning and loading of application components.
- A data synchronization component is needed for updating both the device and server instances of the application with changes to the data on the other sites and to resolve any possible conflicts.

There are difficult architectural and policy issues in the above infrastructure components. A discussion of these issues is beyond the scope of this chapter.

19.3.6 Developing Context-Aware Applications

One can think of a context-aware application as having a triggering aspect and an effecting aspect [19]. The triggering aspect binds to data sources, collects data, analyzes the data, and ensures that the data are relevant to the application. If so, it notifies the effecting aspect, which takes the action corresponding to the trigger. For example, in an application that invokes a computer backup facility when a user Jane is away from her computer, the event that "Jane is at lunch" is the trigger and the act of invoking the backup utility is the effect.

Recall that context-aware applications have three sources of complexity:

1. The heterogeneous nature of data sources.
2. The dynamic nature of context sources.
3. The multiple sources of potentially low-level context data.

Observe that these are all in the triggering component of applications. Current approaches to addressing these issues have focused on creating a reusable infrastructure (middleware or toolkit) that exposes a programming model that hides these complexities [4]. This approach is summarized below.

19.3.7 Source-Independent Context Data

An application obtaining data from heterogeneous sources with inconsistent availability and quality of service should not name a specific source of data. Rather, it should describe the kind of data that is required, so that the underlying infrastructure can discover an appropriate source for the data. This approach, known as *descriptive, datacentric,* or *intentional* naming [20–22] has a number of advantages. It allows the system to select the best available source of data, based on current conditions. If the selected source should fail, the infrastructure can rebind to another source satisfying the same description, thus making the application more robust. New data sources satisfying a description can be introduced, or old data sources removed, without modifying the application; likewise, the application can be ported to an environment in which there is a different set of sources for the described data.

The basic idea of this approach is for an application to specify the desired context data without specifying the exact location and data type of the source, or whether it is coming from multiple sources. These are considerations that will be handled transparently by the infrastructure. In some cases, the infrastructure may discover a data source, such as a device or a web service that directly provides the described data. For example, suppose an application specifies that it is interested in a Boolean value for "Is Jane at lunch?" The infrastructure may discover a data source that directly reports whether Jane's location is the cafeteria. Alternatively, the infrastructure may discover a programmed component, called a *composer* in Norman et al. [4], which computes the described data from other data. In our example, some combination of Jane's calendar, office status, and computer status might be combined by a composer to determine with a degree of certainty whether she is at lunch. A composer may be reusable across multiple applications and may itself be built on top of other composers that handle lower level, more generic, data. For example, the query "Is X at lunch?" could be answered using the answer to a query of the form "Is X located at Y?" and queries of that form might themselves be answered by consulting multiple sources of location data (e.g., active badge, 802.11, or cell tower) with different resolutions, and inferring a composite location with a certain degree of confidence.

Once a composer that can answer the question "Is X at lunch?" is written and added to the infrastructure, it can be reused by all context-aware applications. A composer is itself a data source, just like a sensor, a web service, or a database, and may be discovered by the infrastructure in response to a query for data satisfying a given description.

Some data sources, such as request response web services, are passive or pull-based. Other data sources, such as sensors that trigger alarms, are active or push-based. Flexible infrastructure is capable of discovering both kinds of data sources. An application can then pull the current value from a passive data source or subscribe to be notified each time an active data source generates a new value.

This application development model presents several challenges. One challenge is to define a model for the computations performed in retrieving data from pervasive sources. Another is to provide the application developer with a simple but powerful means for specifying the behavior of a composer. Still another is to devise an appropriate system for describing datasource requirements. The remainder of this section addresses each of these three challenges.

A wide variety of computation models has been proposed. Some systems, such as Tapestry [23] from the Xerox Palo Alto Research Center and Cougar [24] from Cornell University, view sensor data as being added to an append-only database and use Structured Query Language-like models to retrieve the data. In contrast,

NiagaraCQ [25] defines data-retrieval compositions in terms of continuous queries over XML infosets, specified in an XQuery-like language. The Rome system [26] from Stanford University and the Solar system [27] from Dartmouth University specify composer-like entities called *triggers* and *operators*, respectively. Both presume that all data sources are passive. The iQueue computation model [4] from IBM Research allows a composer to obtain input from lower level data sources, including both passive and active sources, and allows the composer itself to act as either a passive or an active source; this model is based on an expression that is evaluated whenever data are pulled from the composer or whenever one of the composer's input sources pushes a new value.

The means for specifying the behavior of a composer depends, of course, on the underlying computation model. For the expression-based model of Norman and coworkers [4], the appropriate specification is the expression itself. The language iQL, described by Norman and colleagues [5], is specifically tailored to the kinds of expressions that are useful in writing composers.

The description of datasource requirements poses a difficult challenge because of the wide variety of data sources. Different kinds of data sources have different interesting attributes and new kinds of data sources are continually being invented. It is untenable to adopt a fixed vocabulary of kinds of data in which current applications are interested, let alone those in which next year's applications will be interested. However, it is feasible to categorize each new data source registered with the infrastructure as belonging to a specified provider kind that can be named in a descriptive query. Some new data sources can be categorized as belonging to providers of an existing kind, although new provider kinds will have to be registered for other data sources. Provider kinds can be categorized in a superkind–subkind hierarchy, such that all attributes of a provider kind are inherited by its subkinds. A query for a provider of kind k can be satisfied by a provider of any subkind of k.

19.4 Conclusions

This chapter has discussed the key difficulties in writing pervasive applications—those that support mobility and context-awareness—and summarized the main approaches that are currently being employed to address these difficulties. The key issue is application development complexity to deal with heterogeneous devices, varying degrees of connectivity, and dynamic data sources. Reuse of application components is the fundamental means of addressing this complexity.

Four basic approaches to enhancing reuse were discussed, based on the well-known MVC application structure:

1. *Device-Independent Views*: These allow an application to capture the basic interaction structure that should be reused across multiple devices and modalities. They should be combined with the ability to fine-tune the presentation when necessary.
2. *Platform-Independent Controllers*: These allow an application to specify the overall control flow across multiple execution platforms, but still allow an application to have different control flow structures for different devices and uses.

3. *Host-Independent Models*: These allow an application to encapsulate the business logic and data in a manner that can be reused regardless of which host a component is instantiated on.

4. *Source-Independent Context Data*: This allows an application to specify the intended context data to be supplied by reusable infrastructure components, which in turn are concerned with the specific data formats, locations, and combinations of physical data sources that provide the actual data.

These approaches have reached different levels of maturity (interestingly, the above order represents the highest to lowest in terms of maturity) in research projects and commercial offerings. Several challenges remain before these approaches can become widely useful.

Acknowledgments

This chapter is a compendium of many ideas that have evolved from projects and discussions with many individuals in the pervasive computing group at IBM, including Jeremy Sussman, Larry Bergman, Rich Cardone, Shinichi Hirose, Andreas Schade, and Apratim Purakayastha.

References

1. Weiser, M., The computer for the twenty-first century. *Scient Am*, 1991; 94–104.
2. Banavar, G., Beck, J., Gluzberg, E., Munson, J., Sussman, J.B., and Zukowski, D., Challenges: An application model for pervasive computing, in *MOBICOM 2000*, Boston, MA, USA, pp. 266–274, 2000.
3. Banavar, G. and Bernstein, A., Software infrastructure and design challenges for ubiquitous computing applications. *CACM*, 2002; 45(12): 92–96.
4. Cohen, N.H., Purakayastha, A., Wong, L., and Yeh, D.L., iQueue: A pervasive data-composition framework, in *Proceedings of the 3rd International Conference on Mobile Data Management*, Singapore, 8–11 January 2002, pp. 146–153.
5. Cohen, N.H., Lei, H., Castro, P., Davis, II J.S., and Purakayastha, A., Composing pervasive data using iQL, in *Proceedings of the 4th IEEE Workshop on Mobile Computing Systems and Applications (WMCSA 2002)*, Callicoon, NY, 20–21 June 2002, pp. 94–104.
6. Krasner, G.E. and Pope, S.T., A cookbook for using the model view controller user interface paradigm in smalltalk80. *J Object Orient Programm*, 1988; 1(3): 26–49.
7. Klyne, G., Reynolds, F., Woodrow, C., Ohto, H., Hjelm, J., Butler, M.H., and Tran, L. (Eds). *Composite Capability/Preference Profiles (CC/PP): Structure and Vocabularies 1.0. W3C Proposed Recommendation*. 2004. http://www.w3.org/TR/CCPP-structvocab/ (15 October 2003).
8. Wiederhold, G., Mediators in the architecture of future information systems. *IEEE Comp* 1992; 25(3): 38–49.
9. Website META Language. http://thewml.org/ (19 October 2002).

10. Kamada, T., Compact HTML for small information appliances. W3C Note. http://www.w3.org/TR/1998/NOTE-compactHTML-19980209/ (9 February 1998).

11. Abrams, M., Phanouriou, C., Batongbacal, A.L., Williams, S.M., and Shuster, J.E., UIML: An appliance-independent XML user interface language. *WWW8/Computer Netw*, 1999; 31(11–16): 1695–1708.

12. Merrick, R.A., Defining user interfaces in XML, in *Proceedings of the POSC Annual Meeting*, London, England, 28–30 September 1999 (http://www.posc.org/notes/sep99/sep99_rm.pdf)

13. Dubinko, M., Klotz, Jr. L.L., Merrick, R., and Raman, T.V. (Eds). *XForms 1.0. W3C recommendation*. 2003. http://www.w3.org/TR/xforms/ (14 October 2003).

14. Microsoft, Mobile Web development with ASP.NET, 2003. http://msdn.microsoft.com/mobility/prodtechinfo/devtools/asp.netmc/default.aspx

15. Bergman, L.D., Banavar, G., Soroker, D., and Sussman, J., Combining handcrafting and automatic generation of userinterfaces for pervasive devices, in *Proceedings of the 4th International Conference on ComputerAided Design of User Interfaces (CADUI 2002)*, Valenciennes, France, 15–17 May 2002, pp. 155–166.

16. Medvivovic, N., Egyed, A., and Rosenblum, D.S., Roundtrip software engineering using UML: From architecture to design and back, in *Proceedings of the 2nd Workshop on Object Oriented Reengineering (WOOR)*, Toulouse, France, September 1999, pp. 1–8.

17. Banavar, G., Bergman, L., Cardone, R., Chevalier, V., Gaeremynck, Y., Giraud, F., Halverson, C. et al., An authoring technology for multidevice Web applications. *IEEE Pervasive Comput*, 2004; 3(3), 83–93.

18. Bergman, L.D., Kichkaylo, T., Banavar, G., and Sussman, J.B., Pervasive application development and the WYSIWYG pitfall, in *EHCI 2001*, pp. 157–172.

19. Sow, DM., Olshefski, D.P., Beigi, M., and Banavar, G., Prefetching based on web usage mining, in *Middleware 2003*, pp. 262–281.

20. AdjieWinoto, W., Schwartz, E., Balakrishnan, H., and Lilley, J., The design and implementation of an intentional naming system, in *Proceedings of the 17th ACM Symposium on Operating Systems Principles (SOSP'99)*, 12–15 December 1999, Kiawah Island Resort, SC, published as *Operating Systems Review*, vol. 33, no. 5, pp. 186–201, 1999.

21. Bowman, M., Debray, S.K., and Peterson, L.L., Reasoning about naming systems. *ACM Trans Programm Lang Syst* 1993; 15(5): 795–825.

22. Intanagonwiwat, C., Govindan, R., and Estrin, D., Directed diffusion: a scalable and robust communication paradigm for sensor networks, in *Proceedings of the 6th Annual International Conference on Mobile Computing and Networking (MobiCom 2000)*, Boston, 6–11 August 2000, pp. 56–67.

23. Terry, D., Goldberg, D., Nichols, D., and Oki, B., Continuous queries over appendonly databases, in *Proceedings of the 1992 ACM SIGMOD International Conference on Management of Data*, San Diego, 2–5 June 1992, pp. 321–330.

24. Bonnet, P., Gehrke, J., and Seshadri, P., Querying the physical world. *IEEE Personal Commun 2000*; 7(5): 10–15.

25. Chen, J., DeWitt, D.J., Tian, F., and Wang, Y., NiagaraCQ: A scalable continuous query system for Internet databases, in *Proceedings of the 2000 ACM SIGMOD*

International Conference on Management of Data, Dallas, TX, USA, 15–18 May 2000, pp. 379–390.

26. Huang, A.C., Ling, B.C., Ponnekanti, S., and Fox, A., Pervasive computing: What is it good for? in *Proceedings of the International Workshop on Mobile Data Management*, Seattle, 20 August 1999, pp. 84–91.

27. Chen, G. and Kotz, D., Context aggregation and dissemination in ubiquitous computing systems, in *Proceedings of the 4th IEEE Workshop on Mobile Computing Systems and Applications (WMCSA 2002)*, Callicoon, NY, 20–21 June 2002, pp. 105–114.

20

Wireless Personal Area Networks: Protocols and Applications

Khaled A. Ali

Hussein T. Mouftah
University of Ottawa

20.1 Introduction

Wireless personal area networks (WPANs) are rapidly evolved short-range wireless communication paradigms. In the near future, such communication systems are expected to play an integral role in our daily activities from simple applications such as controlling a light switch, monitoring the temperature of an object, and so on, to complex applications such as Telehealth monitoring, Smart Grids, Home Automation, Industrial Automation, and so on. From one perspective, the WPANs basic functionalities entail connecting different communication devices together such as a computer and its peripherals. From another perspective, WPANs have another important field of application: sensing and monitoring different physical phenomena in their communication vicinity, processing, to a certain limit, the sensed information, and communicating such information to a remote monitoring and controlling site through other communication media. The other communication technologies which could be involved in the whole communication chain of a WPAN system are: wireless local area networks (WLANs), wireless cellular networks, and/or wide area networks. The internetworking of WPANs with these technologies facilitates the data communication of the short-range WPANs to remote controlling and monitoring locations for further analysis and decision-making processes.

WPANs are designed with a wide vision of applications ranging from low-rate communication systems capable of communicating information of data rates ranging from few bits/s to hundreds of kbits/s, to ultra wideband communication systems capable of supporting multimedia applications of high-quality audio and video streams having data rates ranging from few Mbits/s to a number of Gbits/s. The wide spectrum of such achievable data rates classifies WPANs into low and high data rate communication systems. These classifications place a number of constraints on the system designers which require careful considerations for defining robust, secure, and scalable physical layer and data link layer protocols capable of meeting the design objectives of such systems.

This chapter contains a tutorial-oriented description of WPAN systems such as Bluetooth and ZigBee networks. This tutorial is based on the WPAN standards of the Bluetooth Special Interest Group (SIG) and the *IEEE 802.15 Working Group*. Such standards define the network architecture, the physical layer, and the medium access control (MAC) sublayer of the WPAN systems. The chapter will be arranged into three parts. First, the Bluetooth network architecture and protocols are explained. Then, the IEEE high data rate WPAN system architecture and protocols are studied in the second part followed by a study for the IEEE low data rate WPAN system architecture and protocols. Then, a comprehensive study for a selected number of WPAN-enhanced MAC protocols is provided.

20.2 Bluetooth for WPANs

Bluetooth is a well-defined wireless technology for constructing a mobile ad hoc network. It can be exploited on small scales to build ad hoc WPANs, that is, networks that connect devices such as mobile phones with handsfree headsets, PCs with printers, etc., placed inside a space confined with a circle of 10 m radius [1–4]. Bluetooth was invented in 1994 by Ericsson. The standard is named after *Harald Blåtand "Bluetooth" II*, king of Denmark AD 940–981. The Bluetooth specification is released by the Bluetooth SIG founded by Ericsson, IBM, Intel, Nokia, and Toshiba in February 1998 to develop an open specification for short-range wireless connectivity. The group is now also promoted by 3COM, Microsoft, Alcatel-Lucent, and Motorola. More than 1900 companies have joined the SIG. In 2002, the IEEE 802.15 Working Group has derived the *IEEE 802.15.1* WPAN standard based on the Bluetooth v1.1 specifications. It includes an MAC and physical layer specification. An updated version of this standard, based upon the additions incorporated into Bluetooth v1.2, was published in 2005 citeWPAN05. The objective design of such systems is to have a low-cost, short-range wireless communication ÓÛÓÝËÉ, integrated into a microchip, enabling protected ad hoc connections for voice and data communication in stationary and mobile environments.

The Bluetooth networks are operating in the 2.4 GHz free industrial, scientific, and medical (ISM) band. This makes Bluetooth systems rely on a common transmission medium which requires the coordination of the network nodes' transmission by an efficient channel access mechanism. Such coordination is achieved by means of control information that is carried explicitly by control messages traveling along the medium. Specifically, the Bluetooth MAC protocol behaves as a polling system [2–4]. Hence, the MAC protocol scheduling algorithm by determining the order in which stations are polled is the main element to determine the network efficiency. The Bluetooth specification indicates a *Round Robin* scheduler as a possible solution, that is, each slave is polled in a consecutive order. The SIG is working to impose the Bluetooth technology as the de-facto wireless standard for wireless personal area communication.

20.2.1 Bluetooth Network Topology

A Bluetooth network is composed of a number of electronic devices communicate over a wireless transmission channel through a well-defined set of protocols. Bluetooth devices are generally organized into groups of two to eight devices called piconets, the-fundamental building block of a Bluetooth network, consisting of a single master device and one or more slave devices (Figure 20.1a) The master node of a piconet coordinates the communication with one or more slaves. In such coordination, the master device decides which slave is the one to have the access to the channel. More precisely, a slave is authorized to transmit a single packet to the master node only if it has received a polling message from the master node. A device may additionally belong to more than one piconet, either as a slave in both or as a master of one piconet and a slave in another. These bridge devices effectively connect piconets into a scatternet (Figure 20.1b).

20.2.2 Bluetooth Protocol Structure

The Bluetooth protocols are structured in a layered format as shown in Figure 20.2. This structure comprises Bluetooth radio layer, baseband layer, link management

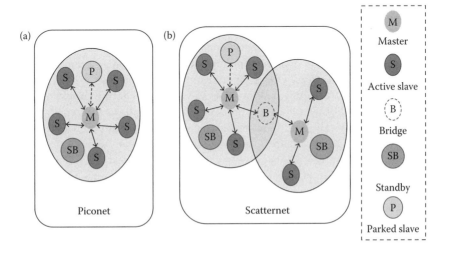

FIGURE 20.1　Bluetooth network topology.

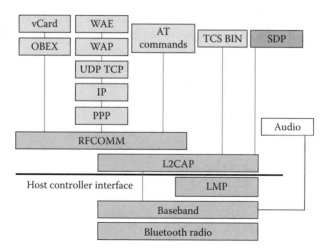

FIGURE 20.2   Bluetooth protocol stack.

protocol (LMP), and logical link control and adaptation protocol (L2CAP). The Bluetooth radio layer provides the physical links among Bluetooth devices, while the baseband layer provides a transport service of packets on the physical link. The LMP is responsible for the setup and management of physical links. The management of physical links consists of several activities such as putting a slave node in a particular operating state (i.e., sniff, hold, or park mode), monitoring the status of the physical channel, and assuring the negotiated quality of service (QoS) profiles such as the node transmission power level, the maximum poll interval, and so on. Also, LMP implements security procedures at the link level.

The upper layer Bluetooth protocols, service discovery protocol (SDP), telephony control protocol specification (TCS), and the RFCOMM protocol, are interfaced with the lower layer Bluetooth protocols through the L2CAP. SDP provides service discovery specific to the Bluetooth environment without inhibiting the use of other SDPs. RFCOMM is a simple transport protocol providing serial data transfer. A port emulation entity is used to map the communication application programming interface to RFCOMM services, effectively allowing legacy software to operate on a Bluetooth device. TCS is provided for voice and data call control, providing group management, capabilities and connectionless TCS, which allows for signaling unrelated to an ongoing call. Both point-to-point and point-to-multipoint signaling are supported using L2CAP channels, although actual voice or data are transferred directly to and from the baseband over synchronous connection oriented (SCO) links. In addition to this protocol layers, the specification also defines a host controller interface that provides a command interface to the baseband controller, linkmanager, etc. The other protocols presented in the figure are application-oriented protocols enabling different applications to run over Bluetooth devices. In the following subsections, we describe Bluetooth radio layer, baseband layer, LMP, and L2CAP in a detailed manner.

20.2.2.1 Bluetooth Radio Layer

In the standard, the physical layer of the Bluetooth is known as Bluetooth radio. Each Bluetooth device consists of a radio component operating in the 2.4 GHz band. This band is divided into 79 different radio frequency (RF) channels which are spaced 1 MHz apart. The frequency-hopping spread spectrum (FHSS) transmission technique is utilized for transmission between the master and slave nodes. The hopping sequence is a pseudo-random sequence of 79-hop length which gives a unique signature for each established ad-hoc network. The establishment of a physical channel is associated with the definition of a channel frequency-hopping sequence (FHS) which has a very long period length; however, it does not show repetitive patterns over a short time interval. This can be achieved by exploiting the actual value of the master node clock and its unique Bluetooth 48-bit device address.

The FHSS system is a robust transmission technique that helps Bluetooth devices coexist and operate reliably alongside other devices operating in the ISM band. Each Bluetooth piconet is synchronized to a specific frequency-hopping pattern. This pattern, moving through 1600 different frequencies per second, is unique to the particular piconet [4]. Each frequency hop is a time slot during which data packets are transferred. A packet may actually span up to five time slots, in which case the frequency remains constant for the duration of that transfer.

In Bluetooth systems, a time division duplex (TDD) transmission scheme is adopted for packet transmission between the master node and its slave nodes. In such scheme, the transmission channel is divided into $f(K)$ time slots, each of 625 μs, and each slot corresponds to different RF hop frequencies. The time slots are numbered according to the master node clock. Herein, the master node packet transmission is permitted in even numbered time slots while odd numbered time slots are reserved for active slaves' packet transmission. The TDD frame structure is shown in Figure 20.3. Basically, the transmission of a packet covers a single slot, but it may last up to five consecutive time slots. For multiple slots, packet transmission, the RF hop frequency for the entire packet transmission is unchangeable which is the one assigned at the first time slot when the packet

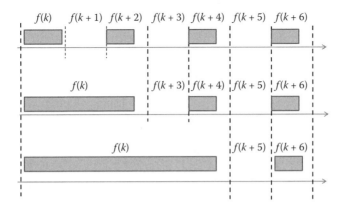

FIGURE 20.3 Bluetooth protocol stack.

transmission has begun. Changing the used RF after transmitting or receiving a packet reduces the imposed interference from other signals coming from other radio systems which are also utilizing the free ISM band. The Bluetooth transmission distance ranges from 10 cm to 10 m. As the device transmission power is increased, such range can be extended up to 100 m.

20.2.2.2 Bluetooth Baseband Layer

The baseband layer shown in the Bluetooth protocol stack is responsible for establishing and releasing the physical connections between the master and slave nodes, traffic transmission and reception over the physical channels, the synchronization of devices belonging to a piconet on master clock, and the management of different power saving states which the device can stay in. In the following, we will describe in depth the Bluetooth connection types and its channel access mechanisms.

20.2.2.2.1 Bluetooth Connection Types

The Bluetooth physical connections can be classified into SCO link and asynchronous connection less (ACL) link which can be established between a Bluetooth master and slave devices. The SCO physical link is a point-to-point, symmetric connection that can be established between the master and a specific slave. It is used for communicating delay-sensitive traffic, mainly voice. In fact, the SCO link transmission rate is 64 kbps. This transmission rate can be maintained by reserving a couple of consecutive slots for master-to-slave transmission and immediate slave-to-master response. The SCO link can be considered as a circuit-switched connection between the master and the slave. On the other hand, the ACL physical link is a connection between the master and all of its slaves that participate in the corresponding piconet. The ACL can be considered as a packet-switched connection between the Bluetooth devices that supports point-to-multipoint transmissions from the master to the slaves. The ACL channel guarantees the reliable delivery of data through the utilization of a fast automatic repeat request.

20.2.2.2.2 Channel Access Mechanism

The channel access mechanism in Bluetooth networks is managed according to a polling scheme. In such scheme, the master node decides which slave node is to have the access permission to the transmission channel by sending to it a packet. The master packet may contain data or can simply be a polling packet. When the slave receives a packet from the master it is authorized to transmit in the next time slot. For SCO links, the master periodically polls the corresponding slave. On the other hand, polling is asynchronous for ACL links. Figure 20.4 presents a possible pattern of transmissions in a piconet with a master and two slaves. Slave1 has established an SCO channel with the master for packets transmission and Slave2 has established an ACL link for communicating with the master node. The SCO link is periodically polled by the master every five slots, while ACL links are polled asynchronously. Furthermore, the size of the packets on an ACL link is constrained by the presence of SCO links. For example, in the figure, the master sends a multislot packet to Slave2, which, in turn, can reply with a single-slot packet only, because the successive slots are reserved for the SCO link.

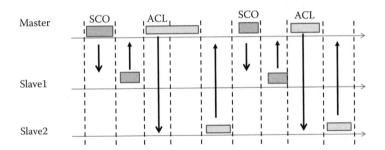

FIGURE 20.4   Bluetooth packet transmission on SCO and ACL channels.

20.2.2.3 Link Management Protocol

Each Bluetooth device has a link manager (LM) that is responsible for link setup, authentication, link configuration, power control, etc. It also discovers other remote LMs and communicates with them via the LMP. The LMP consists of a number of protocol data units (*PDUs*), which are sent from one device to another, determined by the active member address (AM_ADDR) in the packet header. LM PDUs are always sent as single-slot packets and the payload header is therefore 1 byte. To perform such deities, the LM uses the services of the underlying link controller (LC).

LMP packets, which are sent in the ACL payload, are differentiated from L2CAP packets by a bit in the ACL header. They are always sent as single-slot packets and have higher priority than L2CAP packets. This helps ensure the integrity of the link under high traffic demand. The LM translates the commands into operations at the Baseband level, managing the following operations:

- Connecting slaves to piconets and allocating their active member addresses
- Disconnecting slaves from a piconet
- Configuring the link including master/slave switches
- Establishing ACL and SCO links
- Putting connections into one of the low power modes: hold, sniff, and park
- Controlling test modes

20.2.2.3.1 ACL Link Setup

Figure 20.5 demonstrates the LM role in establishing an ACL link between a master and a slave in a Bluetooth network. The demonstration shows the messages involved in setting up an ACL connection. First, the LC layer must establish a link between both devices before LMP messages can be exchanged. This is done by paging the slave through sending an ID packet message from the master. The slave then responds by sending its ID to the master. The master then sends an FHS, so that the slave can adjust its clock accordingly. The slave then sends its ID to the master and now both devices enter into the connection setup state.

After the link establishment between the master and the slave, the master sends a *LMP_host_connection_req* message to the slave. The slave responds by either accepting

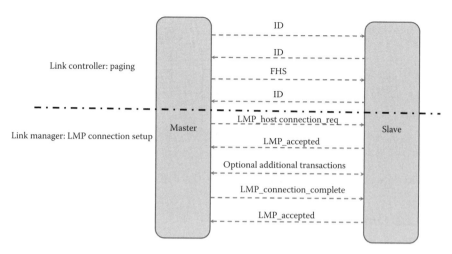

FIGURE 20.5 Link management protocol.

or rejecting the connection request using *LMP_accepted* message. The master and slave exchange more messages if there are any optional transactions remaining. Then the master sends a *LMP_connection complete* to the slave. The slave then responds with an *LMP_accepted*.

20.2.2.3.2 SCO Link Setup

After the establishment of the ACL connection, it can be used to carry the LM PDU for establishing the SCO links between the master and slave devices. As shown in Figure 20.6, the establishment of the SCO can be initiated by either the master or the slave device. To initiate an SCO connection setup phase, both master and slave use a *LMP_ SCO_req* message. However, when the SCO link request is initiated by the master device, the slave device replies a *LMP_Accepted* message which indicates the successful establishment of the requested SCO link (Figure 20.6a) On the other hand, when the slave device initiates the SCO link request message as shown in Figure 20.6b, the master device replies with a *LMP_SCO_req* message. Then the slave device replies with a *LMP_ Accepted* message.

20.2.2.4 Logical Link Control and Adaptation Protocol

The L2CAP provides group management mapping upper protocol groups to Bluetooth piconets, segmentation and reassembly of packets between layers, and negotiation and monitoring QoS between devices. The L2CAP link layer operates over an ACL connection provided by the baseband where a single ACL link, set up by the LM using LMP, is always available between the master and any active slave. This provides a point-to-multipoint link supporting both asynchronous and isochronous data transfer. There are three types of L2CAP channels exist: *bidirectional signaling channels* that carry commands; *connection-oriented channels* for bidirectional point-to-point connections; and *unidirectional connectionless channels* that support point-to-multipoint connections, allowing

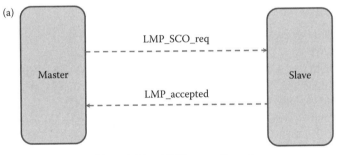

Master initiated SCO connection setup

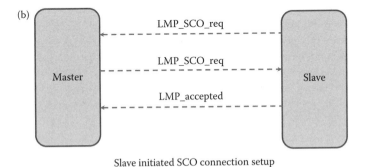

Slave initiated SCO connection setup

FIGURE 20.6 SCO link establishment.

a local L2CAP entity to be connected to a group of remote devices. In a summary, the main features supported by L2CAP are as follows:

- Protocol multiplexing: the L2CAP uses a protocol-type field to distinguish between upper layer protocols.
- Segmentation and reassembly: this feature uses 2 b in the payload header and is required to adapt upper layers packet sizes to the Baseband packet size.
- Group management, groups of addresses, or devices onto piconets.
- QoS management through the exchanging and monitoring QoS parameters.

Bluetooth is currently envisioning almost every electronic device such as mobile phones, digital camera, laptops, and a whole range of other electronic devices. As a result, the market is going to demand new innovative applications, value-added services, end-to-end solutions, and much more. The possibilities opened up really are limitless, and because the RF used is globally available, Bluetooth can offer fast and secure access to wireless connectivity all over the world. However, the power consumption of the Bluetooth devices is considered an obstacle for using such system in other fields such as wireless sensor networks (WSNs) and disposable WPANs. Such limitation has driven the IEEE community to establish a low-power WPAN communication system, which is introduced in the following sections.

20.3 High-Rate WPANs

High-rate WPANs (HR-WPANs) characterized by its high data rate transmission that ranges from 11 Mb/s to 55 Mb/s. It is well suited for many applications which require large bandwidth to sustain their large traffic. These applications may include the distribution of real-time video generated by digital cameras and camcorders that can store multi-megabytes image files and video streams in digital format. Also, a high-quality audio and video can be carried by HR-WPANs. One such popular application is wireless playback feature of a digital camcorder on a TV screen. Another application for HR-WPAN can be found in the area of interactive gaming which is built around 3D graphics and high-quality audio. HR-WPAN can be used to establish the wireless links between multiplayer game consoles and high-definition displays.

The HR-WPANs network topology and the network nodes classifications and functionalities are explained in Section 20.3.1. Then, we explain the layered architecture of the network nodes in Section 20.3.2. The standard HR-WPANs MAC protocol is explained in Section 20.3.3.

20.3.1 HR-WPAN Network Topology

An HR-WPAN topology consists of several nodes that implement a full or a subset of the standard. The HR-WPAN is called a piconet and the network nodes are known as devices (DEVs) [5]. The formation of a piconet requires one node to assume the role of a piconet coordinator (PNC) which provides synchronization for the piconet nodes to the periodically transmitted PNC beacon frames, support QoS, and manage nodal power control and channel access control mechanisms. The HR-WPAN topology is created in an ad-hoc manner which requires no preplanning and nodes can join and leave the network unconditionally.

There are three types of piconet topologies which are defined in the standard:

- Independent piconet
- Parent piconet
- Dependent piconet

The network topologies of these piconets are shown in Figure 20.7, and their characteristics, the objective of their creation, and their node functionalities are detailed in the following.

20.3.1.1 Independent Piconet

An independent piconet is a stand-alone HR-WPAN which consists of a single network coordinator and one or more network nodes. The operation of this piconet type does not depend on other HR-WPANs. The network coordinator manages the network operation through the periodically transmitted beacon frames in which other network nodes use to synchronize with to be able to communicate with the network coordinator as well as performing peer-to-peer communication. The independent piconet topology is shown in Figure 20.7 which can be inferred by its independent PNC. The independent piconet

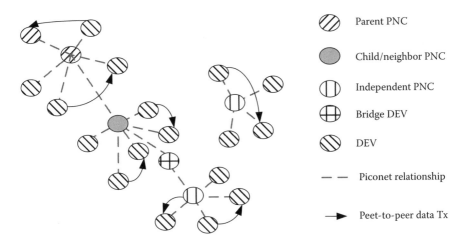

FIGURE 20.7   HR-WPAN network topology.

can communicate with another adjacent independent piconet through a bridge node which is also shown in Figure 20.7.

20.3.1.2 Parent Piconet

A parent piconet is an HR-WPAN that controls the functionality of one or more other piconets. In addition to managing communication of its network nodes, the network coordinator of the parent piconet controls the operation of one or more dependent network coordinators.

20.3.1.3 Dependent Piconet

Dependent piconets are classified according to the purpose of their creation into "Child" piconet and "Neighbor" piconet. A child piconet is created by a node from a parent piconet to extend network coverage and/or to provide computational and memory resources to the parent piconet. The resources allocation such as channel time allocation (CTA) for network nodes in such piconet is controlled by the parent network coordinator node.

The other dependent piconet, the neighbor piconet, is created by the neighbor PNC which is not a member of the parent piconet. The purpose of associating a neighbor piconet with a parent piconet is the lack of a vacant frequency channel for the creation of an independent piconet. Therefore, a neighbor piconet can be created and the radio channel resources of such network will be allocated by the parent network coordinator.

20.3.2 HR-WPAN Node Architecture

The reference model of the HR-WPANs node architecture is shown in Figure 20.8. This model defines the architecture in two parts, namely: node structural blocks part and node management part. The node structural blocks part is mainly composed of three

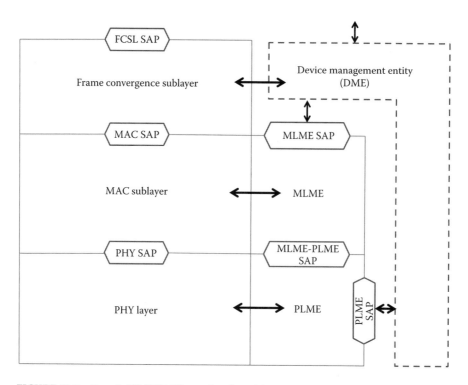

FIGURE 20.8   Generic HR-WPAN layered node architecture.

layers: frame convergence sublayer (FCSL), MAC sublayer, and physical (PHY) layer. The HR-WPANs standard is defined only for the MAC and PHY layers where each layer implements a subset of the standard and offers services to its upper layers and gets services from its lower layers. The FCSL interfaces the MAC sublayer to the upper layers such as the networking layer, application layer, and so on.

The management part of an HR-WPAN node consists of the device management entity (DME). The DME facilitates the functionalities of the MAC and PHY layers and other upper layers. It is a layer-independent entity and can be resided in a separate management plane. The detailed functionality of the DME is not specified by the standard. However, gathering layer-dependent status from the management entities of different layers and setting the values of layer-specific parameters are examples of DME duties.

The DME is interfaced to the MAC sublayer and the PHY layer through designated service access points (SAPs) of the MAC sublayer management entity (MLME) and the PHY layer management entity (PLME), respectively. The DME should be present in each node for correct node operation.

The services provided by the physical and MAC layers are briefly explained below.

20.3.2.1 Physical Layer Services

The physical layer of the HR-WPANs defines the procedures of transmitting and receiving data between two or more network devices through a wireless channel. The physical layer contains two functional entities, namely: PHY function and PLME function (Figure 20.8). The PHY layer services are provided to the MAC sublayer through the PHY's SAP. The main tasks of the PHY layer are the activation and deactivation of the radio transceiver, link quality indication (LQI), clear channel assessment (CCA), and transmitting as well as receiving data packets over the physical medium.

The LQI is a measurement task performed by the physical layer. The LQI is reported for the Trellis-coded modulation coded QAM modes using the estimation of the signal-to-noise ratio. LQI is a characterization of the strength and the quality of a received signal. The use of LQI result is up to the network or application layers.

The CCA is used to evaluate the wireless channel activities before a node can start data transmission. This mechanism is used to minimize packet collisions. Therefore, a node with queued data is required to first sense the transmission medium for a random period of time through the CCA procedure. If the PHY layer informs the MAC sublayer the medium is idle, the node starts data transmission. This process of waiting before transmission is known as backoff process. The backoff procedure is not applied to the transmission of the periodically transmitted beacon frames by the PNC node.

20.3.2.2 MAC Sublayer Services

The MAC sublayer of the HR-WPANs is designed to achieve a set of goals. These goals are: supporting fast connection time, ad hoc networks topology, QoS support, dynamic node membership, efficient data transfer, and secure data communication. The MAC sublayer achieves these goals through two services: the MAC data service and the MAC management service. For data communication, the MAC sublayer communicates with the FCSL through the MAC SAP and being serviced by the PHY layer through the PHY SAP. The management entity of the MAC sublayer; MAC MLME communicates with the DME through the MLME SAPs. The features of the MAC sublayer are beacon management, channel access control through the carrier sense multiple access with collision avoidance (CSMA/CA) scheme, collision-free CTA for management information and data communication, frame validation, acknowledged frame delivery, and node association and disassociation.

20.3.2.3 Node Channelization

Before a PNC can start the formation of an HR-WPAN, it has to select an available channel from the allocated frequency channels defined in the HR-WPANs standard. The HR-WPAN operates on an ISM frequency spectrum ranges from 2.4 to 2.4835 GHz. This spectrum is subdivided into five frequency channels which are grouped into two sets: high-density set of four channels and 802.11b coexistence set of three channels. If there are no WLANs detected in the vicinity of the PNC node, the high-density mode which allows the formation of up to four simultaneous HR-WPANs is selected. On the other hand, since each of the two center channels of the high-density mode overlaps with two WLANs channels, the mphPNC node selects the 802.11b coexistence mode to exist with the detected WLAN. The 802.11b coexistence mode aligns these channels so the effect of the created HR-WPAN operation on the detected WLAN is minimized.

20.3.3 HR-WPANs Standard MAC Protocol

The MAC mechanism for HR-WPANs is based on the MAC protocol detailed in the *IEEE* standard of HR-WPANs [5]. In this protocol, the channel time is subdivided into time slots called superframes. The structure of each superframe is presented in Figure 20.9 which is composed of Beacon Period (BP), an optional CAP, and CTA Period *(CTAP)*. During the *BP*, the coordinator sends a beacon frames for conveying CTA and other management information to other network nodes.

Access to the channel during the CAP is performed through the slotted CSMA/CA scheme. This period is used to transmit commands and asynchronous data. In the CAP, a wireless node is considered in a synchronization mode with the network coordinator when it receives and decodes the network coordinator beacon frames. A synchronized node with buffered packets for transmission first it has to backoff for a random period of time. Then, it senses the wireless channel using the CCA mechanism explained above in the first time slot located at the end of its current backoff period. If the transmission medium has been sensed idle, the node starts transmission in the following time slot. If the channel is busy, the corresponding node has to wait for another random backoff period before it starts reprobing the transmission channel.

The CTAP is the period in which HR-WPAN network nodes exchange data. The CTAP is divided into management CTAs (MCTAs) and generic CTAs. The MCTAs are used for communication between the network nodes and their network coordinator, while the CTA is used by peer nodes to exchange isochronous and asynchronous data.

Whenever a network node needs to start a collision-free multimedia stream transmission; first, it sends a channel time request in the CAP to the network coordinator and specifies the required duration of the requested allocation. The network coordinator then sends an acknowledgement (ACK) to the corresponding node and later sends a channel time response indicating if the request has been accepted or rejected. If accepted, the network coordinator includes also the position of the specified CTA within the CTAP. The node must send an ACK frame in response. Then, all the transmitted packets must be successfully acknowledged by the coordinator.

The Beacon-enabled MAC frame structure is efficient in terms of minimizing nodal power consumption and providing multimedia support in WSN in general and WPANs in particular. The IEEE 802.15.3 MAC standard has been extended for ultrawideband (UWB) systems in which Aloha access mechanism is proposed in addition to the slotted CSMA/CA channel access scheme [6].

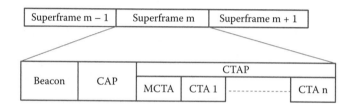

FIGURE 20.9 HR-WPAN frame structure.

20.4 Low-Rate WPANs

The most obvious characteristic of low-rate WPANs (LR-WPAN) is its data throughput, which ranges from 20 to 250 kb/s. Many applications require only limited bandwidth since they are only generating small amounts of data. These applications, such as monitoring and controlling industrial equipment, require long battery life so that the existing maintenance schedules of the monitored equipment are not compromised [7]. Other applications, such as environmental monitoring over large areas, may require a large number of devices in which battery replacement of such devices is impractical. Another promising application for LR-WPANs is in the field of medical application where patient's health conditions can be monitored remotely via an on- or in-body sensor nodes [8,9]. In such applications, health information from a human body can be sensed using a special medical sensors and communicated via LR-WPAN, which is overlied over other communication technologies such as WLANs, to a remote health site where it can be diagnosed by health authorities.

Herein, LR-WPAN architecture as presented in the IEEE standard [10] is described in this section. First, we explain the network topology and the node classifications and functionalities for LR-WPANs in Section 20.4.1. Then, the LR-WPANs nodal architecture is explained in Section 20.4.2. The allocated spectrum for the LR-WPANs is outlined in Section 20.4.3.

20.4.1 LR-WPAN Network Topology

The IEEE 802.15.4 standard has defined two different network topologies to suite different application requirements, namely star and peer-to-peer topologies. The star network topology is shown Figure 20.10a and the peer-to-peer network topology is shown in Figure 20.10b. These network topologies consist of different devices which are classified by the standard into full function device (FFD) and reduced function device (RFD). The

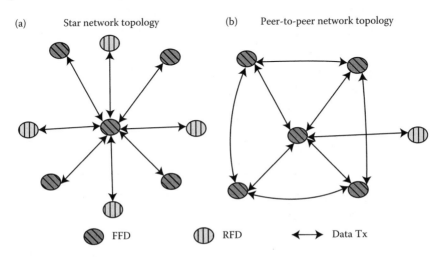

FIGURE 20.10 LR-WPAN network topology.

FFD implements all features of the IEEE 802.15.4 standard and can serve as a regular device or as a network controller known as personal area network (PAN) coordinator. Also, the FFD can serve as a relaying node in an LR-WPAN mesh network topology or as an end node performing only specific duty. On the other hand, RFD implements the minimum required functionalities by the standard and can only function as an end node such as temperature sensor or a light switch which communicates its sensed information only to the PAN coordinator of its network. The characteristics and the targeted applications of each network topology are given below.

20.4.1.1 Star Network Topology

The star network topology is composed of a PAN coordinator and a number of FFD and/ or RFD nodes. When beacon-enabled MAC frame structure is used, network devices synchronize their communication with the PAN coordinator through the periodical transmission of the coordinator's beacon frames and they use slotted CSMA/CA for accessing the transmission channels. Also, the PAN coordinator allocates, upon nodal request, collision-free time slots for supporting mission-critical services. However, when nonbeacon-enabled frame structure is used, there is no beacon frames transmission by the coordinator and the unslotted CSMA/CA is used instead for accessing the transmission channel.

The communication in the star network topology is established between network devices and the PAN coordinator. A network node with an associated application can be the communication initiation or termination point. Also, the PAN coordinator may have a specific application but also can be initiation and/or termination communication point as well as routing communication in the network. A variety of applications can benefit from the star network topology. Some of these applications are healthcare applications, personal computer peripherals, home automation, and games.

20.4.1.2 Peer-to-Peer Network Topology

A peer-to-peer LR-WPAN topology is an ad hoc, self-organizing, and self-healing network topology. It is also formed around a PAN coordinator but differs from the star network topology in which any device may have a peer-to-peer communication with any other device in the network through a single hop or multiple hops. Complex network topologies can be created which facilitate the formation of mesh networking topology. Only nonbeacon-enabled MAC frames with unslotted CSMA/CA channel access mechanism are used in the peer-to-peer network topology. The targeted applications for peer-to-peer network topology are industrial control, monitoring, WSNs, inventory tracing, etc.

20.4.2 LR-WPAN Node Architecture

Figure 20.11 shows a generic LR-WPAN node architecture. The node architecture is divided into a number of structural blocks called layers. Each layer implements a subset of the LR-WPAN standard and offers services to its upper layers and gets services from its lower layers. The layered architecture of each network node comprises PHY layer and MAC sublayer. On top of these layers is the service-specific convergence sublayer which interfaces the MAC sublayer with the logical link control sublayer and other upper layers

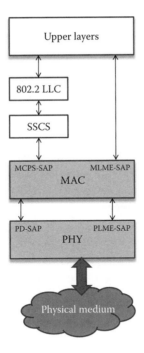

FIGURE 20.11 Generic WPAN layered node architecture.

such as the networking layer, application layer, and so on. The LR-WPANs standards are defined only for the physical layer and MAC sublayer while other layers' specifications are undefined in the standards. The services provided by the physical and MAC layers are briefly explained below.

20.4.2.1 Physical Layer Services

The physical layer provides two services: the PHY data service and PHY management service interfacing with the PLME. The PHY data service enables the transmission and reception of PHY PDUs (PPDUs) across the physical radio channel. The main tasks of the PHY layer are the activation and deactivation of the radio transceiver, channel Energy Detection (ED), LQI, CCA, and transmitting as well as receiving data packets over the physical medium.

The ED measurement estimates the received signal power within the bandwidth of the LR-WPAN channel during both channel scanning phase and packet reception phase on an active physical channel. The measured value is used by channel selection algorithm at the networking layer.

The LQI is another measurement task performed by the physical layer for each received packet. It is a characterization of the strength and the quality of a received packet. The measurement may be implemented using the ED mechanism or the signal-to-noise estimation procedure or a combination of these methods. The use of LQI result is up to the network or application layers.

The CCA task is performed by the PHY layer to evaluate the current transmission channel status: busy or idle. Three modes are used to perform (CCA) task. These modes are:

- *CCA Mode* 1: *energy above threshold:* A busy medium signal is reported when the channel's detected energy level is above the ED threshold.
- *CCA Mode* 2: *carrier sense only:* A busy medium signal is reported upon the detection of a signal with the same modulation and spreading characteristics of the PHY that is currently in use by the device even if the detected signal is below the ED threshold.
- *CCA Mode* 3: *carrier sense with energy above threshold*: A busy medium signal is reported only when carrier sense with energy level above the ED threshold.

20.4.2.2 MAC Sublayer Services

The MAC sublayer provides two services: the MAC data service and the MAC management service interfacing to the MAC MLME-SAP. The MAC data service enables the transmission and reception of MAC PDUs across the PHY data service. The features of MAC sublayer are beacon management, channel access control through the CSMA/CA scheme, collision-free time slots management, frame validation, acknowledged frame delivery, and node association and disassociation.

20.4.2.3 LR-WPAN Node Channelization

Before a PAN coordinator node can start the formation of an LR-WPAN, it has to select an available channel from the frequency channels defined in the LR-WPANs standards. The LR-WPANs are allocated 49 frequency channels in three different ISM frequency bands: 3 channels in the 868 MHz band, 30 channels in the 915 MHz band, and 16 channels in the 2.4 GHz frequency band. Typically, a coordinator node selects one of these channels for the formation of its LR-WPAN and stays on it.

20.4.3 LR-WPANs Standard MAC Protocol

The MAC mechanism for LR-WPANs is based on the MAC protocol detailed in the IEEE standard of LR-WPANs [10]. In this standard, two alternative MAC protocol operation modes are defined: nonbeacon-enabled and beacon-enabled. The nonbeacon-enabled MAC operation mode is explained in Section 20.4.3.1. The detailed beacon-enabled MAC frame structure and operation is explained in Section 20.4.3.2.

20.4.3.1 LR-WPAN Nonbeacon-Enabled MAC Protocol Mode

When the nonbeacon-enabled MAC operation mode is activated, unslotted CSMA/CA protocol is used to control the access to the transmission medium among the competing nodes. Therefore, a node with buffered frames at the MAC queue has to wait for a random backoff period before sensing the transmission channel using the CCA mechanism. If the channel is sensed ideal, frame transmission will commence at the end of the time slot of the current backoff period. Otherwise, the node has to wait for another random backoff period before reattempting sensing the channel condition. The contention-free time slots are not enabled in the nonbeacon-enabled MAC operation mode.

20.4.3.2 LR-WPAN Beacon-Enabled MAC Protocol Mode

The beacon-enabled MAC frame structure specified in the standard is divided into BP, CAP, contention-free period (CFP), and an optional inactive period, where network nodes may enter a sleep mode to reduce power consumption, as shown in Figure 20.12 [10,11]. To form the so-called superframe structure as shown in Figure 20.12, the beacon frame is transmitted periodically by the PAN coordinator without performing a carrier sense operation. The Beacons are also used to synchronize the attached nodes, to define the LR-WPAN, and to detail the description of the superframe structure. The beacon-enabled MAC framestructure is efficient in terms of minimizing nodal power consumption and providing multimedia support in WSN in general and LR-WPANs in particular.

The length of the superframe and the relative length of its active period are configurable and called beacon interval (BI). The BI length is given by: $BI = aBaseSuperframeDuration \cdot 2^{BO}$, and $BO \in \{0, 1, \ldots, 14\}$ is the configurable beacon order (BO). The duration of the active period of the superframe is also configurable and known as superframe duration (SD) which is calculated as: $SD = aBaseSuperframeDuration \cdot 2^{SO}$, where the configurable superframe order (SO) ranges from $0 \le SO \le BO \le 14$. The standard value of the aBaseSuperframeDuration is 960 symbols.

The CAP is divided into equally spaced time slots which immediately commence after the BP. Wireless nodes with packets to transmit are allowed to contend for the transmission medium at the beginning of each time slot according to the above-mentioned slotted CSMA/CA channel access mechanism.

The standard defines three variables, which need to be maintained by the CSMA/CA algorithm for each frame before its successful transmission, namely: number of experienced backoff stages (NB), the current backoff exponent (BE), and the contention window (CW) for defining the number of successful CCA which is required before starting the actual data transmission. The default values of these parameters are defined in the standard [10].

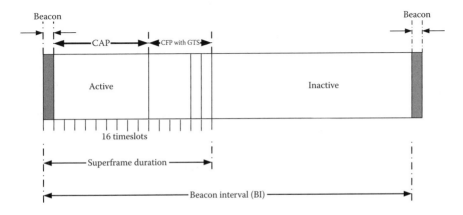

FIGURE 20.12   LR-WPAN frame structure.

TABLE 20.1 HR-WPAN and LR-WPAN Main Characteristics

	HR-WPAN	LR-WPAN
Dara rate	11–55 Mb/s	20–250 kb/s
Transmission range	10–100 m	10–100 m
Battery life	Moderate life	Long life
Topology	Mesh	Star and mesh
Complexity	Low	Low
Types for traffic	Multimedia traffic	Data-centric traffic
Desired frequency band	Unlicensed band (2.4 GHz)	Unlicensed band (2.4 GHz, 868/915 MHz)
MAC frame	Beacon-enabled	Beacon-enabled and nonbeacon

According to the standard, a node with a frame to transmit in the CAP is required to backoff a random number of frame slots and set the CW to two [10,11]. At the end of the backoff period, the node performs the first CCA, and if it is successful, the CW is decremented by one and the node performs the second CCA. If both CCAs were successful, CW is decremented to zero and frame transmission starts at the consecutive frame slot. On the other hand, if one of the CCA failed, the NB parameter is increased by one. If the new NB value is below a predefined threshold, CW is reset to two and another backoff stage is started. Otherwise, the frame is dropped and channel access failure is reported to higher layers. Such channel access mechanism seems to be fair as long as the the generated network traffic has the same priority level. However, when the network traffic has different priority levels, as in multimedia applications, high priority traffic can be reserved collision free time slots in the CFP.

In the CFP, the PAN coordinator reserves a number of time slots, which is called guaranteed time slots (GTS), for a node to perform a collision-free transmission. Each node has to place a request to the PAN coordinator in the CAP for the required number of time slots before it can transmit in the CFP. Then, the PAN coordinator informs the corresponding node of the number of granted time slots, which will be exclusively used for traffic transmission by that node, and the starting transmission slot to be used in the next frame. Therefore, QoS support for different multimedia applications can be achieved.

The CAP and the CFP together form the active portion of the superframe. In the standard [10], the active portion of the superframe is divided into 48 backoff slots where 15 of them are occupied by the CFP. The IEEE 802.15.4 MAC standard also has been extended for UWB systems in which Aloha access mechanism is proposed in addition to the slotted CSMA/CA channel access scheme [6].

Table 20.1 shows a general summary of the main characteristics of HR-WPANs compared with the ones for LR-WPANs.

20.5 WPAN MAC Protocols Performance Evaluation

A selected number of papers from the literature which mathematically or through extensive simulation analyze and enhance the performance of the proposed MAC protocols for the WPANs are reviewed in this section.

A mathematical model for evaluating the performance of the CAP of the IEEE 802.15.4 MAC protocol has been proposed in [12]. The model considers the channel sensing and data transmission for any network nodeover a given time interval as a renewal process by arguing that each node resets its parameters to the default values after each transmission trial or when it senses the medium busy. In such process, access to the radio channel in the CAP is modeled as a three-level renewal process as shown in Figure 20.13.

Each renewal process level is modeled into consecutive cycles. At level$_1$, a cycle starts whenever a tagged node starts stage 0 of its backoff period. The level$_1$ cycle length is X which can be either of type X_1 or X_2 as shown in Figure 20.13. Type X_1 cycle includes no transmission from the corresponding node due to the maximum use of its consecutive failures in accessing the channel which is labeled by the symbol x in the figure. The X_2 cycle contains data transmission by the tagged node after sensing the channel is idle. A level$_2$ renewal cycle Y lies between the ending point of two consecutive X_2 cycles at level$_1$. The Y cycles are classified into $Y1$ cycles in which data transmission of the tagged node collided with another network transmission at X_2 cycle, and Y_2 cycle in the case of successful data transmission at X_2 cycle. At level$_3$, the renewal cycle Z starts at the end of a Y_2 renewal cycle and ends at the end of the next Y_2 renewal cycle. The successful frame transmission of a tagged node in the Z cycle can be viewed as the reward for the level$_3$

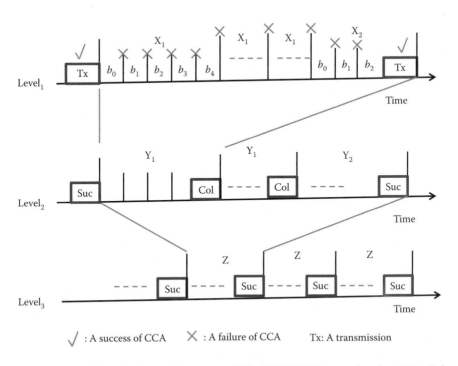

FIGURE 20.13 Three-level renewal process model for WPAN MAC protocol analysis. *Note:* Col: A collision, Suc: Successful Tx.

renewal cycle. Therefore, the tagged node throughput can be obtained as the average reward in a Z cycle.

Closed mathematical formulas for network throughput and average frame service time have been derived from the above-explained model. These formulas are used to evaluate the performance of the proposed model for two different scenarios, namely, single channel sensing, where CCA is performed once, and double channel sensing where CCA is performed twice. The effect of increasing the number of nodes in the network on both network throughput and average frame service time are studied using the default MAC parameters defined in the standard. The study finds that the network throughput decreases and the frame service time increases as the number of network nodes increases. This is due to the increasing probability of simultaneously channel sensing by multiple nodes as the number of node increases. Therefore, most of the network time is wasted in backoff which leads to degradation in the network throughput and increase in the average frame service time.

An enhanced MAC protocol for mesh networking with dynamic bandwidth management named group relay and token solicitation (GRATS) is proposed in [13]. The GRATS protocol allocates channel time at the route level instead of node level CTA allocation proposed by WPAN standard MAC protocol. Also, the source node of each activated route initializes a token and become the token owner. These tokens are identified with the activated route number and generation number. Three options are defined for token relaying mechanism between network nodes in the GRATS protocol. These options are: normal mode, piggyback mode, and solicit mode.

At the CTA for a route, transmission is only allowed from the token owner of the route. After transmitting the preconfigured number of packets or the node queue becomes empty, the token owner passes the token *to* its successor node in the route through piggyback mode. This mechanism is called group relay. When the device next to the destination node finishes its transmission, it returns the token, using normal mode, to the source node directly or by relaying where direct transmission is impossible. Once the token is re-owned by the source node, a new round of group relay begins.

In case of buffer overflow at a relaying node while it receives packets from its predecessor node, the node can solicit the token from its predecessor node by solicit mode. As the relaying node buffering capability becomes available, the solicited token transferred back to the predecessor node.

The GRATS protocol has been analyzed mathematical and through simulation to evaluated the network throughput and packet delay. The obtained results of the GRATS protocol are compared with the results of the standard MAC protocol. The GRATS protocol outperforms the standard MAC protocol. However, the GRATS protocol requires an extensive management of the active routes and their allocated channel times.

The beacon, beacon plus GTS, and nonbeacon modes of the MAC protocol proposed in [10] have been mathematically analyzed in [14]. The analysis considers a star-based network configuration for WPAN. This study shows that beacon-mode operation is suitable under very tight data rate restrictions. Also, the analysis found that GTS has a penalty over beacon operation because of the receiver is required to stay on for a full slot regardless of packet size. Therefore, to minimize power usage, the slot size needs to be selected as close as possible to the packet size. The study has proved that the nonbeacon operation mode provides the best performance since sensor nodes' receivers are not required to power up for receiving the beacon frame.

In [15], an MAC protocol for supporting QoS that modifies the original MAC frame structure is proposed. In such protocol, sensor nodes contend for the access medium in the contention access period (CAP) to only reserve a collision free time slot for data transmission in the GTS period. The contention mechanism in the CAP is based on a slotted Aloha scheme instead of CSMA/CA. The drawback of this proposed MAC protocol is that even nodes with noncritical information have to reserve a collision-free time slot before it can transmit its data.

A distributed queuing body area network (DQBAN) MAC protocol which takes into consideration sensor nodes energy consumption and QoS support for healthcare applications is proposed in [16]. DQBAN utilizes two queues: collision resolution queue (CRQ) and data transmission queue (DTQ). The CRQ controls sensor nodes access requests to the medium, while DTQ schedules sensor nodes collision-free packet transmissions. For providing QoS support, the DQBAN protocol utilizes a cross-layer fuzzy-rule-based schedular for packet scheduling which allows a body sensor node with high priority information to transmit in the next frame collision-free data slot even though not occupying the first position in DTQ. DQBAN requires the management of different queues as well as the fuzzy-logic system implementation in every sensor node.

A scheduling scheme for enhancing the WPAN MAC protocol performance by reducing the average waiting time and meeting the QoS requirements for multimedia traffic is proposed in [17]. The proposed scheduling scheme abstracts the channel allocation scheme to $M/M/c$ queuing model. The network traffic is differentiated into different classes where traffic streams of these classes are having different QoS requirements which need to be met by the network. The proposed scheme considers the saturation condition when all of the transmission slots in the super frame are assigned to different node pairs while high-priority traffic channel time requests continues to arrive to the network coordinator node. In the standard MAC protocol, these requests have to wait their turn before they can be assigned transmission slots. In the proposed protocol, the network coordinator node is allowed to reassign a previously assign transmission slots to a lower class stream to a newly arrive high-priority traffic stream. The average waiting time in the system for different average traffic streams arrival rates and different number of serves is evaluated through simulation for the standard MAC protocol and the enhanced MAC scheme. The obtained results indicates that the newly proposed scheme outperforms the standard MAC protocol.

References

1. Haartsen, J. C. The Bluetooth radio system. *IEEE Personal Commun*, 2000; 7(1): 28–36.
2. Chlamtac, I., Conti, M., and Liu, J. Mobile *ad hoc* networking: imperatives and challenges. *Ad Hoc Netw*, 2003; 5(2): 13–64.
3. Bruno, R., Conti, M., and Gregori, E. Bluetooth: architecture, protocols and scheduling algorithms. *Cluster Comput J*, 2002; 5(2): 117–131.
4. Sairam, K. V. S. S. S. S., Gunasekaran, N., and Rama Reddy, S. Bluetooth in wireless communication. *IEEE Commun Mag*, 2002; 40(6): 90–96.

5. IEEE 802.15 WPAN Task Group 3 (TG3), IEEE 802.15.3: wireless medium access control (MAC) and physical layer (PHY) specifications for high-rate wireless personal area networks (WPANs), in *IEEE*, Septmber 2003.

6. IEEE 802.15 WPAN Task Group 4a (TG4a), *IEEE* 802.15.4a: wireless medium access control (MAC) and physical layer (PHY) specifications for low-rate wireless personal area networks (WPANs), in *IEEE*, August 2007.

7. Lee, J., Chuang, C., and Shen, C. Applications of short-range wireless technologies to industrial automation: a ZigBee approach, in *Fifth Advanced International Conference on Telecommunications*, Venice, Italy, May 2009, pp. 15–20.

8. Kim, D. and Cho, J. WBAN meets WBAN: smart mobile space over wireless body area networks, in *IEEE Vehicular Technology Conference, Fall (VTC 2009-Fall)*, Anchorage, AK, Fall 2009, pp. 1–5.

9. Ali, K., Sarker, H., and Mouftah, H. Urgency-based MAC protocol for wireless sensor body area networks, in *IEEE International Conference on Communications*, CapeTown, South Africa, pp. 1–6, May 2010.

10. IEEE 802.15 WPAN Task Group 4 (TG4). IEEE 802.15.4: wireless medium access Control (MAC) and physical layer (PHY) specifications for low-rate wireless personal area networks (WPANs), in *IEEE*, September 2006.

11. Ramachandran, I. and Roy, S. On the impact of clear channel assessment on MAC performance, in *IEEE Global Telecommunications Conference*, San Francisco, CA, USA, Novemebr–December 2006, pp. 1–5.

12. Ling, X., Cheng, Y., Mark, J., and Shen, X. A renewal theory based analyitical model for the contention access period of IEEE 802.15.4 MAC. *IEEE Trans Wirel Commun*, 2008; 7(6): 2340–2349.

13. Chen, X., Lu, J., and Zhou, Z. An enhanced high-rate WPAN MAC for mesh networking with dynamic bandwidth management, in *IEEE Global Telecommunications Conference*, St Louis, Missouri, USA, December 2005, Vol. 6, pp. 3408–3412.

14. Timmons, N. and Scanlon, W. Analysis of the performance of IEEE 802.15.4 for medical sensor body area networking, in *IEEE Conference on Sensor and Ad Hoc Communications and Networks*, Santa Clara, CA, USA, October 2004, pp. 16–24.

15. Bucaille, I., Tonnerre, A., Ouvry, L., and Denis, B. MAC layer design for UWB LDR systems: PULSERS proposal, in *4th Worksshop on Positioning, Navigation and Communication*, Hannover, Germany, March 2007, pp. 277–2837.

16. Otal, B., Alonso, L., and Verikoukis, C. Highly reliable energy-saving MAC for wireless body sensor networks in healthcare systems. *IEEE J Sel Areas Commun*, 2009; 27(4): 553–565.

17. Zeng, R. and Kuo, G. A novel scheduling scheme and MAC enhancement for IEEE 802.15.3 high-rate WPAN, in *IEEE Wireless Communications and Networking Conference*, Orleans Louisiana, USA, March 2005, Vol. 4, pp. 2478–2483.

18. IEEE 802.15 WPAN Task Group 1 (TG1), IEEE 802.15.1: wireless medium access control (MAC) and physical layer (PHY) specifications for wireless personal area networks (WPANs), in *IEEE*, June 2005.

21

Pervasive Energy Management for the Smart Grid: Towards a Low Carbon Economy

Melike
Erol-Kantarci
University of Ottawa

Hussein T. Mouftah
University of Ottawa

21.1 Introduction

There is an urgent necessity to reduce green house gas (GHG) emissions due to the increasing signs of global warming. The Kyoto protocol, an initiative of the United Nations Framework Convention on Climate Change, has been signed by a large number of countries, where the major goal of the protocol is to enforce some measures to the governments in order to prevent GHG concentrations in the atmosphere to reach at a level that could be dangerous to human life and the planet [1]. GHG is used as a common name for carbon dioxide (CO_2), methane (CH_4), nitrous oxide (N_2O), sulphur hexafluoride (SF_6), and two groups of gases, hydrofluorocarbons and perfluorocarbons, where all of them are translated into CO_2 equivalent (CO_2eq) in GHG calculations. Power, transportation, buildings, agriculture, forestry, cement, chemicals, petroleum & gas, and iron & steel sectors produce the highest volume of CO_2eq emissions.

Among the top GHG-emitter countries, China, United States, and European Union are pursuing new technologies to accomplish their commitment to the Kyoto protocol. Smart grid, electric transportation, and future Internet are few examples of the technologies that are aligned with the low-carbon economy goals. In a recent report of the European Commission, information and communication technology (ICT) is stated as "the engine for sustainable growth in a low carbon economy" [2]. ICT is also expressed as a fundamental concept for the renovation of the electrical power grid in the Energy Independence and Security Act of 2007 by the US government [3]. The term smart grid is used in this statement to denote a power grid that utilizes ICT intensively.

Besides reducing the carbon emissions of the power sector, smart grid also aims to serve as a more reliable and secure grid. In the existing grid, imbalance between the growing demand and diminishing fossil fuels, coupled with aging equipments and workforce, has recently caused problems such as the major blackouts in California and northeast of the United States. Later analyses have revealed that those incidents could have been avoided if efficient monitoring, automation, diagnostic tools, and pervasive communications were available. Smart grid aims to have more reliability using the opportunities that become available with the advances in ICT. Besides reliability, integration of distributed renewable energy generation, distributed storage, two-way flow of information and electricity, and sophisticated energy management at the demand side are among the primary targets of the smart grid [4].

US Greenhouse Gas Inventory Report of 2010 reports that electricity generation is the largest source of CO_2 emissions in the United States, causing 40% of the total emissions across the United States [5]. Therefore, reducing the electricity consumption is important to reach the goal of low carbon economies and it is possible with the integration of renewable sources and energy management at the demand side. The relation between the low carbon economies, smart grid, and residential energy management is illustrated in Figure 21.1. Carbon reduction requires cooperation of the transportation, buildings,

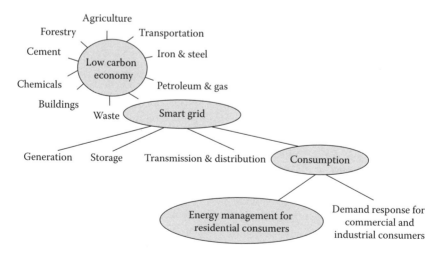

FIGURE 21.1   Low carbon economy and the smart power sector.

agriculture, forestry, cement, chemicals, petroleum & gas, and iron & steel sectors, as well as the power sector.

We focus on the power sector where pervasive communications will play a significant role. We narrow down the scope of our discussion to residential energy management where smart grid and pervasive communication technologies offer diverse energy management applications. Smart grid enables consumers to have more control on their consumption. Pervasive communications enable the adoption of the energy management applications in the daily routines of the consumers while making them personalized and available anywhere/anytime. The key properties of a pervasive computing system have been defined as [6]:

- Sensing and control
- Localization of mobile users
- Context-awareness, that is, recognizing the user's state and surroundings and act accordingly
- Minimum visibility and distraction
- Localized scalability, that is, adjusting the personal computing space based on the interactions and mobility of a person

The primary challenge of pervasive computing is privacy. Without a person knowing or approval, a system may collect private information as a result of continuously monitoring user actions. Even the energy consumption data may yield private information about an individual. In a recent study [7], it has been shown that the absence/presence of an individual, sleeping periods, meal, and shower times could be determined based on the collected power consumption data. Therefore, it is important to have trust between a user and an infrastructure.

In the following sections, we first focus on smart grids and describe its objectives and challenges. Then, we narrow our perspective to pervasive energy management applications for the residential consumers and introduce the recently proposed techniques.

21.2 Smart Grid

The existing power grid has become inefficient to meet the reliability and the availability expectations of the consumers. Consumer markets are filled by a wide variety of electronic devices, and consumers are becoming more and more dependent on those devices. Heating, cooling, lighting, cooking, commuting, entertainment, and telecommunication devices depend on the availability of electricity while the generation of electricity is challenged by the diminishing fossil fuels, and even if the fossil fuel reserves were capable of accommodating the electricity demand for the following centuries, the CO_2 emissions resulting from fossil-fuel-based electricity generation will keep threatening the future of the planet. Besides the growing demand for power, inefficient transmission, and distribution of electricity is another significant factor that is causing losses and increasing the need for more energy generation, while consumer habits and device characteristics accumulate the demand, as well. Therefore, governments are giving primary priority to rethink their way of generating, transmitting, distributing, and consuming electricity and rapidly implementing their smart grids. Note that, smart grid is a generic

FIGURE 21.2   Illustration of a smart grid.

name for the power grid with some level of intelligence; however, implementations of smart grid in different countries may follow different standards and employ different technologies. Figure 21.2 presents an illustration of a smart grid where renewable resources and traditional power generators are integrated, storage is available, and consumers are able to communicate with the utilities.

21.2.1 Objectives of the Smart Grid

Smart grid implementations in different geographic locations may be different; however, the major objectives of the smart grid are similar globally. In this section, we summarize the objectives of the smart grid.

21.2.1.1 Two-Way Flow of Information and Electricity

The existing energy grid is a one-way electricity distribution system where the flow of electricity is directed from the power plants toward the consumer. Energy is generated at power plants which are located close to natural resources, and it is transported by the transmission power lines and the local distribution electricity cables toward the consumer premises. Generation is mostly centralized and energy flows in one direction. One of the major objectives of smart grid is to enable the two-way flow of electricity and information between supplier and consumer, by integrating the advances in ICT.

In the smart grid, advanced metering infrastructure (AMI) of a utility enables communication between the utilities and the consumers where electricity consumption can be monitored by both parties in resolution of minutes. AMI is based on communication with the smart meters which currently have been installed in the majority of the consumer premises.

Primarily, smart meters send information collected from the consumer to the utility and deliver utility-based information to the consumer, in a reliable and secure way. Smart meters also allow the utilities to detect outages, to read meters remotely, and to realize automated energy management, and they enable the use of time-differentiated billing schemes, namely, time of use (TOU), real-time pricing (RTP), critical peak pricing (CPP) which are explained below.

TOU Pricing: Consumer demand (or the load) is known to have seasonal, weekly, and daily patterns. An example of a winter weekday load profile is given in Figure 21.3 for illustrative purposes. The hours when the grid faces high load are called on-peak (or "peak") hours, while moderate and low load durations are called as mid-peak and off-peak hours, respectively.

The existing energy grid is not able to store energy in large amounts; therefore, generation has to be kept in balance with the load. When the demand exceeds the level of the base load, utilities bring in peaker plants into use or purchase electricity from external suppliers. The maintenance of those peaker plants is costly; in addition, they generally use fossil fuels which are expensive and cause high GHG emissions. With the TOU pricing, electricity consumption during peak periods has higher price than consumption during off-peak periods, and the consumers are encouraged to utilize off-peak hours. An example rate chart of an Ontario-based utility is given in Table 21.1 as of 2010 [8]. Note that TOU hours and rates may vary from one utility to another based on the local load pattern and other regional characteristics.

Real-Time Pricing: RTP is also known as dynamic pricing which means the consumers are billed according to the actual price of the electricity in the market. The market price of electricity is generally determined by an independent system operator (ISO) in deregulated markets. ISO arranges a settlement for the electricity prices of the next day and the next hour, based on the load forecasts, supplier bids, and importer bids. The final

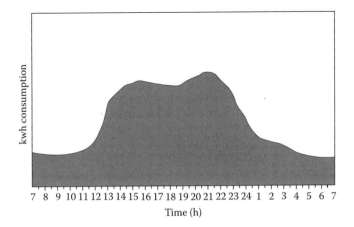

FIGURE 21.3 Daily load pattern of a home in winter.

TABLE 21.1 TOU Rates of a Regional Utility as of 2010

	Period	Time	Rate
Winter Weekdays	On-peak	7:00 a.m. to 11:00 a.m.	9.3 cent/kWh
	Mid-peak	11:00 a.m. to 5:00 p.m.	8.0 cent/kWh
	On-peak	5:00 p.m. to 9:00 p.m.	9.3 cent/kWh
	Off-peak	9:00 p.m. to 7:00 a.m.	4.4 cent/kWh
Summer Weekdays	Mid-peak	7:00 a.m. to 11:00 a.m.	8.0 cent/kWh
	On-peak	11:00 a.m. to 5:00 p.m.	9.3 cent/kWh
	Mid-peak	5:00 p.m. to 9:00 p.m.	8.0 cent/kWh
	Off-peak	9:00 p.m. to 7:00 a.m.	4.4 cent/kWh
Weekends	Off-peak	All day	4.4 cent/kWh

price of the electricity is determined after regulation, transmission, distribution fees, and taxes are added.

Critical Peak Pricing: CPP is applied on several days of a year, for example, very hot days, when the load exceeds a certain threshold. CPP aims to reduce the load of large industrial or commercial consumers on those critical days in order to prevent grid failure. Some utilities reward customers with credits for their corporation on critical peak days which is called as peak-time rebate.

Currently, smart meters have been installed at a large percentage of the consumer premises in North America, Europe, and China and the full installation of smart meters is expected to be completed by 2012. Availability of an AMI is the primary step in implementing pervasive computing applications for the users. While smart meters enable two-way information flow, two-way flow of electricity is also possible where consumers can sell electricity to the utilities.

21.2.1.2 Optimized and Self-Healing Assets

The advances in ICT can be employed in the smart grid to automatically respond to disturbances and failures and to optimize the grid performance. A self-healing grid is desired mostly for security reasons. Under a terrorist attack, the reliability and the availability of the services can be quickly restored by a self-healing grid. The traditional power grid is monitored by the Supervisory Control and Data Acquisition software and its components. In general, human operators handle restoration after a failure, and asset optimization is not realized in real time. In the smart grid, real-time information on the status of the grid can be collected by the advanced sensors, intelligent remote terminal units, and wireless communications. Smart grid will employ those intelligent devices for self-healing and asset optimization, which will be even more critical, especially after the integration of renewable sources and storage.

21.2.1.3 Integrated Renewable Energy Sources

In the existing grid, large-scale energy generation is mainly accomplished by nuclear, hydroelectric, or coal-fired plants depending on the regional characteristics of the power grid. For instance, China's energy generation is primarily based on coal, while in Ontario hydro and nuclear power are dominant and only peaker plants are fossil-fuel-based.

As mentioned before, peaker plants are brought into use by the utilities when the amount of load (demand) exceeds the total capacity of the base plants and the stored energy.

The smart grid will essentially address the environmental concerns. Recently, governments are focussing on renewable green energy sources such as solar, wind, geothermal, and ocean waves to accommodate some portion of the base and the peak loads, in order to reduce their GHG emissions together with their dependency on fossil fuels. In fact, renewable energy sources have been known for a long time; however, they have not been integrated into the energy grid on a wide scale, with an exception of few northern European countries. The reason is mainly due to the intermittent nature of these resources. They are affected by weather conditions and several other factors. Moreover, they are generally available in geographically remote locations, and due to transmission losses, power transportation to urban areas has been challenging.

In the smart grid, renewable generation will be available for large-scale generation as well as small-scale generation such as residential premises. Solar panels and small wind tribunes are already available for home owners. Electricity generated at home can be consumed locally and the excess amount can be sold to the utilities, for example, Ontario Micro-Feed in program allows consumers to sell electricity to the utilities [9]. Managing the use of local resources requires pervasive computing solutions for home owners where they can manage their energy profile from anywhere and at anytime.

21.2.1.4 Improved Energy Storage

In the existing power grid, large-scale, distributed energy storage is not available. Power plants generally utilize inflexible storage methods such as pumped hydro, compressed air, flywheel, flow batteries, and so on. These techniques store energy in the form of potential energy and convert it into electrical power when needed. However, they are not convenient for storing the power generated by renewable energy sources.

Smart grid aims to improve the energy storage in the grid so that energy generated by the intermittent renewable sources can be conveniently integrated. For instance, energy generated overnight by the wind farms may need to be stored until the morning or the evening peak hours. Especially, in a system where generation and consumption can have large variance in real time, energy storage becomes crucial. Otherwise, production from the conventional power plants will have to adapt to these dynamic resources which is complicated and costly.

Recently, the research is directed toward the use of lithium-ion batteries for storage in the smart grid since they will be available in large amounts after the widespread adoption of the plug-in hybrid electric vehicles (PHEVs). PHEVs can provide a distributed storage capacity for the grid with their light-weight and fast discharging lithium-ion batteries. Vehicle to grid (V2G) research focusses on the effective flow of electricity from cars to the grid [10,11]. When V2G and smart charging are combined, PHEVs can be a competitive energy storage solution for the smart grid. Smart charging can coordinate PHEV charging considering the state of the grid or price of electricity. Without coordination, PHEVs can cause several problems, which will be discussed in the following sections.

21.2.1.5 Energy Management

In the power grid generation is controlled by the supplier while the load is determined by the consumers' demands. In the existing grid, utilities use a method called "demand response" which controls the load of the commercial and the industrial consumers during peak periods. Residential consumers have not been considered in the energy management schemes while this is changing with the smart grid. Smart grid is able to reach the residential consumers easily by the help of pervasive communications and offer a rich set of energy management applications. Smart grid handles energy management for those two different consumer profiles, namely, commercial/industrial consumers and residential consumers, in different ways.

Commercial and Industrial Consumers: Commercial and industrial consumers can have a high impact on the load depending on the scale of their business. Demand response term is used for those consumers' reducing their demand following utility instructions, to avoid either high fees or failure of the grid. Demand response is generally handled by the utility or an aggregator where participating consumers are notified by phone calls, for example, to turn off or to change the set point of their heating, ventilating, and air conditioning (HVAC) systems for a certain amount of time to reduce their demand.

In the smart grid, automated demand response (ADR) is considered, where utilities send signals to buildings and industrial control systems to take a preprogrammed action based on a specific signal. Recently, OpenADR standard has been developed by the Lawrence Berkeley National Laboratory and the standard is being used primarily in California [12]. Another well-known data communication standard for building automation and control network is the BACnet. BACnet was initially developed by the American Society of Heating, Refrigerating, and Air-Conditioning Engineers and later adopted by ANSI [13]. BACnet Smart Grid Working Group (SG-WG) is also considering the integration of BACnet to the demand response activities of the smart grid.

Residential Consumers: Residential consumers are the "last-foot" (last-meter) of the electricity grid where most utilities did not have any energy management policies before the smart grid. In fact, without a digitized grid and pervasive communications, residential energy management, which is similar to commercial and industrial demand response, is not scalable. For most residential consumers, energy saving is done with the use of energy-efficient light bulbs and ENERGY STAR qualified appliances. There are also smart homes, although not common, with applications that turn the lights off depending on the occupancy of the rooms or dimming the lights off based on the outside light intensity and shutter positions or adjusting the thermostat based on the outside temperature and sensor measurements.

The smart grid introduces new opportunities for the residential energy management such as energy management based on time-differentiated tariffs where consumers with smart appliances can control their appliances to benefit from lower bills. Especially, for RTP, following the changes in the price of electricity is not practical for consumers while home energy management systems can provide solutions to optimize the energy bills of the consumers. Besides reducing the household energy bills, utilities can benefit from balanced residential loads. In addition, small-scale renewable energy generation can be managed by the residential users.

21.2.1.6 Future-Proof

The renovation of the energy grid is a large-scale project, and it is expected to be completed in the next decade. The complete implementation of the smart grid is desired to be future-proof so that a large-scale renovation will not be required shortly. Therefore, currently installed devices have to be designed with flexibility to allow easy updates while latter installations and technologies have to be backward compatible. The lack of complete regulations and standards, together with the urge in smart grid implementation makes future-proof implementation highly challenging.

21.2.2 Challenges of the Smart Grid

Electricity grid is a highly critical asset and its renovation is challenging from many aspects. Regulations and standardization is one of the major challenges. Currently, various governmental agencies, alliances, committees, and groups are working to provide standards so that smart grid implementations are effective, interoperable, and future-proof. Security is another significant challenge since the grid is becoming digitized and generally using open media for data transfer. Smart grid can be vulnerable to physical and cyber attacks if security is not handled properly. Moreover, the way the grid operates, the economics, and the markets of the power industry are changing, and developing new business models is another challenge for the utilities. Providing widespread consumer participation for the residential energy management applications is challenging as well. Last but not the least, the load on the grid is expected to increase as PHEVs are plugged-in for charging. Uncoordinated charging may cause failures in locations where PHEVs are populated, which introduces another challenge. In the following subsections, we explain these challenges in detail.

21.2.2.1 Regulations and Standardization

Smart grid requires new regulations that are convenient for the power requirements of the next decades. Generally, governmental agencies take the primary responsibility to set up these regulations. In some countries, it is easier to synchronize the regulations due to central governance and fewer players in the electricity market, whereas for some countries the situation is more complex due to higher number of players in the market.

Standardization is another issue that needs to be accomplished so that each new component integrated to the grid follows the same set of rules, and it is interoperable with the other components. Currently, standardization efforts are independently handled in most of the countries that are developing their own smart grids. For example, in the United States, National Institute of Standards and Technology has been given the primary responsibility to develop standards, and it has published the report entitled "Framework and Roadmap for Smart Grid Interoperability Standards, Release 1.0" [14], which presents the available standards and discusses their use in the smart grid applications. In the European Union, the European Commission has defined the European Energy Research Alliance as a key actor for implementing the EU Strategic Energy Technology Plan. Besides standardization, international harmonization of the standards is required to provide consumers the same set of services worldwide. International Electrotechnical Commission, which is an international standardization commission for electrical, electronic, and related technologies, has documented a large number of smart grid standards, already. IEEE has a large number of standards including the IEEE P2030 draft guide entitled "smart grid interoperability of energy technology and information technology operation with the electric power system, and end-use applications and loads." Smart grid has a broad scope; therefore, a large number of new standards are expected to emerge in the following years.

21.2.2.2 Security

Utilities are using computers and software to monitor and control their assets, and they have been experiencing cyber attacks, even before the smart grid. However, the impact and the frequency of those attacks may increase in the smart grid because it will interconnect all the systems digitally. For instance, smart grid is using wireless communications to connect to the smart meters where attackers may eavesdrop or even send fake signals to destabilize the grid. Smart grid may be more vulnerable than the existing grid if security against physical and cyber attacks is not addressed in the first place.

21.2.2.3 Utility Business Models

The operation, economics, and the market dynamics related to the smart grid are quite different from the existing grid where the utilities are challenged with the need for new business models. In the smart grid, consumers can generate energy with renewable sources and accommodate some portion of their consumption through the locally produced electricity or sell electricity to the grid. When these are combined with the novel efficient energy management techniques, zero-energy homes and microgrids become possible, while the costs of the utilities will not be dropping proportionally. For instance, the NOW House project in Toronto, Ontario, has turned several old houses to annually zero-energy homes [15]. On the other hand, meter data management which includes storing and utilizing the high volume of meter data is challenging, as well. To manage the data efficiently, utilities need to invest on green data centers. Among other challenges, the lifetime of the smart grid equipments is shorter than that of the conventional grid components, and their replacement costs will also impact the economics of the utilities. Briefly, utilities need new business models that address the above challenges.

21.2.2.4 Consumer Participation

Consumers prefer easy-to-use, cheap, reliable, and nonintrusive technologies. In the existing grid, electricity has been as simple as plug-and-play, relatively cheap, mostly reliable, and nonintrusive. It is available at a fixed price and consumers do not have to worry about their time of consumption. In the smart grid, time of usage directly affects the consumer bills. From this perspective, smart grid seems to require smart consumers that participate in energy conservation. At this point, home energy management systems that are pervasive and that provide user-friendly and effective energy management for the consumers become crucial.

21.2.2.5 Plug-In Hybrid Electric Vehicles

PHEVs are soon to be on the roads and they may increase the load on the grid and introduce unpredictable load patterns. To prevent PHEVs from instabilizing the grid, smart charging is essential. With smart charging, a PHEV can check the price and the load condition of the grid and charge its battery when the price and the load are low. It can also predict the price of electricity for the following hours and schedule charging to minimize the vehicle operating costs [16]. Smart charging and discharging may also allow PHEVs to be used for distributed energy storage for the smart grid, as discussed in the previous sections.

Smart grid requires the evolution of the power grid in the long term in order to accomplish the objectives that are set forward today. There will be significant challenges during this transition and smart grid needs the help of ICT to overcome these challenges. In the next section, we focus on employment of ICT to the smart grid for pervasive energy management.

21.3 Pervasive Energy Management for the Residential Consumers

Residential energy management applications that are currently available in the market are capable of simple tasks such as turning off lights depending on occupancy of the rooms or dimming the lights based on outside light intensity and shutter positions or adjust the thermostat based on the outside temperature and sensor measurements. These type of comfort-focussed energy management applications initially started with the "smart home" concept several decades ago [17,18], and they can be considered as the primitive implementations of pervasive energy management applications. Today, with the use of wireless sensor networks (WSNs), pervasive computing can penetrate into the energy management applications in the residential premises easier than it used to be; however, smart grid brings in new perspectives to the picture.

Energy management is closely coupled with activity recognition, which is studied extensively in the context of pervasive computing [19]. For instance, the user has a meeting and his/her suit needs to be steamed before his/her meeting; however, steaming in the morning when the other appliances such as kettle or toaster are running will increase the overall power demand of the user. The energy management application can then schedule the running of the appliances based on when the user wakes up, when he/she wants to have his/her breakfast and when he/she wants to dress himself/herself. The appliances can be scheduled to be ready when they are needed and at the same time this can balance the overall use of electricity.

For appliance control, home automation and smart appliances are two important enabling technologies. Regarding the home automation market, there are a variety of solutions with varying capabilities. Intel has recently developed the Home Energy Dashboard that provides a simple interface to consumers to monitor their monthly bills and electricity consumption of their appliances [20]. On the other hand, smart appliances have been a hot topic for appliance vendors for the past several decades where the primary focus has been user comfort and improving the energy efficiency of individual appliances. Appliance to appliance communication or an appliance network is a desired property of the smart appliances; yet due to lack of a common communication technology, appliance networks are not mature. Only the appliances of the same vendor are able to communicate. To overcome this problem, five major Japan appliance vendors recently developed the Echonet standard which has flexible communication options including power-line communication (PLC), low-power wireless, IrDA control, Bluetooth, Ethernet, and wireless LAN [21]. Integration of Echonet to the smart grid and its use to support low carbon economies are reported as the future objectives of Echonet in [21].

Smart grid is a new concept; therefore, energy management applications for the smart grid are rare, and they are being considered very recently. In the following sections, energy management solutions that employ communications and have the possibility to be extended for pervasive energy management in the smart grids are introduced.

21.3.1 Intelligent and Personalized Energy Conservation System by WSN

Intelligent and personalized energy conservation system by WSN (iPower) exploits the context-awareness provided by WSNs to implement an energy conservation application for multidwelling homes and offices [22]. iPower includes a WSN with sensor nodes and a gateway node, in addition to a control server, power-line control devices and user identification devices. Sensor nodes are deployed in each room and they monitor the rooms with light, sound, and temperature sensors. When a sensor node detects that a measurement exceeds a certain threshold, it generates an *event*. Sensor nodes form a multihop WSN and they send their measurements to the gateway when an *event* occurs. The gateway node is able to communicate with the sensor nodes via wireless communications and it is also connected to the intelligent control server of the building. Intelligent control server performs energy conservation actions based on sensor inputs and user profiles. The action of the server can be turning off an appliance or adjusting the electric appliances in a room according to the profiles of the users who are present in the room. Server requests are delivered to the appliances through their power-line controllers. iPower uses Zigbee for WSN communication and X10 for PLCs.

iPower has interactive services as well as personalized services. In the course of interactive services, the intelligent server sends an alarm signal to notify the people in the room that the appliances, lights, and HVAC will be turned off. If the room is unoccupied, those devices are turned off. Otherwise, if there are people in the room, then they can signal their presence by making noise or moving a sensor-attached furniture, and the server does not turn off the devices. In the personalized services, iPower can adjust appliance settings according to the predefined user profiles. In order to recognize the users, iPower requires users to wear/carry identification devices.

iPower is an energy management application that uses a Zigbee-based WSN, and it is similar to the energy management applications in the smart homes. iPower focusses on demand control; however, in the smart grid, users will also have the ability to manage their energy generation and storage. The integration of iPower and the smart grid has not been considered in [22]. In the next section, we will introduce the in-home energy management (iHEM) scheme that interacts with the users to manage their demand considering their generation and storage capacity.

21.3.2 In-Home Energy Management

In-home energy management (iHEM) application uses appliances with communication capability, a WSN, and an appliance manager (AM) to manage the electricity supply for the household demand [23]. In the iHEM application, consumers may turn on their appliances at any time, and considering the availability of the electricity resources (utility grid, solar, wind, and storage), the scheme suggests start times to consumers via LCD displays mounted on the appliances. An illustration of the iHEM network is given in Figure 21.4.

In the iHEM application, an appliance generates a START-REQ packet when it is turned on. This packet is sent to the AM, where upon receiving the START-REQ packet, AM checks the availability of power resources. It communicates with the storage unit of the local energy generator (solar and wind) to learn the available local energy, and it also periodically communicates with the smart meter to receive updated price information from the utility. Price information reflects the state of the grid (e.g., peak, mid-peak, and off-peak). Note that when RTP is used by the utilities it is also possible to attain finer-grained grid state information from the smart meter. AM sends an availability request packet (AVAIL-REQ) to the local resources. When the house is equipped with multiple energy generation devices such as solar panels and small wind tribunes, the amount of energy stored in their local storage units may have to be interrogated with separate messages. Upon reception of AVAIL-REQ, the storage unit replies with the AVAIL-REP packet including the amount of available energy. After receiving the AVAIL-REP packet, AM determines the convenient starting time of the appliance based on the available resources. Then, it computes the waiting time as the difference between the suggested

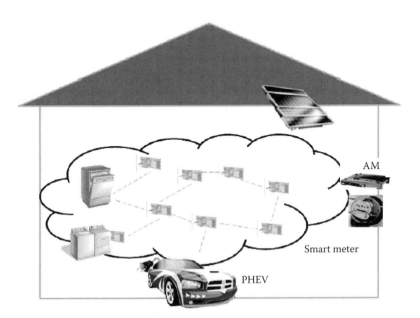

FIGURE 21.4 Smart home with iHEM.

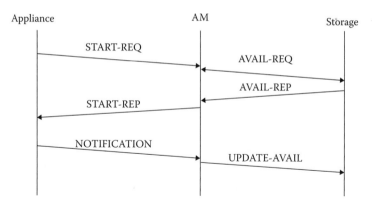

FIGURE 21.5   Message flow for iHEM.

and requested start time and sends the waiting time in the START-REP packet to the appliance. iHEM is an interactive application where the consumer decides whether to start the appliance right away or wait until the assigned time slot. The decision of the consumer is sent back to the AM with a NOTIFICATION packet. This information is required to allocate energy on the local storage unit when it is used as the energy source. In some cases, it is also possible to sell excess energy to the grid operators where some amount of energy needs to be reserved for the appliances that are supposed to run with the local energy. Finally, AM sends an UPDATE-AVAIL packet to the storage unit to update the amount of available energy (unallocated) on the unit. This handshake protocol among the appliances and the AM ensures that AM does not force an automated start time and cause discomfort. The message flow of the iHEM application is shown in Figure 21.5 [24].

The impact of iHEM on the peak load and carbon footprint of the consumers has been analyzed in [25] and [26], respectively. The authors have shown that iHEM has reduced the contribution of the appliances on the peak load, and by reducing the peak hour consumption, iHEM has also decreased the carbon footprint of the household.

iHEM scheme interacts with the users to provide energy management including control over their consumption, generation, and storage. Yet, in iPower and iHEM, the user needs to be at the premises. In the following section, we introduce an energy management application where the users are able to do energy management remotely.

21.3.3 Sensor Web Services for Energy Management

Internet of things (IoT) is one of the enabling technologies of pervasive computing, and IoT is possible with platform-independent mechanisms such as the web services. Web services can invoke remote methods on other devices without necessarily knowing the internal implementation details and enable machine-to-machine communications [27]. Using web services for embedded systems has been considered in the literature, and the challenges employing web services to resource constrained devices such as the sensors have been reported in [27–30]. Groba and Clarke [27] have deployed Web services on the

Sun SPOT sensor platform and analyzed the performance of web services in terms of disk space, message size, response time, and energy consumption. The authors' analyses have shown that sending the same amount of data via web service routines causes almost five times larger packet sizes than sending data with a simple time-stamped format. Moreover, disk space for storing web service routines is shown to vary between 61 bytes and 478 kB for ksoap and ws4d protocol stacks, respectively. Considering the limited memory size of sensors, it is challenging to store the web service routines. It has been shown that delay and energy consumption also increases when Web services are used to deliver the same data. Tiny Web services (TWS) approach [31] aimed to overcome these challenges and offer significant improvements for web services based on a set of modifications to the TCP/IP protocol stack, efficient encapsulation, and compression techniques.

Priyantha et al. [31] have also employed TWS within a home energy management application that they implemented in a volunteer home. TWS utilized power sensors/actors that are able to monitor the current drawn by the appliances and turn on/off the appliances and lights based on occupancy of the home. Occupancy information has been derived from existing motion sensors. Energy management application considered in TWS followed the traditional smart home energy conservation approaches where occupancy-based savings are considered. However, in the smart grid, energy management applications need to interact with time-differentiated pricing, variety of power resources, and the other dynamics of the power grid.

Using sensor web services for energy management (SWSEM) in the smart grids has been recently studied in [32]. The authors assume a smart home containing smart appliances with sensor modules that enable each appliance to join the WSN and communicate with its peers. SWSEM is a suit of three energy management applications. The basic application enables users to learn the energy consumption of their appliances while they are away from home. The current drawn by each appliance is monitored by the sensors on board and this information is made available through a home gateway to the users. Users access the gateway from their mobile devices using web services. SWSEM provides a load shedding application for the utilities, as well. Load shedding is applied to HVAC systems only during peak hours and when the load on the grid is critical. Load shedding is simply turning off an appliance or modifying the set point of the appliance (e.g., HVAC) in order to eliminate or reduce its load. In addition to monitoring and load-shedding applications, SWSEM offers a smart grid compliant application where the amount of energy stored, sold to the grid, or consumed at home can be controlled by the remote user. Since SWSEM application suit is based on sensor Web services, it is a pervasive energy management application where users can access their appliances and manage their consumption profile from anywhere and at anytime.

21.4 Summary and Discussion

Accumulating GHG in the biosphere is considered as one of the major reasons of climate change. Reducing the GHG emissions, before the impacts of climate change becomes severe and threatens the life on earth, has become among the primary concerns of the governments. Leading sectors that contribute to GHG emissions are: power, transportation, buildings, agriculture, forestry, cement, chemicals, petroleum & gas, and iron & steel sectors, where power is the largest contributor. Consequently, integrating clean energy generation, efficient power transmission and distribution, and reduced energy consumption are essential for low-carbon economies. Smart grid aims to achieve those goals, as well as, to deliver secure and reliable power services. Smart grid can reach its targets by applying ICT in every stage of power generation, transmission & distribution, storage, and consumption [33].

In this chapter, we first gave a brief introduction on the objectives and the challenges of the smart grid and then we focussed on the pervasive energy management applications as one of the enabling technologies for low-carbon economies. We introduced three recently proposed schemes, namely iPower, iHEM, and SWSEN. iPower implements an energy conservation application for multidwelling homes and offices using WSNs. In iPower, energy-consuming devices are turned off once a room is detected as unoccupied. Moreover, appliance settings can be modified according to user profiles once a user with an identification device appears in a room. iPower controls a limited set of appliances, and it does not provide control on the energy resources. However, in the smart grid, users can contribute to carbon reduction by choosing the energy resources for their appliances. iHEM is a smart grid compliant application which combines demand management with resource management. iHEM schedules appliances according to the state of the grid and the availability of local resources. It is an interactive application where the user communicates his/her preferences to the scheduler via interfaces on the appliances. Both iPower and iHEM requires the user to be at home in order to interact; however, pervasive energy management applications should also have access from anywhere and at anytime. SWSEN is a suit of sensor Web services application for energy management in the smart grid. It integrates smart appliances with sensor Web services and provides remote energy management capability to users and utilities in the smart grid. Users can monitor their appliances while they are away from home; in addition, they can select from which resource to supply electricity to their appliances.

For the time-being, energy management applications require human interaction and they can sometimes be intrusive. The ultimate goal of the pervasive energy management applications is to execute the best decisions on the behalf of utilities and the consumers based on their needs and preferences. Moreover, energy management applications do not yet integrate PHEVs, which require energy management more than any other appliance since they have the highest energy consumption profile at home. Pervasive computing applications that include PHEVs may yield to rich applications. For instance, PHEV may check the price of electricity and the driver's calendar when it is parked, then calculate the exact amount of charging required to take the driver to his/her destination in the morning, and decide the duration of charging which corresponds to the power consumed by PHEV.

The widespread adoption of pervasive energy management applications is significant for reaching the goals of low-carbon economies; however, for successful market penetration of those applications, privacy issues have to be addressed properly. Data leaving the residential space are critical and strong privacy measures are required to reduce the risk of unwanted data disclosure both for legal concerns and for consumers' adoption of these technologies.

References

1. Kyoto Protocol. [Online] http://unfccc.int/resource/docs/convkp/kpeng.pdf.
2. European Commission Report on Information and Communication Technologies. [Online] ftp://ftp.cordis.europa.eu/pub/fp7/ict/docs/ict-wp-2011-12_en.pdf.
3. Energy Independence and Security Act of 2007. [Online] http://www.gpo.gov/fdsys/pkg/ PLAW-110publ140/pdf/PLAW-110publ140.pdf.
4. Amin, S. M., Wollenberg, B. F., Toward a smart grid: power delivery for the 21st century. *IEEE Power Energy Mag*, 2005; 3(5): 34–41.
5. U.S. Environmental Protection Agency Report on "Inventory of US GHG emissions and sinks: 1990–2008". [Online] http://www.epa.gov/climatechange/emissions/usinventoryreport.html.
6. Satyanarayanan, M., Pervasive computing: vision and challenges, *IEEE Personal Commun* 2001; 8(4): 10–17.
7. Lisovich, M. A., Mulligan, D. K., Wicker, S. B., Inferring personal information from demand-response systems. *IEEE Security Privacy*, 2010; 8(1): 11–20.
8. TOU rates. [Online] http://www.hydroottawa.com.
9. Ontario FIT programs. [Online] http://microfit.powerauthority.on.ca/.
10. Tomic, J., Kempton, W., Using fleets of electric-drive vehicles for grid support. *J Power Sources*, 2007; 168(2): 459–468.
11. Kempton, W., Tomic, J., Vehicle-to-grid power fundamentals: calculating capacity and net revenue. *J Power Sources*, 2005; 144: 268–279.
12. Open ADR. [Online] http://openadr.lbl.gov/pdf/cec-500-2009-063.pdf.
13. BACnet. [Online] http://www.bacnet.org.
14. NIST Framework and Roadmap for Smart Grid Interoperability Standards. [Online] http://www.nist.gov/public affairs/releases/smartgrid 020310.html.
15. NOW House Project. [Online] http://www.nowhouseproject.com.
16. Erol-Kantarci, M., Mouftah, H. T., Prediction-based charging of PHEVs from the smart grid with dynamic pricing, in *First Workshop on Smart Grid Networking Infrastructure in LCN 2010*, Denver, Colorado, USA, October, 2010.
17. Brumitt, B., Meyers, B., Krumm, J., Kern, A., Shafer, S., EasyLiving: technologies for intelligent environments, in *Proceedings of Handheld Ubiquitous Computing*, 2000.
18. Lesser, V., Atighetchi, M., Benyo, B., Horling, B., Raja, A., Vincent, R., Wagner, T., Xuan, P., Zhang, S., The UMASS intelligent home project, in *Proceedings of the 3rd Annual Conference on Autonomous Agents, Seattle*, Washington, 1999.
19. van Kasteren, T., Noulas, A., Englebienne, G., Kröse, B., Accurate activity recognition in a home setting, in *Proceedings of the 10th International Conference on Ubiquitous Computing, Seoul*, Korea, 21–24 September 2008.
20. Intel Home Automation Tool. [Online] http://www.intel.com/embedded/energy/homeenergy.
21. Matsumoto, S., Echonet: a home network standard. *IEEE Pervasive Comput*, 2010; 9(3): 88–92.
22. Yeh, L., Wang, Y., Tseng, Y., iPower: an energy conservation system for intelligent buildings by wireless sensor networks. *Int J Sensor Netw.* 2009; 5(1): 1–10.

23. Erol-Kantarci, M., Mouftah, H. T., Using wireless sensor networks for energy-aware homes in smart grids, in *IEEE Symposium on Computers and Communications (ISCC)*, Riccione, Italy, 22–25 June 2010.

24. Erol-Kantarci, M., Mouftah, H. T., Wireless sensor networks for cost-efficient residential energy management in the smart grid. *IEEE Transactions on Smart Grid*, 2011; 2(2): 314–325.

25. Erol-Kantarci, M., Mouftah, H. T., TOU-aware energy management and wireless sensor networks for reducing peak load in smart grids, in *Green Wireless Communications and Networks Workshop (GreeNet) in IEEE VTC2010-Fall*, Ottawa, ON, Canada, 6–9 September 2010.

26. Erol-Kantarci, M., Mouftah, H. T., The impact of smart grid residential energy management schemes on the carbon footprint of the household electricity consumption, in *IEEE Electrical Power and Energy Conference (EPEC)*, Halifax, NS, Canada, 25–27 August 2010.

27. Groba, C., Clarke, S., Web services on embedded systems—a performance study, in *8th IEEE International Conference on Pervasive Computing and Communications Workshops (PERCOM Workshops)*, Mannheim, Germany, 29 March–2 April 2010, pp. 726–731.

28. Amundson, I., Kushwaha, M., Koutsoukos, X., Neema, S., Sztipanovits, J., Efficient integration of web services in ambient-aware sensor network applications, in *Proceedings of the 3rd International Conference on Broadband Communications, Networks and Systems (BROADNETS'06)*, San Jose, CA, 2006.

29. Gibbons, P., Karp, B., Ke, Y., Nath, S., Seshan, S., IrisNet: an architecture for a worldwide sensor Web. *IEEE Pervasive Comput*, 2003; 2(4): 2233.

30. Delin, K., Jackson, S., The sensor Web: a new instrument concept, in *Proceedings of SPIEs Symposium on Integrated Optics,* San Jose, CA, USA, January 2001.

31. Priyantha, N. B., Kansal, A., Goraczko, M., Zhao, F., Tiny Web services: design and implementation of interoperable and evolvable sensor networks, in *International Conference on Embedded Networked Sensor Systems (SenSys)*, ACM, Raleigh, NC, USA, 2008, pp. 253–266.

32. Asad, O., Erol-Kantarci, M., Mouftah, H. T., Sensor network Web services for demand-side energy management applications in the smart grid, in *IEEE Consumer Communications and Networking Conference (CCNC'11)*, Las Vegas, USA, January 2011.

33. Erol-Kantarci, M., Mouftah, H. T., Wireless multimedia sensor and actor networks for the next generation power grid. *Elsevier Ad Hoc Networks Journal,* 2011; 9(4): 542–551.

Index